D0961611

PROGRAM MANAGEMENT FOR IMPROVED BUSINESS RESULTS

PROGRAM MANAGEMENT FOR IMPROVED BUSINESS RESULTS

Dragan Z. Milosevic
Russ J. Martinelli
James M. Waddell

John Wiley & Sons, Inc.

Published by John Wiley & Sons, Inc., Hoboken, New Jersey.
Published simultaneously in Canada.

For general information on our other products and services please contact our Customer Care Department within the U.S. at 877-762-2974, outside the U.S. at 317-572-3993 or fax 317-572-4002.

Wiley also publishes its books in a variety of electronic formats. Some content that appears in print, however, may not be available in electronic books. For more information about Wiley products, visit our Web site at www.wiley.com.

Library of Congress Cataloging-in-Publication Data:

Milošević, Dragan.
   Program management for improved business results / Dragan Z. Milosevic, Russ J. Martinelli, James M. Waddell.
        p. cm.
   Includes index.
   ISBN: 978-0-471-78354-1 (cloth)
   1. Project management. 2. Project management—Case studies. I. Martinelli, Russ J., 1959- II. Waddell, James M., 1946- III. Title.
   HD69.P75M538 2007
   658.4'04—dc22

                                                        2006025096

Printed in the United States of America.

10 9 8 7 6 5 4 3 2 1

To my wife Dragana and my daughter Jovana

—Dragan Z. Milosevic

To the three most important and supportive people in my life

My wife, Andrea; my daughter Colleen; and my mother, Dolores

—Russ J. Martinelli

To my wife, Susan, and our daughters Lisa, Janine, and Sheri

—James M. Waddell

Contents

Part V: Organizing for Program Management   417

Part VI: Industry Case Examples   491

Final Thoughts on Program Management   547

Preface

Program management has been in practice for decades but isolated to a few industries in which it was developed. It wasn't until the 1980s when it migrated to the commercial sector that it was used more broadly to develop products, services, and infrastructure capabilities. Today, program management is widely recognized as a true management discipline, and institutions like the Project Management Institute (PMI) have finally come to recognize its strategic importance and value to companies. This tells us that the project management community has realized that trying to use project management practices to solve problems that are better resolved by program management practices is an exercise in frustration. Program and project management are both critically important for managing the complex development challenges that face companies today. They are related but are distinctively different disciplines. Program management is strategic in nature and focused on business results, while project management is tactical in nature (management of the triple constraints) and focused on execution aspects of generating project deliverables.

While differentiating between program and project management (as well as other processes and disciplines commonly confused with program management) is one aspect of this book, our intention is to provide a holistic view of the program management discipline. The holistic view describes program management as a critical business function (the business level) that can be used to gain competitive advantage and shows how a program should be managed (the program level) to produce maximum business results. Both the business level and the program level elements covered in this book are based upon real practices by real people in real companies. The concepts presented have been developed through the authors' extensive experience as program management professionals across multiple industries, academic research within companies, and benchmarking activities and interviews with industry representatives.

We believe this is a breakthrough book about a topic that is rapidly gaining interest on a global scale. Historically, there has been little published about program management in the technical press, especially on the business aspects of program management. The research community, with the exception of a small group of researchers in the United Kingdom and Australia, has also missed the business aspects of program management. Fortunately, PMI's publication of a program management standard will spur the development of future literature on the topic. This book aligns well with the new standard and provides necessary details that cannot be included in a generic cross-industry standard.

BOOK PHILOSOPHY AND STRUCTURE

We provide a pragmatic approach by describing the practical aspects of program management. We also provide a blueprint for organizational transition to program management and a how-to approach that is based upon a strong business focus and systems view—both fundamental characteristics of the program management discipline. As stated, this book exclusively focuses on the two levels of program management. The first level is the business level of the company with the purpose of understanding the role of program management in the business model and its links to business strategy and portfolio management. The second level involves using the practices and infrastructure to manage a single program.

To reinforce the practical intent of this book, we include five issue-oriented industry examples, called Program Management In Practice, weaved within the chapters, plus four comprehensive industry examples in the appendices. We chose to use fictitious names for the companies and people presented in the industry examples. We did this to avoid the lengthy process of legal and communications reviews by the companies included.

A significant portion of this book is written in a way that can serve as a blueprint for implementation of program management in companies. It also provides the necessary tools and practices needed by a program manager to manage a program.

When program management is implemented properly, it becomes a primary element of a company's business model. Therefore, this book has a strong focus on business. We present a holistic-systems view of program management as a subsystem in a company's business system.

Additionally, the program management tools included in the chapters are primarily focused on business application.

To represent the robustness of the program management model, we present a diversity of industries throughout the book with a focus on development of products, services, and infrastructure capabilities. The definitions as we use them in this book are as follows:

- *Product:* A manufactured physical good that is new to the firm. This is brought to existence by way of new product development and introduction. [1]

- *Service:* Any primary or complimentary activity that does not directly produce a physical product. It is part of the nongoods transaction between a buyer and seller. [2]

- *Infrastructure capability:* An element of any large-scale technological system that consists of immovable physical facilities that provide essential public or private services through the storage, conversion, and transfer of certain commodities. [3]

Figure P.1 illustrates the overall structure of the content contained in this book.

CONTENT OVERVIEW

In summary, Part I explains what program management is and how it helps the organization accomplish its business strategies. Part II demonstrates the practical aspects of program management, detailing the process of defining, planning, and executing a single program. Program management tools and metrics are daily enforcers of successful execution of programs and are spelled out in Part III. In Part IV we explain what a program manager does and what competencies are needed to effectively manage a program. For executives wanting to pursue a transition to the program management model, Part V provides information necessary to make the transition successful. Finally, Part VI presents real-industry examples of people and companies applying the concepts and principles from the previous five parts.

The details of the book are as follows: Part I begins by providing clarification of the program management discipline and then illustrating how

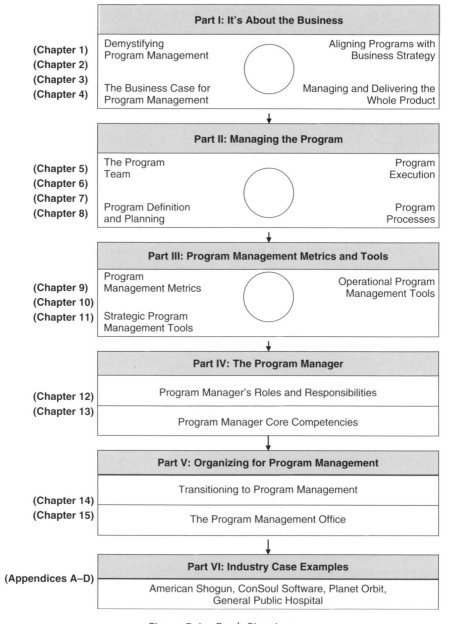

Figure P.1   Book Structure

program management can be implemented as a major element of an organization's business model. Chapter 1 illuminates program management's raison d'etre and compares and contrasts, dimension by dimension, program management to related disciplines and processes such as project management, portfolio management, product management, and management of multiple projects. On that basis, we provide a set of identifying characteristics, or pillars, of program management. To substantiate that a program is really about business success, we present the business case for program management in Chapter 2—specifically, what companies get in return for investing in program management. Chapter 3 illustrates how program management serves to deploy business strategy by aligning the strategy to execution outputs. The purpose of Chapter 4 is to frame the true systems scope and nature of program management to maximize the value proposition of the company by developing and delivering the whole product.

Part II delves into details about how to manage a single program from its inception to end of life. Chapter 5 is about the program team. It emphasizes two critical factors that matter when structuring a program. First, the program team structure must support the highly integrated nature of a program. Second, the program team structure must enable the fundamental elements for team success—from coordination of its activities and interdependent deliverables to collective problem solving. Chapter 6 focuses on program definition and planning, called Define and Plan, phases of the program life cycle. The program definition process involves the steps necessary to integrate business strategy, customer and end user research, and technology research into a viable product, service, or infrastructure concept. The purpose of Chapter 7, program execution, is to focus on execution of the implementation plan to create the whole product, service, or infrastructure and prepare for its launch into the market or end user's environment. Chapter 8 describes the primary program processes that a program manager can utilize to make the operational aspects of a program more efficient, predictable, and repeatable—the program life cycle, schedule management, financial management, risk management, change management, and stakeholder management.

While Part II explains how program managers navigate programs through the phases of the program life cycle using proven practices and processes, Part III explains how performance measures and tools enable the effective execution of the program management processes. In Chapter 9, we shed light on performance measures that we term as metrics for program management. The metrics help focus attention on business results and serve as the critical success factors in programs and projects. Because organizations are both strategic and tactical in nature, we emphasize the need to carefully engineer and install metrics; meaning that they are compatible (or mutually aligned), balanced, consistently implemented, and tiered. Chapter 10 specifies five strategic program management tools that complement one another in implementing the process of strategic planning. The purpose is to enable program managers to fathom the nature and position of their programs in relation to all current and future programs and to companies' business strategies. In Chapter 11, we present five tactical program management tools for program planning and execution. Their aim is to allow the program manager to effectively plan, monitor, report, and control progress of individual programs.

Part IV focuses on the program manager. Specifically, the program manager's primary roles and responsibilities and the competencies that he or she needs to effectively manage a program. In Chapter 12, we begin by explaining that programs are the investment vehicles used to achieve return on investment and other strategic objectives. We also explain why the business manager needs to delegate business responsibilities to a program manager for managing the return on investment for a particular program. To achieve this responsibility, the program manager has the following two primary roles to fulfill: managing the business aspects of the program and leading the program team to success. In Chapter 13, we present the program management competency model. The model contains the necessary knowledge, skills, and organizational enablers to systematically support the recruiting, staffing, professional development, and career planning of the program manager. The competencies and enablers cover customer and market, business, leadership, and process and project management proficiencies.

Part V focuses on the organizational aspects of program management. In Chapter 14, the process of transitioning to program management is explained and a comprehensive plan for implementing the transition is presented. In each stage of the plan we define boundaries for the transition; design an organizational structure, systems, and culture; test pilot the program management model; plan for implementation and political and cultural changes accompanying the transition; implement the transition; and terminate the transition. The final step involved in transitioning to a program management model is continuous improvement.

The program management office, the topic of Chapter 15, provides the infrastructure for continuous improvement. The program management office is also a pillar for managing and controlling multiple development programs. It is often responsible for managing scarce resources across multiple programs, monitoring program status, and taking actions to make sure that deliverables are accomplished in an acceptable and timely manner.

CHAPTER STRUCTURE

As we explain the elements of program management, each chapter follows a similar structure and flow. Figure P.2 graphically illustrates the common elements found in each chapter.

Introduction: This section provides an overview of the topics covered in the chapter. It also illustrates the role of the topic within the broader context of program management, defines the issues of the topic, and explains goals for the chapter.

Concepts: These are the fundamental elements that the program management discipline is built upon.

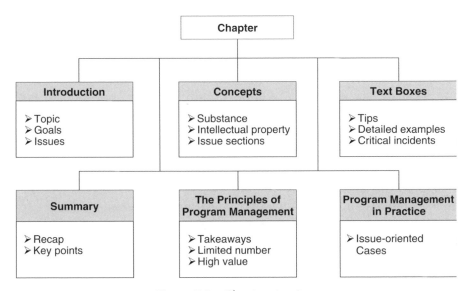

Figure P.2   Chapter structure

Text Boxes: The purpose of the text boxes is to offer readers more depth or breadth of information than the basic text. They appear throughout the book and will provide any of the following:

- Tips
- Detailed examples
- Additional information
- Case incidents
- Background information

Summary: The purpose of this section is to recap the major points of the chapter.

The Principles of Program Management: The principles are guidelines for institutionalizing program management and managing programs. We intentionally limit the number of principles to provide high-value guidelines and concentrated takeaways.

Program Management in Practice: These practices are actually industry examples of limited scope, focusing mainly on one issue but sometimes multiple issues. As a learning device, these examples help the reader see how real companies and people apply the concepts from the chapters. They may be open- or close-ended, and each practice begins with a prologue to help the reader understand the context of the case with respect to the concepts presented.

INDUSTRY EXAMPLES IN APPENDICES

Our goal for the industry examples contained in the appendices is to highlight the application of the program management discipline in real-world examples. In designing this section, we looked at multiple program characteristics such as industry, program size and nature, and program phases. The cases are from different industries, specifically, high-tech manufacturing, software product development, space exploration, and information technology (IT). While one company adapts a hardware product for a new market, another develops a new software product; still, another organization is attempting to identify terrestrial planets in the universe, while the last looks at the design and deployment of an IT solution for the time keeping system in a hospital. The size of the programs vary from 1 million to more than 400 million U.S. dollars.

HOW TO READ AND USE THIS BOOK

Executive managers and program managers are always busy people. Few will have the luxury to read this book cover to cover over a short period of time. Therefore, the book is designed in a modular fashion to address multiple groups of people with different needs, such as the following groups:

- The reader with an interest in a specific chapter—for example, program management metrics—is advised to review the chapter in detail.
- The reader who is interested in a certain part—for example, Part I to understand the strategic role of program management—will be just fine reading only that part.

Whether one opts for out of sequence reading or cover to cover, the major purpose of this book is to help the reader understand and utilize the program management discipline. Further, our goal is to help the following industry professionals:

- *Program Managers:* This book will help program managers improve their knowledge and communicate how program management fits into the overall business model within their companies. For those who are just entering into program management, this book can be used as a practice reference for managing programs. For experienced program managers, this book will help improve their expertise and benchmark their program management systems.

- *Executives and Portfolio Managers:* This book will help executives in charge of developing products, services and infrastructure capabilities to understand business values and competitive advantages of implementing a program management model. In particular, how program management aligns execution output to the business strategies of the company, how program management is an effective practice for managing complexity in products and processes, and how the role of the program manager as the master integrator can be used to better meet customers wants and needs.

- *Project Managers:* With program management as a natural career progression for many project managers, this book will help these individuals understand the differences between program and project management and what core competencies they will need to success-fully function in a program management role.

- *PMP candidates:* For those preparing for the PMP (Project Management Professional) certification exam through PMI, this book is a good reference that provides the detail needed to fully understand the various practices involved in program management.

- *The Textbook Market:* Although this book is designed for the professional market, the lack of quality textbooks for program management as a growing area in colleges makes it an attractive book to fill such a gap. Additionally, this book's numerous case studies and wealth of real-world details are likely to make it appealing as a textbook.

- *Academics and Consultants:* For academics, the book can be used as a research resource and/or a recommended reading for their project management and program management classes. Consultants on assignments to help their clients develop program management processes, train their program professionals, or deploy program management within companies can use the book as a framework for the assignments. In addition, they may find it useful as a text to teach program management principles to students.

Your organizations are in the heat of competitive battles. For them to win, you need to manage programs in a fast, repeatable, and integrated fashion. The way to accomplish this is to equip yourself with the knowledge and fundamental aspects of the program management discipline to develop your products, services, and infrastructure capabilities. This book shows you how.

REFERENCES

1. Kahn, Kenneth B. *PDMA Handbook*, 2nd edition. Hoboken, NJ: John Wiley & Sons, 2005.
2. Evans, James R. and William R. Lindsay. *The Management and Control of Quality*, 6th edition. Mason, OH: Thomson-South-Western, 2005.
3. Jutla, Dawn, Steven Feindal and Peter Bodorik. "KM Infrastructure and Electronics Services with Innovation Diffusion Characteristics for Community Economic Development". *Electronic Journal of Knowledge Management*, Volume 1, Issue 2 (2003), pp. 77-92

Acknowledgements

We would like to thank the many people who have helped in making this book a reality:

Bjoern Bierl, Purasitt Patanakul, Sabin Srivannaboon, and Diane Yates for providing the excellent industry examples used in this book.

Wayne Olmstead and Mary Willner for reviewing our drafts and supplying numerous corrections and insights that made their way into the final manuscript.

Andrea Hayes-Martinelli for her writing contributions to the industry examples and copyediting expertise throughout the book, which greatly contributed to the quality of the final manuscript.

David Churchill, Roger Lundberg, Rick Nardizzi, and Gary Rosen for their invaluable viewpoints, opinions, and insights about program management as it is practiced in industries today.

Tim Rahschulte and Mark Lilly for sharing their collaborative writing model and providing encouragement, motivation, and feedback to Russ and Jim during our early writings.

Our many colleagues and coworkers who have contributed to the concepts presented in this work in many ways.

We are truly blessed to be associated with such a wonderful and supportive community of people!

Advanced Praise for
Program Management for Improved Business Results

"Program Management for Improved Business Results is a great book that provides a refreshing and comprehensive look at how to succeed in enterprise program management. Filled with anecdotes, cases and examples, the authors skillfully pull together the many threads of modern project / program management knowledge and best practice, showing the reader how to apply the tools and techniques, leveraging organizational resources and creating value for the enterprise."

Hans Thamhain, PhD, PMP, Professor, Bentley College, Director of MOT and project management program

---

"The science of Program Management is not about process for process sake. It is about speed, decisiveness and faster time to market. Effective program management delivers profitable growth to the corporation while at the same time building internal alignment and a culture of positive accountability."

David Churchill, Agilent Technologies, Vice-President/General Manager, Network & Digital Solutions Business Unit

---

"With the complexity of challenges in today's environment, having a maniacal focus on program management is absolutely critical for business success. 'Program Management for Improved Business Results' is an insightful book, generous with practical advice, real examples and useful tools to help achieve the necessary focus in managing programs to success."

Diane Bryant, Vice President and General Manager, Server Platforms Group, Intel Corporation

---

"This book explains the fundamentals of program management and how they are applied on programs of varying size and within businesses ranging from small companies to defense industry giants. The authors have collected valuable lessons learned that are useful for the seasoned veteran, for those facing their first opportunity to lead a program, or as a primer for those contemplating a career in program management."

Richard Nardizzi, Senior Program Manager, Lockheed Martin

"Successful organizations use program management to transform great ideas into profitable reality. This book will help program managers and business leaders establish the program management core competencies to improve business value and achieve competitive advantage."

J. Hayden Thomas, Vice President Worldwide Operations, Engenio Storage Group, LSI Logic Corporation.

Part I

It's About the Business

Part I begins by providing clarification of the program management discipline and then illustrating how program management can be implemented as a major element of an organization's business model. This is done in the first four chapters.

In Chapter 1, the unique meaning of program management is identified and described, illuminating its raison d'etre. It explains what program management is and what it is not and compares and contrasts program management with project management, dimension by dimension, clearly distinguishing between the two. Similarly, the program management discipline is set apart from portfolio management and product management. Finally, because many companies often group small- and medium-size projects together to achieve development efficiency, we distinguish the management of multiple independent projects from program management. On the basis of these comparisons and contrasts, we provide a set of identifying characteristics, or pillars, of program management.

To substantiate that program management is really all about business success, in Chapter 2 we present the business case for program management, specifically, what companies get in return for investing in program management—a set of competitive advantages. To illustrate this, we begin with a case of inaction: major business problems and challenges encountered by organizations that do not employ program management. Our intent is to show that instituting program management as a critical business function can solve business problems. For each of these solutions, we show how they create competitive advantages for a company.

To provide more details about this "it's about the business" nature of program management, we chose to elaborate on two of the most critical

1

business problem solutions—aligning programs with business strategy in Chapter 3 and managing increasing complexity through development and delivery of the whole product in Chapter 4.

Chapter 3 argues that an organization can build an integrated management system to increase its success. The primary parts of the system—strategic elements, tactical elements, and program management—are interdependent and must be synchronized to assure the management system is strong as a whole. Acting as an alignment link between the business strategy and execution outputs, program management helps deploy the business strategy, depicting our notion that program management is a major part of a company's business model. An example from information technology in the automobile industry offers a practical application of program management for aligning execution to business strategy.

The purpose of Chapter 4 is to frame the true scope and nature of program management to maximize the value proposition of the company. That value proposition can be visualized through the concept of the whole product, which is a fine example of the increasing complexity of products in the business world. Program management, using the management of highly interdependent projects in a collective and concerted manner, leads the definition, design, production, and delivery of the whole product—a sophisticated way to win the market battle.

Chapter 1

Demystifying Program Management

Even though program management is a widely used and accepted approach to managing product, service, and infrastructure development efforts, the program management practice is not well understood. It has been established as a management discipline for many decades, but uncertainty still exists concerning *what* it is; *why* many companies utilize program management to develop their products, services, and infrastructure capabilities; and *how* it is applied within an organization (see Preface for product, service, and infrastructure capability definitions).

The confusion surrounding program management has a lot to do with its roots, which are in the United States aerospace and defense industries, where it was one of the best kept secrets for decades. Only in the 1980s did program management begin its expansion to the commercial sector, but, even then, the expansion was limited. As people moved from the private sector to the commercial sector, they brought with them program management practices and terminology. Sometimes true program management practices took root in a company; however, other times only the *term* program management took root and was then used to describe *project* management practices. The situation is not much different today than it was in the 1980s. Confusion between program management and other disciplines and processes used to develop products, services, and infrastructure capabilities, such as project management and portfolio management, still exists in many companies, classrooms, and works of literature.

The intention of this first chapter is to remove the mystery surrounding the program management discipline. This is accomplished through the presentation of a concise definition of program management, along with a set of six defining characteristics. We provide a clear distinction between program management and other disciplines and processes that people often confuse—namely project management, portfolio management, and product management. We seek to clarify program management for all members of an organization, from senior executives to individual contributors and help them comprehend the following:

- The definition and set of characteristics of program management
- How program management differs from project, portfolio, and product management
- The link between program management, project management, and business strategy

Understanding these topics is crucial for anyone considering the introduction of the program management discipline within an organization or for anyone needing a better understanding of how to use the program management function within their organization to gain improved business results.

THE "MYSTERY" OF PROGRAM MANAGEMENT

It is safe to say that a fair amount of confusion about program management currently exists in many companies and industries. Take, for example, the following short list of questions we have encountered while discussing program management with practitioners, consultants, academicians, and senior managers:

- What exactly *is* program management?
- Is program management just another name for project management?
- Are program management and portfolio management the same thing because both involve managing multiple projects?
- Isn't a program manager a "super-project manager"?
- Do we need program management if we excel in project management?

Most likely, you have your own set of questions that are—at least in part—motivating you to read this book.

The mystery surrounding program management is perpetuated by a number of factors, including the following:

- Even though many program managers and others familiar with program management have moved to commercial industries from the private sector, the program management knowledge base has remained in the management structure of the original industries that developed it (see box titled, "On Origins of Program Management").

- The term program management has become widely used, or more correctly, misused to define many things such as process improvement, maintenance of business, and continuous and repetitive work activities.

- There is very little literature available that accurately describes the program management discipline. Sometimes the topic of program management is found in modern literature that discusses project management, product development, or infrastructure development, but usually it is only discussed in broad and ambiguous terms.

These factors all contribute to the confusion that exists about program management. In the remainder of this chapter we will provide clarity on the subject, beginning with a concise definition of program management.

On Origins of Program Management

Quality management books commonly state that the Japanese implemented quality and strategic (long-term) programs long before the United States. On the civilian side, it wasn't until the early 1980s that the lack of quality management methods led to difficulty in the United States' ability to compete, which led to the development of quality, project, and program management.

However, the U.S. military argues that they developed and implemented these concepts before the Japanese and that there is evidence of it documented in directives and standards of the U.S. military following the end of World War II. It is believed that these management concepts were used to assist in the formation and organization of the first program office in 1957, then called the Special Project Office (SPO), within the U.S. Department of the Navy. The SPO was established to manage the development of an underwater ballistic missile launch system. Indeed, the structure of the missile launch system program mirrors the program management structures utilized today—a series of interrelated projects (launcher, missile, guidance, installation, navigation, operations, and test) collectively and coherently managed as a program. In the early 1970s, the program management discipline became popular across the U.S. Department of Defense, and the SPO became the first program management office.[1]

On July 1, 1971, the doors of the Defense Management School, later called the Defense Systems Management College (DSMC), opened at Wright-Patterson Air Force Base to admit the first students enrolled in the twenty-week program management course.[2] The original mission of the DSMC was to conduct advanced courses in study of program management and assemble and disseminate information concerning program management. In 1993, the name was again changed to the Defense Acquisition University (DAU) to reflect a new mission and broader scope of academic study and research in program management.[3] Today, thousands of military and military support personnel graduate from DAU annually.

Until the 1980s, the program management discipline and the DSMC that resided within the military and defense industries were well-kept secrets. During this time period, companies that maintained both defense and commercial businesses, such as Boeing, Lockheed, and other aerospace companies, began migrating the program management discipline and management model from their military divisions to their commercial divisions. Program management proved to be very effective in the management of complex product development efforts. Today, the program management discipline and its practices continue to expand throughout many commercial and private industries.

PROGRAM MANAGEMENT DEFINED

A common, universally accepted definition of program management does not exist. If you research the definition through multiple sources, you'll most likely come away with somewhat different definitions—similar in some aspects but different in others. The definition that best describes our practical experiences in managing programs is the following:

"Program management is the coordinated management of interdependent projects over a finite period of time to achieve a set of business goals[4]"

This definition describes the highly effective and well-proven model of program management as a primary business function by which new products, services, or infrastructure capabilities are conceived, developed, and brought to market.

The key words in the definition of program management stated above are *coordinated management, interdependent projects, finite period*, and *business goals*. Bringing a new product to market, or a new infrastructure capability on line, requires the work of many functions—such as hardware engineering, software engineering, mechanical engineering, marketing, manufacturing, and testing. Programs, therefore, are organized into a program core team (PCT) and a set of highly cross-discipline project teams. **Coordinated management** of multiple projects means that the activities of each project team are synchronized through the framework of a common life cycle executed at the program level by the PCT. Steven Wheelwright and Kim Clark properly articulated the need for effective cross-functional management, as follows:[5]

> Outstanding development requires effective action from all of the major functions in the business.... From engineering one needs good designs...from marketing, thoughtful product positioning, solid customer analysis, and well-thought-out product plans; from manufacturing, capable processes.... But there is more than this. Great products and processes are achieved when all of these functional activities fit well together. They not only match in consistency, but they reinforce one another. In short, outstanding development requires integration across the functions.

For program management, cross-functional coordination and integration has to be extended to include *cross-project* coordination and integration. Each program is made up of multiple projects, each of which is most likely cross-functional in nature. Mary Willner, a validation manager for Intel said,

> With one set of desired business results for the program, coordination extends beyond just schedule coordination; it also requires coordination to ensure the stated business objectives are met. If compromises are required (for example, cost, feature, schedule), its resolution is managed as a coordinated effort across the interdependent projects.

As the term implies, **interdependent projects** are those that have a mutual dependence on the output of other projects to achieve success. Commonly, the interdependencies come in the form of deliverables that are the tangible outputs from one project team that become the input to another project team or teams. Program management ensures that the dependencies between the multiple projects are managed in a concerted manner.

A **finite period** means that a program is a temporary undertaking, having a point of beginning and a point of ending. A program is of limited duration, a one time venture that begins with clearly defined business objectives and ends when the objectives are attained. The finite period concept in our definition is very important because some other definitions of programs imply that programs have an ongoing nature.

Accomplishment of the stated **business goals** is the overriding objective of a program and the ultimate responsibility of the program manager.

For example, in product development, a key goal of a program is usually to deliver the product to the market on time. In a competitive environment, time to market is arguably the most closely tracked metric by both the program manager and senior management. We don't dispute that delivery of the right product at the right time is critical, especially because we have had plenty of personal experiences in which that was the primary measure of our success; however, delivery of the product is only the mechanism to realize the true business goals—such as capturing additional market share; increasing profit through sales and gross margin growth; and strengthening brand value through quality, features, and customer support.

> **Program Management Definition**
> **Program management** is the coordinated management of interdependent projects over a finite period of time to achieve a set of business goals.

PROGRAM MANAGEMENT CHARACTERISTICS

Now that the definition of program management has been addressed in detail, we will present the six primary characteristics, or pillars, that help describe the true nature of program management as a unique business function:

- Program management is strategic in nature.
- Program management provides a focal point for ownership and accountability for business results.
- Program management aligns functional objectives to business objectives.
- Program management is cross-project and multi-disciplined.
- Program management enables horizontal collaboration.
- Program management requires a capable business leader—the program manager.

Program Management Is Strategic in Nature

The program management discipline helps to ensure that a program is closely aligned to, and directly supports, the achievement of a business's strategic objectives (see Chapter 3). In effect, it is used to direct the

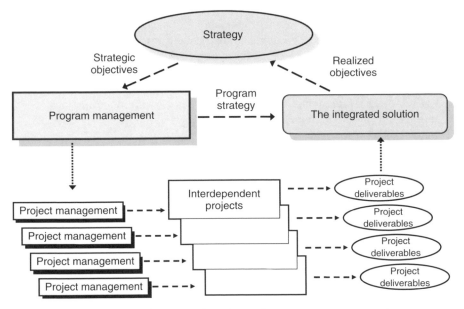

Figure 1.1   The strategic nature of program management.

activities involved in the implementation of strategy (see box titled, "Turning Strategy into Action at Intel"). Figure 1.1 illustrates the link between program management and business strategy.

The program management function links execution to strategy by integrating the deliverables and work flows of multiple interdependent projects to develop and deliver an integrated product, service, or infrastructure capability. This integrated solution becomes the means by which the strategic objectives are achieved.

Turning Strategy into Action at Intel

Intel Corporation identified a strategic objective to converge computing and wireless communication technologies into a single product solution. Legacy solutions involved a microprocessor to handle computing and a separate component, or add-in card, for wireless communication for its personal laptop computer. Intel's strategy to achieve this technology convergence objective involved the development of a new family of microprocessors that combine the two technologies. The market now knows the resulting product as the Centrino™ family of microprocessors.

Intel uses the program management discipline to direct the activities involved in implementing strategy. In the example above, a program manager is responsible for the development and launch of each new Centrino microprocessor into the market. In doing so, he or she is responsible for executing the strategy in the form of a product development and launch, which in turn is the means to achieve the strategic objectives of the business. Therefore, development and delivery of the product becomes the means to achieve the business results intended. Centrino, in fact, was also the means to achieve another Intel strategic objective—to expand beyond the microprocessor and deliver platform solutions to its customers. Centrino was Intel's first commercialized and branded platform product and its first move toward platform program management.

Program Management Provides a Focal Point for Ownership and Accountability for Business Results

In many organizations that do not utilize the program management model, ownership and accountability of the product, service, or infrastructure development effort is shared by the functional managers of the business as the product moves through the development life cycle. Generally, project ownership and accountability passes from research during the concept phase to marketing during the feasibility phase, to engineering during the planning and prototyping phases, to manufacturing during the production readiness phase, and, finally, to marketing for product launch. Passing of the ownership baton can work in a perfectly conceived, planned, and executed project but quickly breaks down when problems begin to surface and personal accountability is required on the part of one or more of the functional managers. With a program management model, there is no debate or subjectivity about who owns and is accountable for the business success or failure of the program; the program manager assumes the full responsibility throughout the development life cycle.

Program Management Aligns Functional Objectives to Business Objectives

Each functional organization within a company normally has a set of objectives to achieve as an organization. But what happens if these functional objectives do not support, or worse yet, are in direct conflict with the strategic business objectives of the company? This dilemma is a difficult problem facing many businesses today and is known as

agency theory.[6] Agency theory occurs when functional managers design objectives that provide the greatest benefit for their organization but consider the strategic objectives of the company secondary.

The program management discipline can be used to reduce the effects of agency theory by aligning functional objectives to corporate or business unit objectives—remembering that products, services, and infrastructure capabilities are the *means* to achieve business objectives. The functional objectives become a crucial part of the overall success of programs, which, in turn, are a crucial part of achieving the overriding business objectives of the firm.

Program Management is Cross-project and Multi-disciplined

Programs, by design, are cross-project in nature, as multiple projects are coherently and collectively managed to achieve the program output. Additionally, the projects that make up a program are normally centered on a single discipline within an organization, such as software development, hardware development, customer support, and manufacturing. To reconcile the cross-project, multi-discipline nature of programs, many organizations employ a matrix structure to span the various functions needed to effectively develop a product, service, or infrastructure capability. The program management discipline is the link that sews the matrix together and enables the cross-project teams to perform cohesively. Organizationally, program management provides the opportunity to manage development efforts across the traditional line structure of an organization, contributing to faster decision making and improved productivity.

Program Management Enables Horizontal Collaboration

A new model has emerged where knowledge work is digitized, disaggregated, distributed across the globe, produced, and reassembled again at its source.[7] Team collaboration can now occur in real time and without regard to geographical boundaries or distances. Companies that are thriving in this new business model are the ones that are successfully integrating horizontal collaboration of work. A key learning that has emerged is that program management is an effective business model for managing the horizontal collaboration, and for integrating the output of specialized knowledge workers into total solutions. The program management discipline enables this horizontal collaboration.

Program Management Requires a Capable Business Leader—The Program Manager

Managing a program is a complex undertaking. It requires much more than planning, tracking, and controlling the work of a cross-functional team. The program manager serves as the catalyst for converting ideas into products, services, and infrastructure capabilities that, when delivered or implemented, become the means to achieve a set of business objectives. The program manager is someone who thinks and acts like a general manager (GM), or a CEO of a small company. In doing so, the program manager has two primary roles, as follows: to manage the business on his or her program and to lead a set of highly interdependent project teams throughout all phases of the program life cycle (PLC) (see Chapter 12). Companies that use the program management discipline as intended understand that these roles require a unique set of core competencies, skills, and personality traits. In Chapter 13 we describe the skills needed within the four program management core competency areas, which include business and financial, market and customer, leadership, and process and project management acumen.

DIFFERENTIATING BETWEEN PROGRAM AND PROJECT MANAGEMENT

Two distinct trends have played a key role in the emerging need to succinctly distinguish between program management and project management. First, there is a recognized need within business management to improve the link between business strategy and operational execution. Second, there is an increasing trend toward larger and more complex product, service, and infrastructure development efforts. These trends are fully comprehended in the program management model and give rise to its increased usage as a critical business function.

Program management and project management are related but distinctly different disciplines. It is important for everyone within an organization to understand the distinction between the two to link project output to business strategy and integrate the efforts of multiple project teams to achieve a common set of business goals.

Summary of Program and Project Management Differentiation

Table 1.1 provides a high-level summary of the *important* differences between program and project management. The primary differentiator is the core area of focus. Program management is strategic in nature and focused on the business success of the program, while project management is tactical in nature and focused on the successful execution and delivery of one subsystem, or element, of the integrated solution. All other factors in the summary (alignment, responsibility, management dimension, risk management, work effort, processes, skills, and capabilities) are subfactors of the primary differentiator.

We refer to project management as tactical in nature based on the Project Management Body of Knowledge (PMBOK) and the dominant industrial practices. PMBOK is a very respectable standard—de jure U.S. national standard and de facto global standard. Per PMBOK, project management is about management of a single, individual project, whose primary focus is accomplishment of the triple-constraint goals (time, cost, and quality).[8] We use this view as a benchmark to compare program and project management.

Alignment of Objectives—Strategic Versus Tactical

Program management is *strategic* in nature and focused on *business* success; however, project management is *tactical* in nature and focused on *execution* success. More importantly, the program manager must ensure that, from concept to launch, the program remains in alignment with, and in support of, the strategic objectives set forth by senior management. This includes alignment with the organization's strategic plan, its product portfolio and road map, and the business-related objectives such as financials, market penetration, and technology advancement. The project manager, in turn, is responsible for ensuring the work and resulting deliverables of the project team are in alignment with and in support of the program objectives.

Table 1.1   Program and project management differentiation summary

Differentiating Factor	Program Management	Project Management
Strategic vs. Tactical	Strategic in nature, focused on business success	Tactical in nature, focused on execution success
Alignment of Objectives	Alignment of execution to business strategy	Alignment of deliverables to triple constraints (time, cost, and quality)
Scope of Responsibility	Successful delivery of the right product, service, or infrastructure capability at the right time	Successful delivery of project deliverable(s) per triple constraints
Vertical vs. Horizontal Responsibility	Manages horizontally across the functional projects involved in the program	Manages vertically within a single project
Work Effort	Assures the cross-project work effort remains feasible from a business standpoint	Assures work effort generates desired deliverables on time, within budget, and at required performance levels
Management of Risk	Concerned with cross-project risk affecting the probability of program and business success	Concerned with single-project risk affecting the probability of project and technical success
Life Cycle Involvement	Involved in all phases of the development life cycle, from definition to end of life	Primarily involved in the planning and implementation phases of the development life cycle
Process Orientation	Ensures consistent use of common processes by all project teams	Ensures effective and efficient implementation of processes on a single project team
Skills and Capabilities	Breadth of business, leadership, customer/market, and project management skills	In-depth project management and functional specific technical skills

Scope of Responsibility

On a broader scale, the program manager must assume the responsibility for the attainment of the combined objectives from each of the functional project teams used to deliver the product, service, or infrastructure capability. This may include marketing, hardware development, software development, mechanical development, manufacturing, validation, testing, and customer support.

In a nutshell, the program manager's job is the successful delivery of the right product, service, or infrastructure capability at the right time. This requires management of the interdependent issues across the multitude of projects. For example, if the hardware development project team encounters a quality issue that will impact the timing of its deliverable to the manufacturing project team, the program manager must determine if it's better to delay the deliverable (and the work of the manufacturing project team) or reduce the quality target. This is a cross-project issue to be solved at the program level. In contrast, a project manager is responsible for the scope of work within his or her project only.

Vertical Versus Horizontal Responsibility

Figure 1.2 demonstrates the concept of vertical project management and horizontal program management; both program and project managers are responsible for the effort and deliverables but in different dimensions. The project manager directly manages the effort and work flow

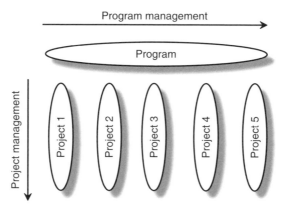

Figure 1.2   Horizontal and vertical management of a program.

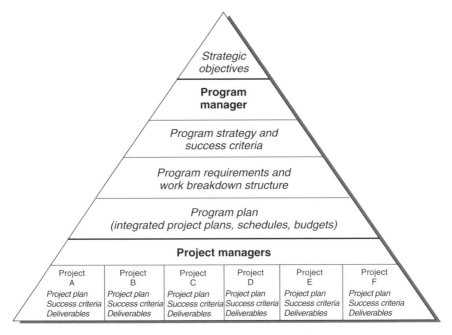

Figure 1.3   The program hierarchy.

within his or her project team. This is called **vertical responsibility**. Project managers are directly held accountable for the plans, schedules, objectives, deliverables, risks, and quality levels as they pertain to their respective projects. In contrast, the program manager manages horizontally across the functional projects involved on the program and is held accountable for the integrated plans, schedules, deliverables, risks, and overall quality output of the multiple projects.

Work Effort

The program manager assures that the cross-project work effort remains in alignment with the strategic objectives and is feasible from a business standpoint by focusing *across* the functional groups to ensure that the deliverables, timing, and other interdependencies between the groups are met in accordance with the overall program plan and schedule (Figure 1.3). By contrast, the project manager assures that work effort generates deliverables on time, within budget, and at required performance levels.

Risk Management

Both the program and project managers are responsible for identifying and managing risk on a development program but do so differently. Project risk management involves identifying and managing risks that may affect the technical success for a single functional project. Program risk management involves identifying and managing cross-project risks that may affect the overall business success of the development program.[9] (See Figure 5.2 in Chapter 5 for an illustration of this concept.)

Life Cycle Involvement

Life cycle in this context pertains to all of the phases a product, service, or infrastructure capability will transition through from the time of its inception to its eventual phase out. By virtue of the program management model, the program manager is involved in all phases of the life cycle. This includes the definition, planning, implementation, launch, and sustain phases. Project managers are typically involved only in the planning, implementation, and, occasionally, the initial launch or go-live phases of the life cycle.

Process Orientation

The distinction between program and project management comes in how the processes and procedures are established for and executed on a program. The program manager is responsible for ensuring that company processes and procedures are established for the program and that they are consistently used by all project teams. The project manager is responsible for effective and efficient implementation of the processes and procedures established by the company, as well as those established by the managers of functional organizations.

Skills and Capabilities

The breadth and depth of skills and capabilities is also a differentiating element between program and project management. Project managers must have in- depth knowledge of the domain they represent and experience in project management. In comparison, program managers must have a working knowledge of the intricacies of each of the functional projects involved with the program, such as marketing or software development. Program management core competencies must also include business, leadership, customer/market, and project management skills to effectively lead a development effort (see Chapter 13).

DIFFERENTIATION BETWEEN PROGRAM AND PORTFOLIO MANAGEMENT

At times, confusion also exists between program management and portfolio management. One of the causes of this confusion may be that they are both commonly broadly defined as the management of multiple projects. But this is where the similarity ends. In the following section, we provide a brief characterization of portfolio management for readers who are not familiar with the process. We then describe the key distinctions between portfolio management and program management.

Characterizing Portfolio Management

The senior management team of an organization utilizes the portfolio management process to synthesize current and future collective intelligence of the organization. They use it to select, prioritize, fund, and resource the portfolio of products, services, or infrastructure opportunities that will best achieve the attainment of the strategic objectives. In synthesizing the intelligence of the organization, various key factors about the business and business environment must be analyzed to obtain the right mix and number of opportunities. Such factors may include the following:[10]

- Company strategic objectives
- Customer wants, needs, and usage requirements
- Competitive intelligence
- Current and future technology capability of the enterprise
- Risks and potential rewards
- Resources and other assets available to plan and implement the portfolio of products, services, or infrastructure capabilities

The portfolio management process, by necessity, crosses all company disciplines that are pertinent to the successful development of the portfolio for products, services, or infrastructure capabilities. The objective of the portfolio effort is to ensure that the company is working on the opportunities that offer the highest probability for attractive financial and strategic returns at the lowest possible risk. Opportunities are ranked and prioritized based upon a set of criteria that represents *value* to the organization. Resources are then allocated to the highest value and most strategically significant products, services, or infrastructure opportunities. Low-value opportunities must be cut, returned for redefinition, or put on hold until adequate resources are available.

Summary of Program and Portfolio Management Differentiation

Table 1.2 provides a high-level summary of the important differences between program management and portfolio management. The primary differentiator is that portfolio management is a decision-making *process,* while program management is a key management *function* within an organization. All other factors in the summary (determining and obtaining value, management of risk, and management of resources) are subfactors of the primary differentiator.

Process Versus Function

Senior management of an enterprise utilizes the portfolio management *process* to evaluate, prioritize, select, and resource new products, services or infrastructure ideas that will best contribute to the attainment of the strategic objectives of the business. The program management *function* is used to determine the business and execution feasibility of a selected idea; the idea then turns into an actionable plan that is successfully executed and delivered as a tangible product, service, or infrastructure capability.

Determining and Attaining Value

The heart of the portfolio management process is the ability of the senior management team to determine the business value of a product, service, or infrastructure opportunity. Therefore, the portfolio management process identifies the critical factors that determine opportunity value.[11] Common factors may include the following:

- Alignment to strategic objectives
- Technology and commercial risk

Table 1.2   Program and portfolio management differentiation summary.

Differentiating Factor	*Program Management*	*Portfolio Management*
Process v. Function	A management *function* utilized to determine the business and execution feasibility of a selected idea. The idea then turns into an actionable plan that is successfully executed and delivered to the customer	A *process* utilized to evaluate, prioritize, select, and resource new ideas that best contribute to the attainment of the strategic objectives of an organization
Determining and Obtaining Value	Focused on ensuring that the business value is attained for a single opportunity throughout the development and market introduction process	Focused on determination of the business value of all existing opportunities of the organization
Risk Management	Management of risk across all disciplines involved in the development of a single product, service, or infrastructure capability	Determination of the business and technical risk of each opportunity concept, balancing risk and return for the aggregate portfolio of opportunities
Resource Management	Staffing the PCT, ensuring the project teams are adequately staffed throughout the development life cycle	Aligning resources to opportunities that provide the greatest strategic value to a business

- Financial reward or return
- Estimated market segment share
- Technology advancement

Once the business value is determined for an opportunity within the portfolio and it is selected for funding and resource allocation by the senior management team, the opportunity (in the form of a product, service, or infrastructure concept) is assigned to the program management function within the enterprise. The program managers are then responsible for turning each of the portfolio ideas into a tangible product, service, or infrastructure capability and *delivering* the value.

Managing Risk

The senior management team manages portfolio risk from both macro and micro perspectives. Macro-level risk management of a portfolio involves determining the overall risk level of the aggregate opportunities within the portfolio and then determining the right balance based upon the risk tolerance of the organization. A key element of the portfolio management process is balancing the portfolio risk against the potential reward.

Figure 1.4 illustrates an example of portfolio risk versus reward. Risk is assessed as high, moderate, or low, and financial value is assessed in terms of return on investment (ROI) of the program. The size of the bubbles represents the relative development budget of the investment for each capability. The senior management team must also balance

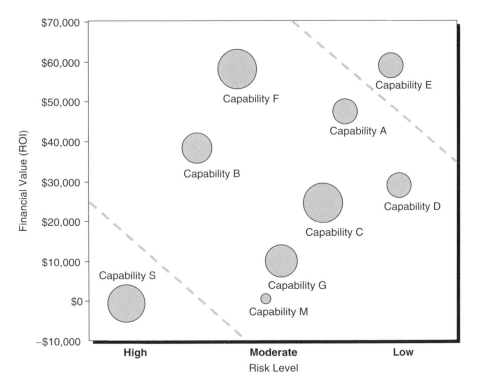

Figure 1.4   Portfolio risk versus reward.

the potential risk exposure for each opportunity against the determined business value on a micro level.

Once a product, service, or infrastructure opportunity is approved and funded, the program manager becomes responsible for the risk/reward ratio of that single opportunity.

Resource Management

Businesses typically have more ideas than human and non-human resources available to carry them out. As a result, resources become overcommitted and weighed down by an overwhelming list of opportunities to pursue. Portfolio resource management involves aligning resource demand to capacity and assigning resources to products, services, or infrastructure opportunities that provide the greatest value to a business. The end result of a well-executed portfolio management process is a balance between high-value opportunities and the number of available resources to execute the opportunities.

Efficient and effective resource management is needed for the development of an opportunity throughout its life cycle. This becomes the responsibility of the program manager and the functional managers of the organization. For the value of an opportunity to be realized, the program designed to deliver it must be adequately staffed.

DIFFERENTIATING BETWEEN PROGRAM MANAGEMENT AND PRODUCT MANAGEMENT

The terms *program management* and *product management* are sometimes used interchangeably, which leads to confusion between the roles of the program manager and product manager. Indeed, they are both responsible for the success of a program; the difference being that the product manager ensures the product remains viable from a market perspective while the program manager ensures the entire program is successful in achieving the business objectives.

Summary of Program and Product Management Differentiation

Table 1.3 illustrates that program and product management are symbiotic but are two distinctly different disciplines. The primary differentiator is

Table 1.3   Program and product management differentiation summary.

Differentiating Factor	Program Management	Product Management
Business v. Product Champion	Business champion responsible for achieving specific business objectives	Product champion responsible for identifying the customer needs that the product is meant to satisfy
Organizational Affiliation	A program manager is part of the general business or program management office organization	A product manager is a member of the marketing organization
Area of Focus	Product definition and feasibility, program planning, execution, launch and post-launch support	Market and customer research, new product planning, new product launch, and post-launch support
Specialist v. Generalist	Cross-discipline generalist providing strategic and business guidance to the development team	Marketing specialist providing marketing guidance and direction to the development team

that a product manager is the product champion, while the program manager is the business champion on a development program. All other factors in the summary (organizational affiliation, primary areas of focus, and specialist versus generalist) are subfactors of the primary differentiator.

Organizational Affiliation

The primary role of the product manager, who is normally part of the marketing organization, is to size, segment, and target key market opportunities. He or she does this by utilizing his or her unique understanding of the customer to identify unfulfilled needs and new market opportunities.[12] The program manager, by contrast, usually reports to a GM or the program management office and often is viewed as the bridge between marketing and development. The program manager utilizes the information generated by the product manager to ensure the right product is developed and delivered to the markets that the business services. In most instances that we have observed, the product manager is a member of the PCT, which the program manager leads.

Primary Areas of Focus

The main areas of focus for the product manager include market and customer research, new product planning, new product launch, and post-launch product management.[13] Probably the most critical direct interaction between the program manager and the product manager is during the product definition and product proposal phase. During this phase, the product and program manager, along with engineering, define a viable differential value proposition for the product. It is the responsibility of the program manager to ensure that this proposed value proposition has a high probability of achieving the strategic objectives for the business. The product manager, in turn, is responsible for ensuring the value proposition meets customer needs. When the product manager and the rest of the team are confident in the value proposition, the program manager will proceed with submitting the product proposal to senior management for approval.

Specialist Versus Generalist

The product manager is the marketing specialist and product champion with the primary role of identifying the customer needs that the product is meant to satisfy. He or she provides marketing guidance and direction to the development team,[14] typically in the form of a market requirements document, which is the foundation for the creation of the product requirements document by the engineering specialists. The product manager may serve as the functional project manager who represents the marketing function on a program.

By contrast, the program manager is a generalist whose key role is the business champion for a product development program. The program manager is ultimately responsible for the overall success of the new product in the market and for achieving the specific business objectives pertaining to the program. He or she ensures success through the efforts of the product manager and other members of the PCT.

DIFFERENTIATING BETWEEN PROGRAM MANAGEMENT AND MANAGEMENT OF MULTIPLE PROJECTS

In this section, we compare and contrast program management as the management of *interdependent* projects with the management of multiple *independent* projects. Interdependent projects within a program are aimed toward the achievement of a common business objective, where the successful completion of deliverables from one project is needed for the success of the other projects—if one project fails, they all fail. Program management ensures the dependencies between the multiple projects are managed concertedly.

Multiple independent projects, however, do not share a common objective; rather, they are each aimed at achievement of separate business objectives. Each project is stand-alone and can be managed by a common project manager with no apparent impact on one another—the success or failure of one project does not affect the other projects.

To illustrate the difference between managing multiple projects and managing a program, let's look at two training course development examples. The first scenario is the development of multiple "how to" remodeling courses by a home improvement retail center. The second is the development of a systems administration curriculum by a for-profit university. In both scenarios, the work performed to develop the courses is similar and can be accomplished through good project management practices. The difference is in *how* the collective set of courses is managed.

The primary objective for the how-to remodeling courses is to increase merchandise sales by strengthening the homeowner's competency and confidence in this area. The courses to be developed are the following:

- Plumbing basics
- Electricity basics
- Ceramic floor installation
- Wallpaper hanging

In this case, the courses are entirely independent of one another because each course is focused on a unique area, and there are no interdependencies between them. Therefore, each course development can be set up as a stand-alone project that independently contributes to the business objectives. Figure 1.5 illustrates the independent nature of these four projects. Development of the four courses can be managed separately with different project managers.

The primary business objective for the systems administration curriculum is to increase enrollment revenue by offering compelling, certified, and competitive courses for customers to obtain an associate's degree

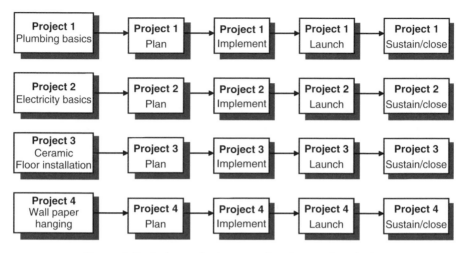

Figure 1.5   Independent course development projects.

in systems administration. The program consists of multiple courses, including the following:

- Introduction to networks
- Network administration
- Database design and development
- Visual basic programming

In this case, the courses are highly interdependent. The curriculum cannot achieve certification without the successful development of all courses, and the program cannot generate enrollment revenue without certification.

Because of the interdependent nature, each course development effort should be managed as a single project within a larger curriculum development program. The program should be managed by a program manager who is responsible for achievement of the business objective—increased enrollment revenue through deployment of the systems administration curriculum.

Managing multiple interdependent projects requires the integration of project planning, execution, and sustaining activities at the program level, as shown in Figure 1.6.

As illustrated, if even one project team does not deliver their respective course, the cross-project integration that occurs at the program level will

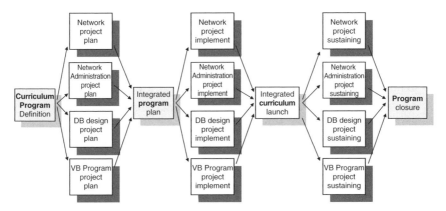

Figure 1.6   Example of a curriculum program with interdependent projects.

be unsuccessful. When this occurs, the program as a whole will fail and the business objective driving the need for the program will be unmet.

THE "MYSTERY" OF PROGRAM MANAGEMENT REVISITED

This chapter begins by stating that there is a fair amount of confusion about program management and also provides a short list of questions from practitioners, consultants, academicians, and senior managers. We revisit these questions below and include answers based on the context of this chapter.

What exactly Is Program Management?

Program management is the coordinated management of interdependent projects over a finite period of time to achieve a set of business goals.

Through the coordinated management of the interdependent projects and focus on achieving business results, program management provides the following business value:

- Ensuring that project execution is closely aligned to and supports the achievement of a business's strategic objectives
- Providing better development efficiency and decision making within the program team to achieve rapid time to money
- Improving customer satisfaction with well-defined communication channels and communication messaging with customers

Is Program Management Just Another Name for Project Management?

Program management is *not* just another name for project management. The primary differentiator between program and project management is the core area of focus, as follows: Program management is strategic in nature and focused on the business success of the program, while project management is tactical in nature and focused on the successful execution of tasks and deliverables within the classic triple constraints of time, cost, and quality.

Are Program Management and Portfolio Management The Same Thing because Both Involve Managing Multiple Projects?

Program management and portfolio management are *not* the same. They differ in that portfolio management is a decision-making *process,* while program management is a key management *function* within an organization. A business will have a portfolio of programs that will be selected, prioritized, and resourced by senior management. The program managers and their teams then deliver the intended business value for each program within the portfolio through the definition, planning, implementation, and launch of products, services, or infrastructure capabilities.

Isn't a Program Manager a "Super-Project Manager"?

A program manager is *not* a "super project manager." A program manager is a business leader who needs business, leadership, customer/market, and process and project management skills (see Chapter 13). Program management offers a single point of accountability for the business results of a development program. A project manager is tactically focused and needs in-depth project management and functionally specific technical skills.

Do We Need Program Management If We Excel in Project Management?

The answer to this question is business specific and depends upon how project management is used within the business. We offer the following guidance:

- If your development efforts are tactical and focused on execution, then you *do not* need program management.
- If your development efforts are low in complexity, with few interdependencies, then project management methods will suffice.
- If your development efforts are tactical and you need to focus more on strategic and business success, then program management is a viable business model to pursue.
- If your development efforts are growing in complexity and you are struggling with the management of many cross-team interdependencies, then you *do* need program management.

When considering these options, keep in mind that the complexity of the company's products, services, and infrastructure capabilities are important factors in the decision to use program management. As we cover in Chapter 4, industry benchmarking, research, and personal experience shows us that a project management-only approach is only sustainable to fairly low levels of project complexity. In cases of higher complexity, program management brings value by integrating the work output of multiple highly interdependent projects to deliver products, services, or infrastructure solutions that contribute to the bottom line of a company.

SUMMARY

The intent of Chapter 1 is to identify and remove the mystery surrounding program management. The chapter begins with an explanation of why program management gets confused with other disciplines, with a concise definition of "the coordinated management of interdependent projects over a finite period of time to achieve a set of business goals." To shed more light on what program management is and is not, we describe the distinguishing factors between program management and project management, portfolio management, management of multiple projects, and product management. This comparison is focused on the six primary characteristics or pillars.

Understanding and deploying the six pillars of program management will help program managers to improve business results through the achievement of a business's strategic objectives, positive contribution to the bottom line, and increased customer satisfaction.

The Principles of Program Management

▼ Similarly aligned projects are linked into programs that are tied to the business strategy of the organization to realize the power of program management

▼ Program management is a focal point for ownership and accountability of business results

▼ Program management is strategic in nature and focused on business success; project management is tactical in nature and focused on execution success

▼ The program manager manages horizontally across the functional projects involved with the program, while the project manager manages vertically within a single functional project.

▼ Program management aligns functional objectives to business objectives through the development process—products, services, and infrastructure capabilities are the *means* to achieve business objectives

▼ Program management is cross-project and multidisciplined

▼ Program management requires a capable business leader whose core skills go beyond technical aspects and include business, leadership, and program and project management process competencies to effectively lead new development programs

▼ Portfolio management is a planning and decision-making process to select the optimum portfolio value, while program management is the function that ensures the portfolio value is attained

▼ A product manager is a marketing specialist whose primary role is product champion, and a program manager is a generalist whose primary role is business champion and master integrator

REFERENCES

1. Ashie, Ibrahim A., Department of the Navy Strategic Systems Programs Office, *http://www.dau.mil/pubs/pm/pmpdf94/ashie.pdf*
2. Defense Acquisition University Press. *U.S. Department of Defense Extension to a Guide to the Project Management Body of Knowledge*. Fort Belvoir, VA: Defense Acquisition University Press, June 2003.
3. Summers, Wilson. "Before DSMC, There Was DWSMC." *The Program Manager* (January-February 2000).
4. Martinelli, Russ and Jim Waddell. "Demystifying Program Management: Linking Business Strategy to Product Development." *PDMA Visions Magazine* (January 2004): pp. 20–23.

5. Wheelwright, Stephen C. and Kim B. Clark. *Revolutionizing Product Development: Quantum Leaps in Speed, Efficiency, and Quality*. New York, NY: Free Press Publishing, 1992.

6. Pearce II, John A. and Richard B. Robinson. *Strategic Management: Formulation, Implementation, and Control*. New York, NY: McGraw-Hill Publishing, 2000.

7. Friedman, Thomas L. *The World is Flat*. New York, NY: Farrar, Straus and Giroux Publishing, 2006.

8. Project Management Institute. *A Guide to Project Management Body of Knowledge*. Drexell Hill, PA: Project Management Institute,2004.

9. Martinelli, Russ and Jim Waddell. "Managing Program Risk." *Project Management World Today* (September-October 2004).

10. Cooper, Robert G., Scott J. Edget, and Elko J. Kleinschmidt. *Portfolio Management for New Products*. Cambridge, MA: Perseus Publishing, 2001.

11. Cooper, Robert G., Scott J. Edget, and Elko J. Kleinschmidt. *Portfolio Management for New Products*. Cambridge, MA: Perseus Publishing, 2001.

12. Moore, Geoffrey A. *Crossing the Chasm*. New York, NY: HarperCollins Publishing, 1991.

13. Life Cycle Strategies. "Three Fundamentals for Effective Product Management." *Life Cycle Strategies Website (www.lifecycle.com / pm /)*, 1999.

14. Haines, Steven. "Help, I'm a new product manager!." *Sequent Learning Networks*, February 2003.

Chapter 2

The Business Case for Program Management

In this chapter we present the case for program management as a primary business function within firms that develop products, services, and infrastructure capabilities. In doing so, we evaluate program management on the basis of how it adds value by creating competitive advantages for a company. In creating these advantages, program management itself becomes a source of competitive advantage that a company can use to outplay rivals.

Eight common critical business problems that plague companies today are presented first. Then, explanations about how the program management discipline has been successfully implemented to overcome each of the problems are provided. These explanations encompass the eight elements of the business case for program management. Finally, the value proposition of program management is summarized in the form of advantages, concluding that it is a relevant way to build a great company.

This chapter seeks to help executives in charge of program management, and practicing program and project managers understand the following:

- Why a project management-only approach creates business problems when applied in some critical business situations
- How program management helps as an integrating mechanism for business-model deployment by aligning execution activities with business strategy
- How program management helps tame the fuzzy front end of development by aligning market and technology research

- How program management can be used to manage increasing complexity and accelerated time-to-money goals, mitigate business risk, and manage the business's return on development investment
- How program management helps a company create a competitive advantage to outplay business rivals

NEW DAY, NEW PROBLEMS

The message of "it's about the business" is repeated throughout this book. Program management is a primary business function that coordinates and aligns the execution work of multiple functions in multiple interdependent projects toward the achievement of a firm's strategic business objectives.

Over the past few decades, companies have invested an enormous amount of time and resources on improving their project management capabilities to gain products, services, or infrastructure development efficiency. Most senior managers within these companies agree that the investment in project management methods, tools, and practices has had a moderate to significant effect on improvement in the planning and execution of development projects. However, many managers and executives whom we have spoken with expressed several significant problems with using a project management-only approach to develop new capabilities in their organizations. The following list describes their concerns:

Lack of business integration: A company's business model deals with whether the revenue, cost, and profit target economics of its strategy demonstrate the viability of the enterprise as a whole.[1] The crucial point then is the realization of strategy. Since we know that only 10 percent of all concocted strategies are realized and 90 percent fail, it is clear that the real difficulty is with the implementation of the strategy, not the development of the strategy.

Implementation of strategy involves many business functions and processes that need to be coordinated and integrated into a synchronized business action that will hit the desired revenue, cost, and profit target. Project management was engineered to be a coordinating and integrating mechanism. The trouble is that project management has become tactically focused on the triple constraints—time, cost, and quality—and many times fails to serve as the business integrator focused on the implementation of strategy.

Misalignment between strategy and execution: In a large number of organizations, a misalignment exists between companies' strategic objectives and the corresponding abilities to effectively identify, manage, and execute the projects targeted to deliver the objectives.[2] There is often a chasm between business objectives and project management activities. Projects may be efficient and "on target" with respect to time, cost, and quality but fail to achieve anticipated business results such as increased market share or increased worker productivity.

Misalignment of market and technology research: A related problem to the issue addressed above is the poor track record of companies successfully and consistently integrating market and technology research to produce compelling concepts that are focused toward the attainment of desired business objectives. The early definition phase of the development life cycle is commonly called the fuzzy front end and is a complex and difficult stage for any company. A firm's ability to successfully identify the right product, service, or infrastructure capability to develop is critical to the future lifeblood of an organization. To frame the problem succinctly, if a company is developing the wrong product, service, or infrastructure solution, it doesn't really matter if the project management processes are efficient and effective.

Poorly managed ROI: With multiple development efforts in various stages at any one time, GMs of business units are unable to manage the investments in all development budgets and resources. This responsibility falls on the project managers. However, most project managers are challenged to adequately manage the business aspects of products, services, or infrastructure development efforts because the bulk of their training and experience focuses on operational execution of projects.

Increasing complexity: Due to the amount of complexity required to meet customer demands for performance, features, and customization, development efforts are often beyond the scope of a single project.[3] The days of stable, slow to evolve designs are a thing of the past. Today, multiple projects with tightly linked activities and deliverables are required to deal with the complexity of solutions demand from customers. This requires the simultaneous management of multiple, highly interdependent projects, which a project management-only approach is challenged to provide.

Slow time to money: Acknowledgement by firms that time to money can be a significant competitive advantage has made historical methods of product, service, or infrastructure development obsolete.[4] The project hand off or waterfall approach in which project management ownership is transferred (or sometimes thrown over the wall) from one functional project team to the next is too slow to gain time to money competitive advantage. A purely concurrent development method in which functional development occurs in parallel is also inefficient because of the high potential for significant rework late in the project. Eventually, the efforts have to synchronize to reach an integrated solution. This synchronization many times occurs just before or in the early part of final validation and testing. When the interfaces between the concurrent development efforts are not properly defined, communicated, or managed, rework is required and time to money advantage is lost.

Unmitigated business risk: Business risk encompasses all unknown and uncertain events that may prevent an enterprise from executing its strategies, meeting its performance goals, and achieving its business objectives. It also includes anything that may negatively impact the well-being of the enterprise itself. Project-oriented risk-management practices, methods, and tools are effective in managing technical risks and providing valuable information for scope, budgets, and schedule trade-off decisions. However, they are not very effective in identifying and managing the greatest threats to the business success of a company. Current project management curriculums often fall short on providing the necessary business skills that project managers need to fully comprehend the link between strategic business objectives and the output of projects. As a result, many project managers do not possess the breadth of knowledge and experience necessary to identify and manage risk across multiple interdependent projects which can threaten the success of the enterprise.

Ineffective global collaboration: The world is flat. This is how Thomas L. Friedman, in his highly acclaimed book titled *The World is Flat*, describes the phenomenon that began in the 1990s and continues today whereby knowledge work can be digitized, disaggregated, distributed, produced, and reassembled across the globe.[5] The flattening of the world has enabled the remote development of products, services, and infrastructure capabilities, allowing people in countries like India, China, and Russia to participate in knowledge work with partners across the world. This has created a new business model in which horizontal collaboration is required across the globe to remain competitive. However, many companies have historically operated under a silo, or stove-piped, business model, in which horizontal collaboration across the operating functions that make up the stove pipes is foreign, let alone collaboration across the globe. One by one these companies are being forced to adopt a horizontal, cross-functional business model. Most are struggling with this because of a lack of knowledge and required capabilities.

The problems outlined in the last section are *business problems*. They are *not* project execution problems, and the project management-only approach has shown little capacity to solve them. Organizations that have instituted a program management model to develop products, services, or infrastructure capabilities have done so to address some or all of the problems described above. Therefore, program management is meant to be viewed and deployed as a *primary business function* within the organization.

BUILDING THE CASE FOR PROGRAM MANAGEMENT

How do organizations that have deployed a program management model address the business problems described in the previous section? We focus on the answer to this question in the following section by presenting the program management value proposition for each of the business problems and, in the process, building the business case for implementing the program management model within an organization.

Integrating the Business Model Elements

A company's business model deals with whether the revenue, cost, and profit (or efficiency, cost, and profit) economics of its strategy demonstrates the viability of the enterprise as a whole. To elaborate, a business model is the mechanism by which a company generates revenue and profits and serves its customers through the deployment of both strategy and implementation. In particular, a business model includes the following elements.[6]

- Selecting customers
- Acquiring and keeping customers

- Creating value for its customers
- Presenting to the market through promotion and distribution strategies
- Defining and differentiating its products, services, or infrastructure offerings
- Defining and managing the tasks to be performed
- Utilizing its resources effectively
- Capturing profit

What is the role of program management in the deployment of the business model? Program management is the mechanism by which the work of the various operating functions within a company is integrated to create an effective business model. For example, consider the business functions of marketing, engineering, manufacturing, and finance. Each function has its own language and jargon. For instance, marketing language talks about the four Ps (product, price, place, promotion), finance discusses discounted cash flow, engineering discusses baud rate, and manufacturing is interested in yield. To say that experts from different functions often don't understand one another is an understatement (horror stories about inter-functional misunderstandings abound). A program manager, however, can understand and speak all functional languages.

Program management integrates a company's functions through the use of a PLC that coordinates the functional plans into one synchronized program plan. It integrates all actions into a synchronized program that smoothes out the interfaces between project teams.

Here is what David Churchill, vice president/general manager of Network & Digital Solutions Business Unit for Agilent Technologies, said about the role of program management:

> Most firms have the right product ideas, technical talent, and marketing capability to support their business strategies. Organizations many times have difficulty performing to expectations when they cannot turn their strategies into successful execution. Program management is strategic to the firm because it provides the ability to convert business plans into actions that will achieve the intended objectives; it helps bridge the gap between strategy and execution.

Hence, we see program management as a key part of a company's business model. When it is properly conceived and implemented, it helps deploy the business model and execute business strategies to give a business competitive advantage, as illustrated in Figure 2.1.

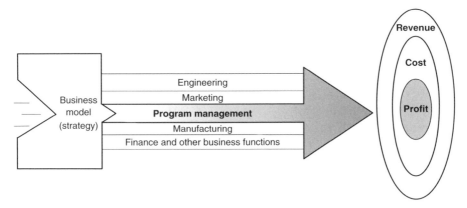

Figure 2.1   Program management delivers business results.

Aligning Business Strategy and Execution

The second aspect of the business case for program management is the alignment of business strategy and operational execution. Many organizations engage in yearly strategic planning activities that focus on the identification of long-range business objectives, as well as high-level plans on how to achieve them. Good strategic management practices identify *what* an organization wants to achieve (strategic objectives) and *how* they will be achieved (strategies) over the time horizon, which is typically three to five years. Strategic objectives may include revenue growth, market share increase, decreased cost of goods sold, or technology advancement. For product development companies, strategy consists primarily of a collection of product ideas that, when turned into tangible products, contribute to the achievement of the primary business objectives. For service-oriented companies, strategy consists of a set of services that collectively contribute to the achievement of the business objectives. And for infrastructure development organizations, strategy consists of a suite of infrastructure capabilities that collectively contribute to the attainment of the strategic business objectives.

As we explained in Chapter 1, program management is strategic in nature, but project management is meant to be applied more tactically. Practitioners and researchers have observed that a project management-only approach to development often leads to a misalignment between a company's strategic management process and its output. Plainly stated, the products, services, or infrastructure capabilities that the company delivers may not contribute to the attainment of the strategic objectives that it has targeted. Chapter 3 discusses this identified gap between strategic elements of the development process and project execution.

We view program management as the organizational glue that can translate strategic business objectives into actionable plans that, if managed properly, can be used to achieve the desired business results. The gap between strategic elements and project execution is then effectively eliminated—a clear competitive advantage.

Taming the Fuzzy Front End

Management teams in successful and innovative companies fully understand that some of the greatest opportunities reside in the fuzzy front end, or definition phase of the PLC.[7] The ability to accurately forecast future customer and market needs and then integrate those needs with leading edge technologies is critical for companies to survive in their respective industries.[8] This work is never simple and the fuzzy front end of product, service, or infrastructure development has served as a test in frustration and a lesson in patience for many. The high failure rates associated with a weak definition have led to negative connotations toward the fuzzy front end.[9] However, this phase is a natural part of every development program and, if managed effectively, can provide great business opportunity and competitive gain. There is no shortcut to creating a strong definition; therefore, companies looking to successfully produce competitive products, services, and infrastructure capabilities on a consistent basis must learn to effectively harness the innovation and research engines involved in the front end of the development process.

The key to realizing the opportunities that the fuzzy front end presents is to "tame" the fluid and ambiguous nature of this phase. This can be accomplished by establishing a framework and well-defined targets to provide focus and employing effective leadership to cut through the ambiguity and competing agendas that characterize the fuzzy front end.[10] The program management model provides both of these things.

To converge on a viable product, service, or infrastructure definition as rapidly as possible, the program manager can establish a framework of targets to be used as motivators and to focus the effort of the cross-discipline team. The targets must concentrate on both time and business. Time-focused targets consist of regularly scheduled review meetings with senior management to demonstrate continued convergence progress in the development of the concept. Business-focused targets include the relevant objectives that drive the need for the solution (for example, time-to-market window, key technologies, and cost targets) and shape the final concept. These targets are the early success criteria that form the basis of the program strategy that provides the guiding direction to achieve business success.

By employing program management principles in the fuzzy front end, an organization will realize the following three key benefits: the leadership necessary to effectively integrate customers, technology, and business perspectives; a framework to enable rapid convergence toward a viable product, service, or infrastructure concept; and a business champion and leader who ensures the product, service, or infrastructure definition fully supports the strategic objectives set forth by senior management.

Managing the ROI

Within a development organization, the general manager makes decisions concerning investments (in the form of products, services and infrastructure) that will help achieve the anticipated business results. The investment normally comes in the form of monetary funding and human and non human resources. The GM nearly always invests in multiple programs that are managed concurrently but with varying life cycles. The total investment budget is divided among the various programs within the organization's development portfolio (see Figure 2.2).

Once a product, service, or infrastructure capability is delivered, it begins generating the intended return (for example, revenue, market

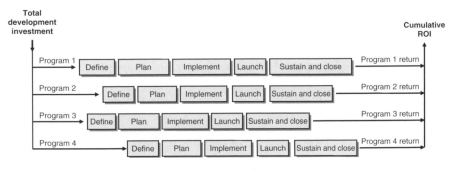

Figure 2.2   ROI for a portfolio of programs.

share, increased productivity). The cumulative return of each program represents the organization's ROI.

The difficulty and challenge for the GM comes in managing the investment of *all* development efforts of the organization, in addition to the existing products, services, or infrastructure capabilities already launched and operational. Unless the company is a start-up with a single focus, it is not possible for one GM to manage the business aspects of a development program. He or she must hand this responsibility off to the program manager. In effect, the program manager serves as the *GM Proxy* for his or her piece of the business investment. A primary role of the program manager, therefore, is to manage the business on the program to ensure the ROI is achieved.[11]

Managing Complexity

In many aspects of life, humans have a tendency to push the norm or current status quo. Our ever-increasing wants and desires drive our collective environment toward more challenging and exciting ends. This is true in our careers, relationships, activities, and especially in the products and services we utilize.[12] Consumer demand for customized solutions is a continual cycle; once a new product or service becomes status quo, the opportunity arises for a company to develop the next generation solution. This translates into a decrease in a product, service, or infrastructure's life cycle but an increased opportunity for continual refresh. This can keep a company's revenue stream constant and growing, if it continually provides competitive solutions to the markets and customers it serves.

Along with the opportunities described above, the increase in product, service, or infrastructure complexity caused by the demand for customized solutions presents some significant challenges. Complexity in development manifests itself in the following areas: designs have become more complex as features and integrated capabilities increase; the process to develop and manufacture the solutions; the ability to integrate multiple technologies with end user wants; and the current global, multinational approach to development.

Companies that are succeeding in managing these issues are doing the following two key things: adopting a systems approach to development and adopting the program management model to effectively integrate complex solutions. Early adopter companies in the automotive, aerospace, and defense industries continue to utilize the systems and program management approach for their products under development. More recently, companies such as Intel, Tektronix, and Hewlett-Packard have found great success in utilizing systems and platform concepts coupled with the program management model to develop their products, services, and infrastructure capabilities.

Accelerating Time to Money

Besides demanding increasing complex solutions, consumers also want accelerated delivery of new technologies. It is a well-known fact that in today's highly competitive world, time to money is a critical factor in gaining an advantage over one's competitors. For most companies, gaining time to money advantage means decreasing the cycle time required for developing a product, service, or infrastructure capability. This requires the adoption of new development models. Historically, the two most dominant development approaches were the project hand-off and concurrent methods. A brief overview of each approach is included in the following section, along with an example from the medical product industry that chronicles the company's struggle to find a development model that meets accelerating time to market demand.

In the project hand-off approach (also called the relay race), each functional team sequentially works on their element of the project, then hands both the work output and project ownership over to the next functional team in line (see box titled, "The Perils of the Project Hand-off Method"). This concept is illustrated in Figure 2.3.

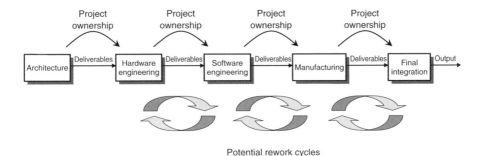

Figure 2.3   Project hand-off method.

The primary problem with this approach is that the development work accomplished at each hand off occurs within a single function. Errors introduced upstream have to be reconciled downstream, resulting in multiple rework cycles that consume time to money advantage.

The Perils of the Project Hand-off Method

Hospi-Tek is a medical equipment manufacturing company who has historically used a project hand off approach to develop its products. They are currently under intense time to market pressure from their primary competitor, forcing senior management to reevaluate their approach.

Under the project hand-off method, the Hospi-Tek product development effort began with the architectural team who developed an architectural concept and derived the high-level requirements of the medical device from the work of the product marketing team. The architectural concept and specification were then handed off to the hardware engineering team who assumed ownership of the project. The engineering team developed the hardware requirements, engineering specifications, and the product design, which were then handed off to the manufacturing team who assumed ownership of the project. The manufacturing team developed the manufacturing processes, retooled the factory, and produced the physical product. The product and project ownership were then handed off to downstream engineering teams, such as the software development team. The software team developed the software stack, then handed the combined hardware/software product, as well as project ownership, to the validation and test team. Finally, the validation and test team performed product and component-level testing to ensure the product achieved the functional, quality, usability, and reliability requirements.

Management of the project was accomplished through a project management-only model, with multiple project managers in control of the project as it progressed through the development life cycle. Thus, a project manager with the functional expertise specific to the phase of development the product was currently in assumed ownership of the project.

This method of development is common in smaller, less mature, and technically focused companies in which true project and program management value is usually not well understood and the engineering function reigns king. Unfortunately, this method is not scalable, and as a company begins to succeed and grow, product and process complexity requires the management team to look at alternative methods to structure and manage its product development efforts. This was the case with Hospi-Tek.

The concurrent development method was created to decrease cycle time—an improvement over the project hand-off method.[13] It involves the various functional teams working simultaneously to deliver their

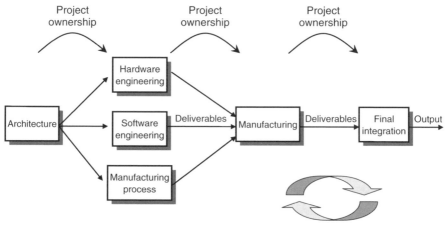

Figure 2.4   Concurrent development method.

elements of the product, service, or infrastructure capability under development (see Figure 2.4). The concurrent development approach, however, has been found to be less than optimal in reducing development cycle time because of what we call the "big bang event" that inevitably has to occur in concurrent development.

The big bang occurs when the concurrently developed elements are integrated. Great inefficiencies can occur due to rework caused by poor requirements definition, lack of change management in the functional development efforts, poor communication between the concurrent development teams, and so on. As a result, it is common for development teams to spend as much or more time performing integration and testing as on concurrent development of the elements (see box titled, "Little Gain with Concurrent Development"). Depending on the extent of any misalignment of the functional outputs being developed, the concurrent method may actually lead to an increase in cycle time over a hand-off approach.

Little Gain with Concurrent Development
Following the advice of a well-known consulting firm, Hospi-Tek moved to a concurrent method to develop their medical equipment products faster. As depicted in Figure 2.4, much of the sequential project work associated with the project hand-off method was pulled forward and performed in parallel with

the early hardware design work. One can see that the opportunity for cycle time reduction is significant because all elements, including the production processes, are developed simultaneously. However, the opportunity for cycle time reduction is still at great risk as the development enters the integration stage. This is the "big bang" stage in which all elements of the product are integrated for the first time. Hospi-Tek struggled with this stage and the significant misalignment that occured between the product elements (hardware, software, and manufacturing) very late in the process. This misalignment caused rework that, in many cases, was significant. Hospi-Tek came to believe that the integration process takes nearly as long as the design and development phases and began planning their projects accordingly. The real source of the problem is not in the integration phase but in the earlier design and development phase. The concurrent development method contains a major flaw; it leads people to believe that because early design and development efforts occur in parallel, cycle time is automatically reduced. However, unless the early concurrent development teams are also communicating, sharing critical cross-discipline information, and collaborating on producing the concurrent elements, significant rework usually awaits downstream.

As a result, Hospi-Tek utilized the concurrent development process for two product generations and achieved about a 10 percent reduction in product development cycle time—still insufficient to remain competitive.

So the question becomes how do companies achieve a significant reduction in development cycle time? The best in-class companies that we've seen, worked for, and have been engaged with accomplish it through an integrated development approach, which is driven by a true program manager.

In the integrated approach, the cross-discipline teams work shoulder to shoulder through the development life cycle, from conception to end of life. The work of each team is coordinated with and reinforced by the work of the other teams. To increase the probability of time-to-money efficiencies, the historically back-end functional activities such as manufacturing, validation, and testing are pulled forward in the development life cycle. Additionally, the front-end activities such as architecture and marketing are involved longer in the development cycle to ensure seamless integration. The big bang phenomenon of integration is replaced by continuous, iterative-design, develop, and integrate cycles (Figure 2.5). The work of all functions involved in the development effort is comprehended in the concept, business case, implementation plan, implementation activities, launch, and support activities. The integrated

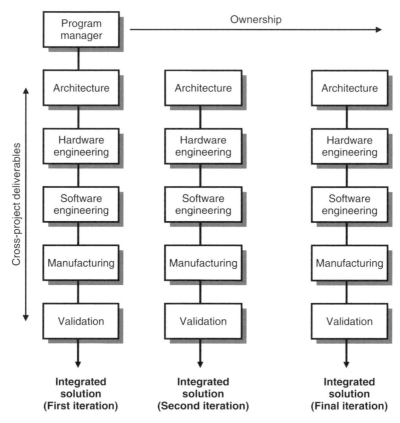

Figure 2.5   The integrated development method.

team takes responsibility for joint consideration of alternatives, decisions that will have consequences to multiple functions, and ownership of the success of the development program.

For the integrated development process to achieve its full potential, the cross-project, multi-discipline team is led by a program manager who can work across the disciplines to ensure their work is indeed fully integrated and aligned to the business objectives. Achieving time-to-money advantage requires tight management of the interfaces, shared risks, and open communication channels between the teams. This involves orchestrating, coordinating, and directing the work of the various functional teams. The program manager brings the required general management, business, and leadership skills required to effectively lead the integrated, cross-project development team (see box titled, "Applying Integrated Development").

Applying Integrated Development

Fortunately, the horizon appears brighter for Hospi-Tek. The company now understands the limitations of the concurrent development method and of the capabilities of its program managers. They have hired a director of program management who, at the time of this writing, is evaluating the merits of utilizing a fully-integrated approach to develop their medical devices and full use of their program management structure and capability. To give their program managers the foundation in knowledge, skills, and capabilities, they will need to fulfill their role, program management training and benchmarking activities are under way.

Mitigating Business Risk

Informed risk-taking can provide a means to gain competitive advantage. However, risk taking does not mean taking chances. It involves understanding the risk/reward ratio and then managing the risks, or uncertainties, that are involved in each development effort. Failure to understand the business risk involved can lead to substantial loss for the enterprise. As the business champion for the program, the program manager is responsible for managing business risk.

Good risk management practices allow program teams to move from concept to launch as quickly as possible by removing potential barriers well ahead of the point in which they become impediments to time-to-money goals.[14] To be successful, the program manager must manage risk across multiple, interdependent projects. This requires a different perspective and technique than managing risk for a single project (see Chapter 8); if one of the projects fails, the entire program will likely fail.

Navigating the Flat World

The flattening of the world has created opportunities for companies that have learned to take advantage of this new phenomenon. It enables complex systems to be broken down into specialties that are then worked on, reassembled, and integrated into a total solution.[15] Additionally, a company can now take advantage of worldwide resources and alternative methods (such as open-source development) to develop the various elements of the total solution.

However, many of the companies that have adopted forms of distributed horizontal collaboration are struggling to manage the work that takes place around the globe and the clock. Companies that utilize a program management model are consistently succeeding in the management of distributed development efforts that employ outsourcing, open sourcing, in sourcing, and offshoring. The reason? It takes a special type of business model and skill to mold the work of specialists through a high degree of horizontal collaboration to create an integrated solution. As explained in Chapter 1, this is exactly why program management was developed as a business model in the defense and military industries several decades ago—to integrate the work of highly skilled specialists to create and deliver a holistic, complex solution.

PROGRAM MANAGEMENT AS A COMPETITIVE ADVANTAGE

The essence of business strategy lies in creating competitive advantages to outplay rivals.[16] Companies have successfully deployed program management to address and resolve the business problems identified in the previous sections. In doing so, they have created competitive advantages over their business rivals who have not deployed program management. To emphasize this point, we evaluate the elements of the program management business case from this perspective:

Integrating the business model elements: Program management is the mechanism by which the work of various operating functions within a company is integrated to create an effective business model. When program management is properly conceived and executed, it helps deploy the business model and strategies much more effectively than an uncoordinated or ad hoc approach.

Therefore, rivals can be outplayed in the long run by a company's ability to execute accurately, timely, and repeatedly through program management practices that integrate and synchronize the work of the operating functions, while focusing on intended business objectives. Program management is very effective in breaking down the functional barriers that can prevent efficient product, service, or infrastructure development.

Aligning business strategy to execution: A company that translates good strategy into successful programs and execution is bound to outplay a rival that has only good strategy. Program management can be viewed as the organizational glue that translates strategic business objectives into actionable plans and then manages the tactics to achieve the desired results. When the program management model is implemented, the gap between strategic elements and project execution is effectively eliminated and the company gains a clear competitive advantage.

Taming the fuzzy front end: The inability of companies to control the fuzzy front end of products, services, or infrastructure development has contributed to a high rate of failure for many development endeavors. However, if managed correctly by employing sound program management practices, the fuzzy front can lead to great opportunity and competitive gain for a business. Program management can establish a sound framework within the fuzzy front end to project accurate forecasts of future customer and market needs and then integrate those needs with leading edge technologies, which creates competitive challenges for business rivals.

Managing ROI: Once a product is launched into the market, or a service or infrastructure capability goes live, it begins generating the return intended. The cumulative return of each product, service, or infrastructure capability represents the organization's ROI. A skilled and competent program manager is the primary business manager for a program and is responsible for focusing on and ensuring that the ROI is met. Program management practices put a continual focus on the business aspects of developing products, services, and infrastructures that lead to better use of investment resources.

Managing complexity: Program management was originally conceived for managing highly-complex development undertakings. It puts a systems structure in place and provides a framework which disaggregates the complexity into manageable elements. By contrast, companies not employing program management practices can and do become consumed by the complexity.

Accelerating time to money: By using an integrated development approach that is driven by the program management function, time-to-money goals are optimized. The program management model is built on the development, management, and delivery of interdependencies between the functional elements of the program throughout the development life cycle. By incorporating and managing the cross-project deliverables through an iterative and integrated development process, a limited rework scenario develops. This translates to faster time-to-money possibilities—an advantage that brings an extended sales cycle, premium prices, higher profits, and faster learning over companies that do not use program management practices.[17]

One company leader we talked with who agrees with this assessment is Gary Rosen, vice president of Engineering for Varian Semiconductor Equipment. Rosen describes the role of program management in accelerating time to money:

> If you have a strong program management function, products are closer to what the customers want, and the team spends less time iterating late in the program to meet customer expectations. A program manager adds clarity for the engineering team by balancing market requests with engineering capabilities', therefore, setting realistic customer targets. This results in more efficient use of resources, which allows a program team to deliver what the customer wants the first time and then move on to the next product development program.

Mitigating business risk: Risk taking is necessary in many industries, if a company is intent on being a market leader. Products, services, or infrastructure capabilities that don't push the risk envelope probably aren't worth the development investment required—unless the company is intent on being a market follower, which is a valid strategy for many companies. Market leaders understand they must run directly toward risk to put distance between themselves and their competitors.

Program risk management methods flush out the business risk involved in a development program, then effectively manage the risk at both the program and project level to maximize the reward/risk ratio.

Navigating the flat world: Companies that utilize a program management model for product, service, or infrastructure development have succeeded in managing highly distributed development efforts that the flattening of the world has made possible. These companies are now able to take full advantage of worldwide specialty resources by effectively managing the horizontal collaboration between the multiple project specialists that are highly dispersed.

Now, back to the question: Can program management be a competitive advantage? We believe that the eight elements of the business case for program management prove that the answer to this question is a resounding yes.

SUMMARY

Even though companies have invested an enormous amount of time, money, and resources trying to improve their operational and project management capabilities, many of these companies still face serious problems associated with their development process. The most serious ones identified include the following: misalignment between strategy and execution, poorly managed ROI of development dollars, increasing complexity of design and process, and slow time to money. The program management model is an effective approach companies use to confront these business problems in the development of their products, services, and infrastructure capabilities.

This chapter reviews the following eight elements of the business case for program management: integrating business model elements, aligning business strategy and execution, taming the fuzzy front end, managing ROI, managing complexity, accelerating time to money, mitigating business risk, and navigating the world. These eight elements of the business case for program management create a competitive advantage for companies that employ them. In many instances, they are the "secret sauce" that helps a company transition from good to great.

The Principles of Program Management

▼ Program management is a primary business function that is focused on the business aspects of the organization

▼ Program management provides the missing link between organizational strategy and project output

▼ Program manager functions as the GM Proxy for his or her development program.

REFERENCES

1. Thompson, J., Arthur A. and A. J. Strickland III. *Strategic Management: Concepts and Cases*, 13 ed. McGraw-Hill/Irwin, 2002.
2. Morris, P. W. G. and A. Jamieson "Moving from Corporate Strategy to Project Strategy". *Project Management Journal* 36(4), 2005: 5–18.
3. Cleland, David I. *Field Guide to Project Management*, 1st edition. New York, NY: Van Nostrand Reinhold, 1998.

4. Smith, Preston G. and Donald G. Rinertsen. *Developing Products in Half the Time: New Rules, New Tools*, 2nd edition. Hoboken, NJ: John Wiley & Sons, 1998.

5. Friedman, Thomas L. *The World is Flat*. New York, NY: Farrar, Straus and Giroux Publishing, 2006, pp. 439–440.

6. Wikipedia, *The Free Encyclopedia*: Wikipedia, (*http://en.wikipedia.org/wiki/Business_model*):

7. Smith, Preston G. and Donald G. Rinertsen. *Developing Products in Half the Time: New Rules, New Tools*, 2nd edition: Hoboken, NJ: John Wiley & Sons, 1998.

8. Koen, P. A., G. M. Ajamian, et al. (2002). Fuzzy front end: Effective methods, Tools, and Techniques. The PDMA ToolBook for New Product Development. P. Belliveau, A. Griffin and S. Somermeyer. New York NY:, John Wiley & Sons: 5–35.

9. Cooper, Robert G. *Winning at New Products: Accelerating the Process from Idea to Launch*, 3 ed. Cambridge, MA: Perseus Books, 2001.

10. Martinelli, Russ. "Taming the Fuzzy Front End". *Project Management World Today* (July–August 2003).

11. Martinelli, Russ and Jim Waddell. "Program Manager Roles, Responsibilities and Core Competencies". *Project Management World Today* (November–December 2004).

12. Gharajedaghi, Jamshid. *Systems Thinking: Managing Chaos and Complexity*. Woburn, MA: Butterworth-Heinmann Publishing, 1999.

13. Thamhain, Hans. J. *Management of Technology*. Hoboken, NJ: John Wiley & Sons, 2005.

14. Martinelli, R. and Jim Waddell. "Managing Program Risk". *Project Management World Today* (September–October 2004).

15. Friedman, Thomas L. *The World is Flat*. New York, NY: Farrar, Straus and Giroux Publishing, 2005, pp. 439–440.

16. Hamel, Gary and C. K. Prahalad. *Competing for the Future: Breakthrough Strategies for Seizing Control of your Industry and Creating the Markets of Tomorrow*. Harvard Business School Press, 1994.

17. Morris, P. W. G. and A. Jamieson "Moving from Corporate Strategy to Project Strategy". *Project Management Journal* 36(4), 2005: 5–18.

Chapter 3

Aligning Programs with Business Strategy

Historically, the primary managerial functions and processes of product, service, or infrastructure development have been defined and viewed as independent entities, each with its own purpose and set of activities. For example, executive management normally performs the strategic processes that set the course of action for the organization. Portfolio management and project selection are commonly thought of as senior and middle management responsibilities. Program planning and execution processes are performed by the program manager and core team, while project managers and team leaders are responsible for project planning and execution processes. Each of these functions and processes is executed separately by a different set of people within the organization. At best, the strategic element feeds the portfolio element, the portfolio element feeds the program management element, and the program management element feeds the projects and team execution that delivers the products, services, or infrastructure outputs. In many cases, this still results in projects that are not tied directly to either the business strategy or the organization's portfolio.

Companies invest much time, money, and human effort into refining and improving each of their independent functions and processes, only to come to the inevitable conclusion that they are not coming any closer to effectively and efficiently turning their ideas into positive business results. Increasingly, this fact is leading business leaders to the realization that their independent variables cannot remain independent. Rather, they must be transformed into a set of interdependent elements that form a coherent development system.

This chapter uses the systems concept to explain and demonstrate how program management, and other critical managerial functions and processes within an organization, must be defined and executed as elements of an integrated management system. It explains how an organization can be viewed as a holistic, coherent system that is composed of critical managerial elements and processes. In doing so, this chapter will help senior executives, program managers, and project managers accomplish the following:

- Build an integrated management system model
- Deploy the strategic *and* tactical elements of the integrated management system
- Effectively align the strategic and tactical elements of the integrated management system

By taking a systems approach through an integrated development model, an organization can realize improved business alignment between execution output and business strategy. As we demonstrate in this chapter, the program management discipline plays a pivotal role in aligning the tactical work output of multiple project teams with the mission and strategy of an enterprise.

THE SYSTEMS PERSPECTIVE

We define a system as an assemblage of interrelated elements or subsystems comprising a unified whole. It can be visualized as a simple entity consisting of four primary elements: inputs, output, interdependent subsystems, and an environment within which the system operates. Figure 3.1 illustrates the generic systems model.

The nature of the system elements are unique to each problem that the system resolves. The subsystems are highly interactive and interdependent on one another. The role of the subsystems is to utilize the inputs provided by the environment to produce a desired output. Both the inputs and the output are heavily influenced by the particular environment within which the system exists.

THE INTEGRATED MANAGEMENT SYSTEM

When one looks at an entire business from a systems perspective, a holistic view of the enterprise emerges. Within the business enterprise system,

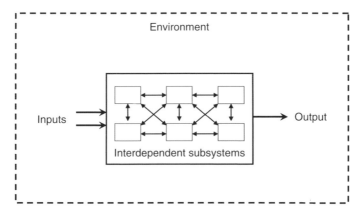

Figure 3.1   The generic system model.

key subsystems such as the corporate mission, strategic objectives, organizational functions, organizational structure, critical processes, and programs exist to effectively and efficiently convert the business inputs into the desired outputs—technological advantage, cost value leadership, or profits. Like any system, the subsystems are highly interdependent upon one another. For example, the mission of the business enterprise influences the strategic business objectives defined; moreover, the objectives define the functions that are needed to achieve the objectives, as well as how the enterprise is organized. The strategic objectives, functions, and organizational structure all have a direct influence on the selection of the critical processes and tactics utilized to convert inputs to outputs.

Finally, the business enterprise operates within, and is influenced by, a dynamic environment. Examples of environmental factors that have an impact on the mission, structure, operation, and output of the business enterprise system include shareholder expectations, domestic and world economic conditions, technology trends, customer usage models, and competitor actions.

The heart and engine of growth for the business enterprise system is the development component that consists of the management functions and critical processes needed to convert inputs into an output.[1] We refer to the set of functions and critical processes as the integrated management system. The integrated management system, shown in Figure 3.2, is the mechanism from which new products, services, or infrastructures are conceived and developed to realize the mission and strategic objectives of the business.[2]

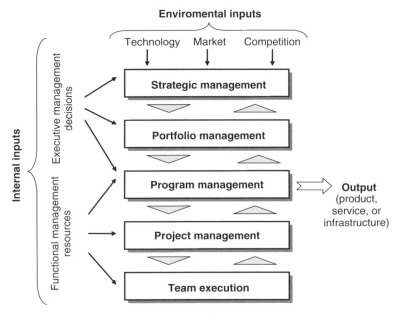

Figure 3.2   The integrated management system.

It is common practice to view the primary subsystems of the integrated management system within a business as independent entities, each with its own set of activities, processes, and tools. Additionally, in practice, there tends to be a real separation between the strategic and tactical subsystems of the development system. This leads to misalignment between the business objectives and the work output of the organization, which may result in an unfulfilled mission and strategic objectives (see box titled, Three Beer Drinkers as a System).

Three Beer Drinkers as a System

This is a bizarre story from a program management workshop in which the instructor mightily struggled to explain the concept of a system in plain language. Help came in an unexpected way when an attendee named John stood up and said, "I have an example from a systems workshop I attended that I can share. Last year I had a high school reunion. It was the same as earlier reunions—just plain boring. So three of us, Paul, David, and I, decided to have our own reunion party. The rule we set was simple: When one of us stopped drinking, the party was over. One hour later, after just four glasses, Paul fell fast asleep (he never really drinks), and the party was over. David and I were really disappointed because the party ended just when it started to be fun.

A few weeks later, I attended the system workshop, and the instructor asked us to think about our own experiences with systems in our everyday lives. Our reunion party occurred to me. The party was the system, and the three of us were each subsystems. The weakest subsystem controlled the whole system. Paul's beer drinking capacity actually determined how much beer I had."

We understand how the subsystems of the integrated management system (Figure 3.2) work together as a holistic system, but the lesson from the beer party tells us that the weakest subsystem also controls the entire integrated management system. For example, no matter how good strategic management is, if the portfolio process is poorly executed, the whole system will operate at a suboptimum level. The moral of the story is that for an integrated management system to work at a high level, all subsystems must work at the same high level.

Contrary to this common practice, it is important to view the integrated management system as a collection of interdependent subsystems. In doing so, a holistic perspective to product, service, or infrastructure development becomes possible. Among the advantages in viewing the functions as an integrated system—instead of a set of loosely dependent elements—is that the system becomes more flexible and receptive to improvement when an organization is looking to gain additional efficiencies. By taking the ad hoc approach, an organization finds itself chasing and attempting to cure a set of symptoms, instead of taking a systems approach that will most likely lead to root-cause analysis and cure of overriding problems.

This is analogous to actions one may take if his or her automobile is performing poorly and exhibiting problems such as fuel inefficiency or rough idling. By taking an ad hoc approach to the problem, one may try to treat the poor fuel efficiency by adding an oxidization additive or replacing the filter. To treat the poor idling problem, one may change the spark plugs or adjust the engine timing. Any or all of these actions may yield an improved performance for a period of time, but they will not solve the root problem if the automobile is in need of an overall tune-up. By viewing the engine and ignition functions of the automobile as a system, a holistic approach to diagnosing and resolving the root problems becomes possible.

The subsystems of the integrated management system are categorized into the following three types: the strategic subsystem, the tactical subsystem, and the program management subsystem. The strategic subsystem consists of the strategic management and portfolio management processes. The tactical subsystem includes the project management and team management functions. The program management subsystem is the organizational glue that translates strategy into actionable plans and tactics that achieve the desired business results. The following sections will look at each of the three subsystems that make up the integrated management system.

THE STRATEGIC SUBSYSTEM

The strategic subsystem of the integrated management system is comprised of two primary processes normally performed by the senior management team of an organization—strategic management and portfolio management (see Figure 3.3).

The strategic management process involves definition of the company mission, analysis of the internal and external environment, identification of the strategic objectives, and definition of strategic options to fulfill the objectives. The portfolio management process includes the review and selection of the strategic options that will be implemented and

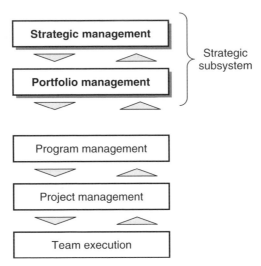

Figure 3.3   Strategic elements of the integrated management system.

also evaluates the success of the current process. The following sections explore the strategic and portfolio management elements of the strategic subsystem in detail.

Strategic Management

Strategic management is defined as a set of decisions and actions that result in the formation and implementation of plans designed to achieve a company's objectives.[3] It includes the following three primary elements: the business mission, strategic objectives, and strategy. The mission statement is a broadly framed statement of intent that describes *why* the company exists in terms of its purpose, philosophy, and goals. Strategic objectives define *what* the business wants to achieve over a specific time horizon, which is typically three to five years. Strategy is the business's game plan that reflects *how* it will accomplish the objectives. It is imperative that all three elements of strategic management be in place to properly frame the direction of the organization. Figure 3.4 shows a generic strategic management process flow.

Environmental Analysis

Understanding the current and future environment that the firm will be operating in is critical to the strategic management process. The senior management team must have a comprehensive knowledge of the following: the future direction of the economy in the company's

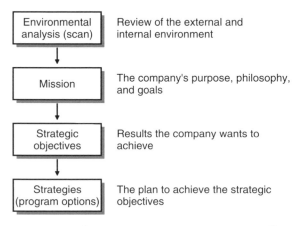

Figure 3.4   The strategic management process flow.

market, the impact of potential technological breakthroughs, local and foreign political climates of countries in which it operates or wishes to operate, the size and stability of its supplier base, and the firm's own resource capability and limitations. This knowledge of the current and future state of the external and internal environments will influence the company's marketing scheme, as well as its choice of strategic objectives and strategies.

Company Mission

Some have argued that a company mission is of little use because big visions are rarely realized.[4] However, without a mission, a company lacks the overriding statement of purpose for the business; its philosophy toward its customers, employees, and competitors; and its goals such as survival, growth, and profitability. Most importantly, without a mission, there is no basis to judge the relevancy of opportunities, threats, and program options presented to a company's management team. All possibilities would have equal value.

In our view, defining the company mission is perhaps one of the most important responsibilities of the senior management team. The mission statement defines why the business exists, what the strategic intent of the company is, what its core values are, and how the management team measures success. The company mission statement should be designed to accomplish several outcomes, as follows:

- Ensure that everyone within the organization understands the company purpose
- Provide the basis for allocating the organization's resources
- Establish the guidance to translate company objectives into programs, projects, and work elements
- Specify the means for which attainment of company objectives can be assessed and assets controlled.

The mission statement should describe the company's products, markets, and technological areas of emphasis in a way that demonstrates the values, priorities, and goals of the management team.[5] For most companies, financial and economic goals greatly influence the strategic direction of the enterprise. This may be either explicitly or implicitly stated in the company mission statement. Take, for example, these two mission statements. The first, from Proctor & Gamble, explicitly includes sales and profits as part of its mission statement. The second, from Merck, implicitly implies profit as a superior rate of return.

Here is Proctor & Gamble's mission statement.

> We will provide branded products and services of superior quality and value that improve the lives of the world's consumers. As a result, consumers will reward us with leadership sales, profit, and value creation, allowing our people, our shareholders, and the communities in which we live and work to prosper.[6]

Here is Merck's mission statement:

> The mission of Merck is to provide society with superior products and services by developing innovations and solutions that improve the quality of life and satisfy customer needs, and to provide employees with meaningful work and advancement opportunities, and investors with a superior rate of return.[7]

A good mission statement promotes a sense of shared expectations among all members of the organization; provides common purpose, direction, and goals for the company; and defines a company's intent for shareholders, employees, customers, suppliers, and the community in which it operates.

Strategic Objectives

As stated earlier, the strategic objectives define what the company wants to achieve within a multiyear period. To achieve long-term prosperity, management teams commonly establish strategic objectives in the following seven areas[8]:

1. *Profitability*: The ability to survive and achieve a company's other strategic objectives depends greatly on its ability to meet its profitability objectives. Profitability objectives may include increased level of profits, return on assets or investments, and profit growth normally expressed as earnings per share for publicly traded enterprises.
2. *Productivity*: These objectives normally focus on how efficient an organization is in creating output from input and may focus on decreasing cost of goods produced, reduced throughput or cycle time, minimization of factory defects, or increased reuse of components or software.

3. *Competitive position*: A business's relative dominance in the marketplace as compared to its competitors; it is normally measured in total sales and/or percentage of total market share. Examples of competitive position strategic objectives include increasing market share by one percent per year, increasing yearly sales by opening new markets, or increasing quality to strengthen brand image.
4. *Employee development*: Career growth and new opportunities for the employees of the enterprise, primarily to foster the development of future leaders and reduce overall employee turnover.
5. *Employee relations*: The purpose is to increase company loyalty and employee productivity; it includes improved working environment and conditions, implementation of rewards and bonuses based on company performance, and consistent access to senior management.
6. *Technology leadership*: If a company chooses to position itself as a technology leader, it must establish aggressive objectives to provide technology improvements and changes in its products or services. Objectives may also include investment in improved manufacturing technologies to lower production costs, integration of several discrete technologies into a single solution, or funding research activities to increase security and safety.
7. *Public Responsibility*: Businesses have learned that company image translates to brand value. Brand value provides a company the opportunity to charge premium prices—relative to their competitor's—for their products or services. Objectives may include increased adherence to land, water, and air contamination to decrease pollutants, increased funding for community charities and public education programs, and sponsorship of community activities and events.

It should be pointed out that a business normally does not strive to identify strategic objectives in all seven areas presented above, but rather only in those areas that align with and fully support attainment of the corporate mission. Additionally, strategic objectives should possess a number of attributes for effectiveness. A strategic objective must be specific in nature by stating what will be achieved, when it will be achieved, and how it will be measured. An objective must be challenging enough to raise the bar on corporate performance but attainable to prevent employee frustration and lack of motivation. Finally, it must be flexible to adapt to changes in the organization's operating environment.

Strategies

Strategy defines *how* a business will achieve the strategic objectives it has established. Strategy consists of the portfolio of ideas that, when fully developed, will contribute to the attainment of the strategic objectives[9]. Superior products, services, and infrastructure capabilities are the means to build technology leadership advantage over competitors, expand market share, increase revenue, decrease operating costs, and strengthen brand value (see box titled, "If You Don't Know Where You're Going, any Road Will Get You There").

To be credible and achievable, the organization must manifest the various strategies into programs that the company intends to fund and resource. The sum of the program outputs will provide the combined means to achieve the desired objectives. Business, technology, and market strategies must culminate in a program strategy to be executed.

If You Don't Where You're Going, Any Road Will Get You There

In a business-to-business software company, we witnessed incomplete success measures for programs that led to misalignment between program execution and business strategy. This is our conversation with a program manager.

Q: What kind of company are you, and how many programs do you do per year?

A: We are Terra Software, a market leader in its niche, $60 million in annual revenues, doing three to four programs per year.

Q: What mechanism do you use to plan for the program success measures?

A: We use the program brief to plan the entire program strategy, including its success measures.

Q: Who issues the brief and to whom?

A: The executive committee issues it to the program manager.

Q: What kind of success measures do you plan for?

A: The release date and the annual sales the software will make.

Q: Does the brief include the planned profits, market share, etc., for the program?

A: No. I have been a program manager here for many years and, frankly, I don't remember a program brief that defined such measures.

Q: Don't you care about profits and market share?

A: I never thought about that but now it makes sense. However, the program managers at Terra Software are not trained to deal with such stuff.

Obviously, Terra Software lives by the mantra: Since you are not given strategic success measures (if you don't know where you are going), any program result is good (any road will get you there).

Portfolio Management

The second strategic element of the integrated management system is the portfolio management process. Portfolio management is defined as a dynamic decision process, whereby a business's list of active programs is constantly updated and revised.[10] As is usually the case, organizations have many more products, services, or infrastructure ideas than available resources to execute them. As a result, resources become overcommitted, and an organization must find a way to broker competing demands for its limited resources. Portfolio management is an effective process for identifying and prioritizing products, services, or infrastructure programs that best support attainment of the strategic objectives. Programs are ranked and prioritized based upon a set of criteria that represents value to the customer and the organization. Resources are then assigned to the highest value and most strategically significant programs. Low-value programs must be cut, returned for redefinition, or put on hold until adequate resources are available.

According to the model developed by Cooper, Edgett, and Kleinschmidt, the three fundamental principles of portfolio management are as follows: maximize the value of the portfolio of programs in terms of company objectives, achieve a balance of programs based upon a number of vectors, and establish a strong link between the portfolio and organizational strategy. Accomplishing these principles should also result in matching the number of selected programs with the available resources. This is sometimes referred to as **resource capacity planning**.

Maximizing the Value of the Portfolio

The foundation for an effective portfolio management process is based on the ability of the management team to determine the factors that constitute program value for their organization. The team must also establish a prioritization system to evaluate one program against the others. Once it gets beyond this hurdle, which is sometimes monumental, the organization will have accomplished the following two things: a real understanding of how value is defined and a prioritized portfolio of programs that clearly differentiates between those that provide the most and least value to the business. Thus, maximization of the portfolio is established. The punch line of portfolio management is now possible. The management team can allocate and concentrate its resources to the programs that provide the most value to the organization', therefore, achieving maximum output from a limited input.

Figure 3.5 is an example of a prioritized portfolio of programs from a manufacturer of Internet communication products. Resource allocation occurred in priority order, with priority one programs receiving full allocation of resources, priority two programs receiving full allocation of resources once priority one programs are fully staffed, and priority three programs receiving full resources once priorities one and two programs are fully staffed.

Achieving a Balanced Portfolio

Creating a balanced portfolio involves making thoughtful decisions at the macro level about which types of programs the company chooses to invest in and the levels of investment. This is analogous to the decisions one makes concerning personal financial portfolios. One decides which types of investment to make (for example, stocks, bond, mutual funds,

Program portfolio (unranked)	NPV ($M)	Evaluation Criteria			Risk x NPV	Ranked Programs	Priority
		Likelihood to succeed				Risk x NPV ranking	Resource priority
		Tech	Mrkt	Overall			
Program A	$38	20%	70%	14%	$ 5.3	Project D	1
Program B	$15	15%	50%	8%	$ 1.2	Project F	1
Program C	$13	35%	75%	26%	$ 3.4	Project E	1
Program D	$13	80%	80%	64%	$ 8.3	Project A	2
Program E	$19	45%	70%	32%	$ 6.0	Project C	2
Program F	$9	90%	85%	77%	$ 7.0	Project H	2
Program G	$9	50%	45%	23%	$ 2.0	Project G	3
Program H	$5	95%	60%	57%	$ 2.9	Project B	3

NPV = net present value

Figure 3.5 The prioritized portfolio.

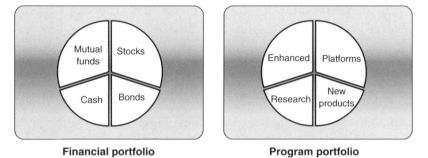

Financial portfolio   Program portfolio

Figure 3.6   Comparison of financial and product portfolios.

cash, and so on) and at what percentage. Similarly, companies decide which programs to invest in—platforms, new products, iterations, and research—and at what percentage (see Figure 3.6). The objective is to balance investment so that both the current and future value propositions to customers are secured and the strategic objectives of the company are funded.

The **investment buckets** approach to balancing a portfolio is the most common method we have seen in our experience. The selection and structure of the investment buckets are unique to each business and are highly influenced by a company's strategic objectives, market segmentation, and product, service, or infrastructure types. Cooper, Edgett, and Kleinschmidt offer seven possible dimensions of how to define an organization's investment buckets, or portfolios, as follows:[11]

- Strategic objectives
- Product line
- Market segment
- Technology type or platform
- Program or project type
- Familiarity matrix
- Geographical region

A financial industry software company we worked with, for example, defines its investment buckets from a strategic standpoint. Its portfolio structure was established on the basis of funding the company's mission. The mission was defined as follows: Drive material impact on (our) margin, growth, and influence by bringing to market software solutions and services and delivering and influencing technologies that enhance

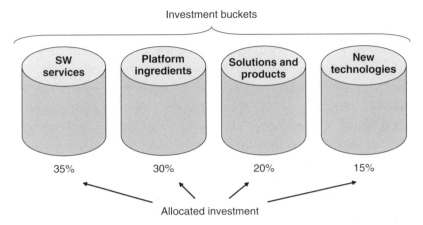

Figure 3.7   Software development portfolio example.

the value of (our) platforms. The investment buckets (or portfolio of programs) and allocated funding levels for this organization are shown in Figure 3.7.

Balancing the portfolio achieves the following two primary things for the business: It allocates funding based on strategic intent, and it dedicates funding and resources to specific portfolios. As a result, the company's strategic objectives are adequately funded in priority order, and resources are focused within the investment buckets defined, eliminating cross-portfolio resource contention.

Establishing a Strong Link to Strategy

The third principle of portfolio management establishes the fundamental imperative that resource allocation must be driven by strategic intent. Strategic objectives must influence how the company spends its investment money, which is normally in the form of a research and development (R&D) budget. As Cooper et al. so eloquently explains, "Well meaning words on a strategy document or mission statement are meaningless without the funding and resource commitments to back them up."[12]

Ensuring strategic intent can be accomplished through implementation methods for maximizing the value of the portfolio and establishing balance. When strategic elements are included in the program prioritization criteria, the value of each program becomes apparent and the results are greater. Also, by utilizing strategic objectives and the mission to define the investment buckets of the enterprise, the overall structure of a company's portfolio solution becomes aligned to its strategic objectives. Then, by allocation of funds and resources to the investment buckets, an organization can be confident that its resources are assigned to programs that provide the greatest strategic value.

Executing good decisions that give the portfolio a robust set of programs that reinforce the business strategy requires strong leadership from executive management. Executives must ensure that the process is effective and that, at the end of the day, the tough choices are made. Without strong leadership, an organization will continue to overcommit its resources and underachieve attainment of its strategic objectives. The box titled, "Do You Agree with This Portfolio Verdict?" is an example of such a scenario.

Do You Agree with This Portfolio Verdict?

The first thing Peter, Star Tech's new software development manager, did was to convene a meeting to review the status of the software development team's programs. There he learned that programs were taking far too long to complete. His feeling was that the pipeline was clogged, with too many programs trying to get through it—28 in total. There were some obvious good programs, but what he was perplexed by was the large number of low-value programs. No two of them were in the same product performance/market segment area. The approach was a strategic scattergun. Practically, there was no glue to bond all of the programs together and a lack of strategic focus was more than evident.

All of the programs held a promise of high reward, which was obviously a good sign. The problem was that all of them were high risk, with a reasonable probability to fail technically or commercially. Additionally, almost all programs were long term. The absence of "quick hits" to balance the long-term development programs to bring in near-term revenue was apparent. Also, almost all of the programs were in the programming phase, with no programs in the requirements definition or closure phase.

Okay, Peter thought to himself. *Now the issues seem clear. To talk to my direct reports, I need to summarize this into a few words, describing the current state of our portfolio of programs.* The following is a list of points that Peter wants to relay to his team:

- Our portfolio is unfocused.
- There are too many programs, causing resources to be spread too thinly.
- We are taking on too much business risk.
- Our portfolio is poorly balanced.
- Our portfolio does not support our business strategy.

Do you agree with Peter's assessment? What we see in this example is the all too common scenario plaguing product, service, and infrastructure development companies. Their portfolio of programs is too large, they are not well balanced, and they are not strategically focused. The result is predictable, programs take too long, cost is too high, and the business objectives of the company are not achieved.

THE PROGRAM MANAGEMENT SUBSYSTEM

At the heart of the integrated management system is the program management function, as shown in Figure 3.8.

The program manager does *not* create the mission or strategic objectives. This is the role of senior management. The program manager does *not* plan and execute the project deliverables. This is the role of the project managers and project team members. The program manager *does* ensure the attainment of the value proposition identified by the portfolio management process by delivering an integrated solution through the collaboration and coordination of multiple interdependent projects. If the value proposition is attained, alignment between business strategy and project execution is achieved.

So then, what does it mean to say a company's programs are aligned with its business strategy? What elements of program management align

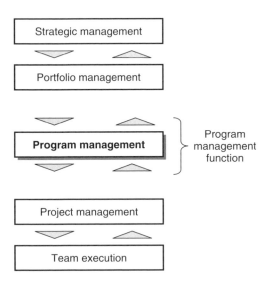

Figure 3.8   Program management function of the integrated management system.

to which elements of business strategy? And, how do you align them? The following section addresses these questions.

First, what is alignment? Alignment is the degree to which priorities of an organization's program management practices are compatible with that of its business strategy. Second, what gets aligned? The focus and content of the program management elements are aligned within attributes of the business strategy. Let's first determine what the program management elements are and then define their focus, content, and attributes.

Program Management Elements

These following six elements make up the program management subsystem:

1. Strategy
2. Organization
3. Processes
4. Tools
5. Metrics
6. Culture

Program strategy is defined as the approach, position, and guidelines of what to do and how to do it to achieve the highest competitive advantage and the best value from the program.[13] As previously mentioned, business, technology, and market strategies must culminate in a program strategy to be executed, as illustrated in Figure 3.9.

The program strategy manifests itself into four important elements of the program:

- The product, service, or infrastructure definition
- The program business case

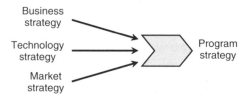

Figure 3.9 Program strategy focuses on business, technology, and market strategy.

- The program success criteria
- The integrated program plan

Every development program relates to a product, service, or infrastructure capability. The development process begins with ideas and conceptualization. Developing the product, service, or infrastructure definition includes synthesizing end user, technology, and market research to establish a conceptual idea to address the business, technology, and market strategies of the organization. These elements are instantiated in the definition.

The business case is the most vital part of the strategy. It establishes the degree of likelihood that the concept will help to achieve the strategic objectives of the organization. It spells out rationale as to why the customer will buy or use the product, service, or infrastructure capability and what makes it better than other alternatives. The business case also presents the competitive advantage. For example, the advantage may be a combination of product attributes, functionality, performance, and quality.

The program success criteria establishes the business criteria with which to define and measure program success, and, in effect, set the tactical targets to achieve the strategic objectives. The criteria makes it clear in advance how the program results be evaluated. Success and failure criteria will be outlined in detail and the different success dimensions will be judged, such as business, technology, cost, and timing. The business perspective is the dominate criteria that, depending on the type of program, may include the growth pattern of sales, market share, profitability, or increased worker productivity during a period of several years.

The integrated program plan sets the program scope and defines how the objectives will be obtained via the delivery of a product, service, or infrastructure capability. Scope defines the program boundaries, the final deliverables on the program level, and the work that will and will not be done on the program. It also includes an estimated resource profile and overall budget needs. Therefore, the integrated program plan defines the program deliverables and shows the integrated time, resource, and budget needs.

Program organization consists of the structure that is needed to successfully deliver the intended results. The most common error companies make in implementing program management is failing to understand the difference between a program structure and a project structure. Chapter 5 will describe in detail the two critical factors to consider when structuring a program, that is, the structure must support the highly integrated and interdependent nature of the projects within the program, and the structure must support the fundamental elements of team success.

Program processes such as the PLC, schedule management, financial management, risk management, and so on, help to make the operational aspects of the program more efficient and effective (see Chapter 8). Most importantly, program processes help to ensure that the work of the multiple project teams is being managed consistently and in a coordinated manner.

Program metrics (or performance measures) gauge how well the program strategy works, how much progress is being made toward selected goals, or how well programs are performing to date (for example, profitability index and staffing levels).[14,15,16,17] The intent is to focus attention on the business aspects of the program (see Chapter 9).

Program tools are the procedures and techniques that process the data involved in a program and extract the information necessary to determine progress toward achievement of the program objectives and support important program decisions (see Chapters 10 and 11).[18]

Program culture refers to the behavioral norms and expectations shared by members of a program.[19] The essence of culture is to have a sense of identity with the program organization and to accept investing both materially and emotionally in the success of the program (see Chapter 14).[20]

Aligning Program Management Elements with Business Strategy

Figure 3.10 illustrates how program management elements should be driven by the business strategy of a company. A company chooses a strategic direction by selecting competitive attributes of its strategy (for example, time to market, quality, cost leadership, or product feature set).

These attributes, in turn, are used to drive the focus and content of the program management elements for a particular program. For example, if the competitive attribute of time to market is chosen, the primary strategic focus of the program becomes schedule driven, and the elements

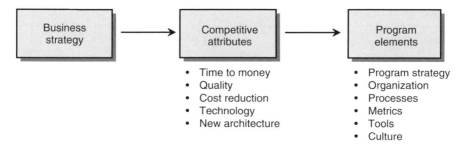

Figure 3.10   Aligning program management elements to business strategy.

are configured to support this strategy. Examples of program management element configuration used by different companies are shown in Table 3.1.

To extend the alignment process further, the focus and content of the program management elements shape the project management elements of the tactical subsystem of the integrated management system.

Table 3.1   Aligning program management elements to strategy

Strategy		*Program management*
Competitive attributes	*Program focus*	*Content of program management elements (examples)*
Time to market	Schedule driven	(Strategy) Dropping features if necessary in a trade-off situation, spending additional money to recover projects if they slip
		(Organization) Building a flexible structure to facilitate the speed of project execution
		(Process) Overlapping and combining phases, milestones, activities
		(Metrics and tools) Focusing on schedule
		(Culture) Building schedule-oriented program culture (for example, rewarded speed)
Superior quality	Quality driven	(Strategy) Delaying schedule if necessary in a trade-off situation
		(Organization) Building a flexible structure to ensure the quality level of the product
		(Process) Having a sequential, iterative process
		(Metrics and tools) Focusing on scope/risk
		(Culture) Building quality-oriented program culture (for example, rewarded quality)

Table 3.1   Continued

Strategy		Program management
Competitive attributes	*Program focus*	*Content of program management elements (examples)*
Cost reduction	Process-improvement driven	(Strategy) Focusing on process improvement (Organization) Having a flexible structure and adapting it to changes in process improvement, with the ultimate goal of saving costs (Process) Having a standardized process with templates (Metrics and tools) Focusing on schedule/cost (Culture) Building cost-oriented program culture (for example, rewarded cost)
Innovative feature and cost	Feature and cost driven	(Strategy) Focusing on features with the major concerns of program cost and schedule (Organization) Having a flexible structure to ensure the innovative features at minimum cost (Process) Having a standardized but flexible process (Metrics and tools) Emphasizing feature tracking and cost (Culture) Building an innovative and cost-oriented program culture (for example, rewarded innovation)
Superior quality and cost	Quality and cost driven	(Strategy) Focusing on quality with the major concerns of cost and schedule (Organization) Having a flexible structure to ensure the best quality at minimum cost (Process) Having a standardized process with the aim of minimizing cost and ensuring superior quality and service (Metrics and tools) Focusing on quality, cost, and schedule (Culture) Building quality- and cost-oriented program culture (for example, rewarded quality and cost)

Table 3.1   Continued

Strategy		Program management
Competitive attributes	Program focus	Content of program management elements (examples)
Science and cost	Science and cost driven	(Strategy) Focusing on science with the major concerns of cost (Organization) Involving technical teams to ensure that the science goals are met at minimum cost (Process) Having standard guidelines with the aim of checking that the science goals are met within the expected cost (Metrics and tools) Focusing on technical, cost, and schedule (Culture) Building science- and cost-oriented program culture (for example, rewarded science goals)

Project management elements (organization, processes, tools, and culture) must follow the same configuration as the program management elements to gain consistency across the projects. Through this process, systematic alignment between business strategy and tactical execution is achieved.

However, once a company achieves alignment between strategy and tactics, it does not mean that it no longer has to focus attention on the alignment. The focus and content of the program strategy, as well as the program organization, processes, metrics, tools, and culture, need to be periodically reviewed and checked against the competitive attributes and strategy of the business.

The operating conditions (the actual conditions of program implementation) may or may not be the same as those assumed in the planning phase of the program because of changes in the environment. The actual operating conditions may differ significantly from the original (or intended) business strategy and need to be changed to remain realistic. This process provides strategic feedback and helps a new strategy emerge.[21] An example of what can happen when this step is not followed is provided in the box titled, "Asleep at the Wheel."

Asleep at the Wheel

Conventional wisdom says that when a program is well aligned with business strategy, it will be successful. The following true story demonstrates a different picture.

The Tripdale program was on the path to success. The program strategy, especially the two elements of product definition and business case, were meticulously thought out. The product feature set was chosen in direct negotiations with the lead customer. This got the program started off on the right foot. Detailed program planning was conducted and the phase gate reviews were performed timely and thoroughly but without customer representation. The strategic alignment of the program was frequently discussed in the PCT and with the vice president in charge of new product development. All parties agreed that it was a well-aligned program on its way to be a money maker.

However, near the end of the program, the lead customers were contacted to announce the program completion and product availability. The customers reviewed the product details and stated that they were looking for a different feature set. What a surprise! The product suddenly had no market.

Because there was no regular, periodic contact with the customers, the alignment disappeared at some point, and the PCT did not know that they were actually misaligned. *The impact was $800,000 of lost development funding.*

This could have been avoided if the PCT was frequently in contact with the customers and reviewed program alignment during the periodic phase gate reviews. The alignment index tool presented in Chapter 10 is an excellent aid in determining program alignment.

The moral of this true story is that a program manager always has to be aware of the alignment of his or her program to the strategies of the business.

TACTICAL SUBSYSTEM OF THE INTEGRATED MANAGEMENT SYSTEM

Tactical elements of the integrated management system include project management and team execution, as illustrated in Figure 3.11. Project management and team execution form the basis for planning, implementing, and delivering the interdependent elements of products, services, or infrastructure capabilities.

This is where the real hands-on work gets completed. The functional project managers are responsible for the detailed planning and execution of the deliverables pertaining to their respective operating functions (for example, hardware engineering, software engineering, and marketing). Each functional project manager, along with his or her respective team

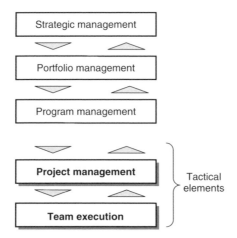

Figure 3.11   Tactical subsystem of the integrated management system.

members, develops a specific project plan and schedule. The results of this planning will be further refined and combined between the program manager and the functional project managers for the overall program. They will ensure that all of the interdependencies, gaps, and overlaps are appropriately addressed and integrated into the master program plan. Once the program plan is approved, each project team executes their part within the program structure, with the oversight of the program manager.

Project management and team execution are the subjects of volumes of books. Thus, we will not attempt to summarize the details of project planning and execution in this book. However, the key differences between project management of an independent project versus one within a program are worth noting. The projects within a program are highly interdependent; therefore, a significant amount of cross-project collaboration is required of the project managers. Obviously, this is not the case for independent projects. Additionally, project managers for a program are directed by a program manager who has full responsibility, authority, and accountability for the success of it and all the projects.

PROGRAM MANAGEMENT AS A COMPETITIVE ADVANTAGE

Program management is an effective means to bridge the chasm between organizational strategies and project execution. It is a core business function that integrates the work of multiple project teams and focuses that work on the attainment of the company goals. Throughout the development life cycle, program management methods measure the work output against the goals of the organization.

A company that has good business strategy, programs, *and* execution is bound to outplay a rival that has good strategy but cannot execute. Effective use of the program management element of an integrated management system creates a competitive advantage.

Gary Rosen, vice president of engineering for Varian Semiconductor Equipment, describes the competitive advantage of program management in the following way:

> *Good program management goes right to the bottom line, it improves a company's P&L* (profit and loss). *A company that delivers more products, better products, and does so faster wins the competitive race. Program management makes better, faster, and cheaper a reality.*

SUMMARY

Program management as a business function has a central role in aligning programs with business strategies. The role of program management is the mechanism from which new products, services, or infrastructure are conceived and developed to realize the mission and strategic objectives of the business. The integrated management system includes the following three subsystems: strategic, program management, and tactical.

The strategic subsystem, including strategic management and portfolio management, concocts the business strategy, clearly identifies the strategy's competitive attributes, and chooses a set of programs to accomplish the strategy. The program management subsystem elements (program strategy, organization, process, tools, metrics, and culture) are then aligned with the competitive attributes of the business strategy. Lastly, the tactical subsystem, including project management and team execution, is aligned to business strategy through the program strategy and master program plan.

Program management links and aligns the projects within a program to the business strategy of the company. This enables a company to consistently execute its strategy and create a competitive advantage over its rivals.

The Principles of Program Management

▼ Program management is executed as an element of an integrated management system

▼ The value proposition identified by the strategic elements are attained by integrating the work output of the tactical elements through the program management model

▼ The program strategy consists of the following three key elements: the product, service, or infrastructure idea, 'the business case', and the strategic success criteria

---

Program Management in Practice

LorryMer Information Technology
Sabin Srivannaboon and Dragan Z. Milosevic

Prologue

This industry example focuses on the information technology (IT) department within LorryMer Corporation, a leader in a specialized motor vehicle industry in North America. LorryMer sales began to suffer substantially in the early 2000s due to the recession in the U.S. economy. As a result, the company was forced to change its primary strategic objectives to focus on cost efficiency. In particular, cutting operating costs to improve the company's competitive position.

First, the example shows how LorryMer's IT programs were aligned with the company's strategic goals and business values. We show how the alignment process was tailored for LorryMer's new business strategy. In doing so, we present an effective alignment tool used by the company. The tool is simply a chart designed to present the alignment between business goals, business values, and the IT programs.

This case demonstrates that alignment between business strategy and program execution begins at the top, where strategy drives the desired business results. Effective program management practices then deliver the business results.

In Troubled Waters

LorryMer specializes in designing, developing, and manufacturing a complete line of technologically advanced motor vehicles. The company has been in business for more than six decades and now operates seven major vehicle manufacturing plants, with one plant in North America. With more than 14,000 employees, its mission is to provide the highest standard of technological innovation and premium quality to its customers. LorryMer remains committed to the highest degree of innovation and quality and is dedicated to meeting its customers' needs.

Saul McBarney, LorryMer's IT program management officer, has been with the company for 25 years. McBarney is proud of the company's history, strategy, and background. "We are unique in terms of listening to the customers. We find out what customers' business needs are first and then develop products for them that meet their needs."

However, 2001 was a painful year for LorryMer. As a result of economic hardship in the motor vehicle industry, the company's sales and profits dropped, and its losses exceeded $1 billion. In September 2002, the *Refrigerated Transporter* reported that depressed-used automobile prices caused many carriers to extend trade cycles because the sale prices on the automobiles were less than what was owed.

As a response to the crisis, the company embarked on a major cost-saving initiative in late 2001, which later produced a number of cost-saving programs. James Ostar, a LorryMer business value account manager and a member of the business advisory group, said, "It was about cost, cost, cost, and nothing else."

Nevertheless, LorryMer continued to experience substantial business challenges in the following years. With the economy still suffering, LorryMer's automobile production in 2002 did not meet 2001 production levels, which were significantly lower than 2000. Funding for any new programs was minimal, especially during 2003. LorryMer needed to change to survive.

Overcoming Strategic Obstacles

Bill Mennon, chief information officer for LorryMer, said, "The business environment for all automobile manufacturers remained depressed and was not expected to regain its historically high levels until late 2004. We knew we had to make some changes to be ready to compete." LorryMer focused on reducing costs quickly. As a result of relentless cost cutting programs, LorryMer's operating costs were dramatically reduced and breakeven profit was achieved in 2003.

"However, we knew those changes were tactical and not sufficient," Mennon said. "We had to make additional changes." Therefore, during 2005, LorryMer modified its strategy by focusing on five new core values, as follows: providing market leadership and brand coverage; pursuing technological innovation; partnering with operators for maximum productivity; focusing on the needs of its customers, employees, communities, and the environment; and being an advocate for their industry.

Aligning Information Technology Programs to Business Strategy

During the past several years, the IT strategy at LorryMer has been to maintain legacy systems. Most of the new IT investments were dedicated to e-business websites for after-market sales and marketing. Ostar said, "Eighty-five percent of what we sell is a complete commodity available from thousands of other people. So the only thing we really have to sell to increase profit is service." The remaining incremental IT spending supported enhancements to keep the legacy systems compliant with government regulations and provided some basic level of additional functions.

To improve the company's competitive position, radical changes were needed to create an IT environment that was better positioned to support all five of the new core values. As part of the improvement efforts, the IT department was reorganized and made much more agile in 2005. That was done by replacing the functional structure with a matrix organization, consisting of application (IT application developers and service providers) and program management (program managers and business system analysts) competency centers. Then, a business advisory group concept was introduced. It consisted of value account managers who provided IT focus in business units and acted as the IT voice of the customer, which was something LorryMer lacked previously. Also, a joint strategic planning session of business units and IT was initiated in the middle of 2005. Its purpose was to determine the business needs to be supported by the replacement of legacy IT systems. McBarney commented, "This was previously unheard of. In the old system, IT goals were either left to IT or imposed on us by the company. Now, our strategic goals and direction were determined by the end users in the joint strategic planning."

To ensure IT strategy and business strategy alignment, IT activities and programs had to complement the needs of the business. The IT team used an alignment chart tool to accomplish this (see Figure 3.12). Mennon explains, "The alignment chart provides a strategic mapping of the business goals, the business values (initiatives), and the IT programs. As part of the strategic changes, we were tasked by the company to design the alignment process, part of which was accomplished by the alignment chart. The chart was designed to help us visualize the alignment between business goals, business value and the programs."

Ostar offered an example of how to read the chart. "A bubble on the intersection of business goal 4 and business value 3 means that they are aligned. Further, a bubble indicates that business value 3 intersects with program 3, meaning they are aligned. In summary, program 3 delivers business value 3, which achieves business goal 4. The alignment is all about IT programs producing business value by helping us achieve our business goals. That is the language we want everyone to speak."

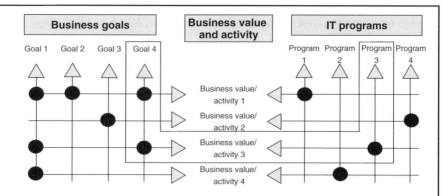

Figure 3.12   An example of an alignment chart tool.

The alignment chart is a useful mechanism that not only helps businesses visualize the alignment efforts but also builds a standard language in the company in which every IT investment is aligned to an organizational need (for more on the alignment charts and tools see Chapter 10).

Alignment Begins at the Top

On paper, the alignment chart is straightforward, but the execution of the steps in the alignment process is more complex. LorryMer has a mix of formal and informal processes for ensuring proper IT strategy and business strategy alignment. The processes are internally embedded within the business strategy formulation and throughout the PLC.

The six steps in the alignment process are as follows (see Figure 3.13):

1. Strategic planning
2. Informal portfolio management
3. Envisioning
4. Planning
5. Development
6. Deployment

In step 1, strategic planning, the strategic plan document is developed to help accomplish the goals of IT in support of the company's business strategies and goals for the three-year-planning horizon.

According to Mennon, "Our challenge within IT has always been to take a look at how we treat the goals, whether they're well defined or not, and determine what our strategic plan is to help accomplish those goals. Our approach is an evolutionary process. Once you know where you want to go, you take it one step at a time and one program at a time, and each program gets you closer to the goals."

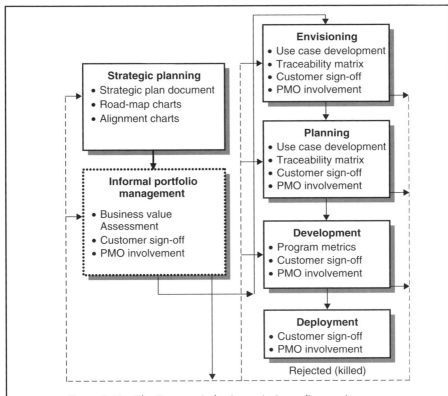

Figure 3.13 The Program to business strategy alignment process.

Then, business value, called strategic initiatives and programs, are formulated based on the goals of the business strategy. These are done by the business advisory group and Program Management Competency Center. Tools at the strategic level, which are used to ensure the quality of the alignment, include the strategic plan document, road map charts, and alignment charts, as follows:

- *Strategic plan document* is a tentative capital plan that shows the snapshot of all program summaries, program types, estimated head count and payback. It is based on the business strategy and goals of different business units that IT supports.
- *Road map charts* are used to define where the company wants to be in IT in the next three-year time frame. They address all of the business goals of different business units that IT attempts to support and their time frames.
- *Alignment charts* are the mapping of the business goals, the business value, or strategic initiatives, and the IT programs.

Step 2, the informal portfolio management process, is a significant part of LorryMer's alignment process. Once the portfolio management process is completed, it produces information about a set of candidate programs, their investment opportunities, and program priorities. In short, the process reveals the most viable programs and possible risks that could occur in the envisioning, planning, development, and deployment phases of their life cycles. McBarney said, "Portfolios help to assess and monitor programs that are the best investment." Currently, this assessment is done in the portfolio process through business value assessment (BVA). BVA consists of a set of criteria prepared by value account managers. The program management office (PMO) is asked to rank the criteria regarding business-value weight, which is expressed on a 1–10 scale with 1 being the lowest value and 10 being the highest value. Second, it assesses business risk, which is also expressed on a 1–10 scale with 1 being the highest risk and 10 being the lowest risk. An example of BVA criteria is shown in Table 3.2.

Programs are then evaluated and selected based on BVA, program type, program size, priority, and business area. Once programs are selected, they go through the standard life cycle phases of envisioning, planning, development, and deployment. A program manager is usually assigned at the end of the portfolio process or the beginning of the envisioning phase, which is step 3 of the alignment process.

McBarney commented, "Ideally, we like to have a program manager on board when envisioning is just ready to begin. That allows enough time to develop a close relationship with the business system analyst, who is responsible for detailed planning and execution of the program. Sometimes we engage a program manager immediately at the time the portfolio process starts. Therefore, the program manager is responsible, along with the value account manager, for the business value of the program and is solely responsible for achieving the program's other requirements."

"In the envisioning and planning stages (steps 3 and 4), the ball is in the program manager's court", Jeff Barrison, a program manager, said.

Use case development and traceability matrixes are used to develop the program plan. A program manager maps the details of the program with the use case, describing a series of tasks that users will accomplish from using the software, which also includes the responses of the software to user actions. The program manager also traces program actions back to the initial requirements (traceability matrix) to ensure that the plan is still aligned with the requirements. In addition, the business system analyst and the program manager meet with the business members and end users to get sign off, as part of a mechanism to make sure that the program is still in line with the business goals and requirements.

Table 3.2   LorryMer business value assessment

Business-Value Weight	*1-Lowest 10-Highest*	*Business-Risk Weight*	*1-Highest 10-Lowest*
Generates at least $1 million in revenue per year		Regulatory requirement	
Cost savings of at least $500,000 per year		Necessary cost of doing business; key business processes will cease if not done	
Improves employee communication, culture, or morale, resulting in a 10% increase in the next annual employee survey results		Required to support a new product	
Product quality that directly affects JD Powers measurement		Replaces legacy systems that inhibit required business process changes	
Improves dealer/customer experience, resulting in an increase of 1/2 market share		Synergy with ZBZ	
Cost avoidance of at least $500,000 per year		Affects existing product build capability	
Payback in less than 18 months		Affects existing motor vehicle and parts sales capability	
		Current customer satisfaction level will decrease as measured by JD Powers	
		Competitive threat will result in the loss of market share and/or net profit	
TOTAL*		**TOTAL***	

*Note: the higher the total, the better

After the program plan is approved, the PCT starts to implement step 5, or the development stage, of the alignment process and, later, the deployment stage (step 6). According to Barrison, program metrics are used as a mechanism to track progress to plan. At the end of each program life cycle phase, a program manager is required to present the program status to the PMO and get customer sign off to be able to proceed from one phase to the next. Therefore, customer sign off and PMO involvement are the other control mechanisms to make sure that the program is still in line with the expectation throughout the program life cycle.

The program status report is used as a vehicle to communicate between the program team, PMO, and customers. When a program does not meet the requirements at each of the phase gates, adjustments are required (if the program is not killed). This mechanism is referred to as a corrective approach, in which a new strategy or action emerges from a stream of managerial decisions through time.

Closing

Aligning programs to business strategy is new to LorryMer. Mennon remarked, "We are learning the alignment process, and I am happy with its results. For the first time, we are aligning IT programs with the business strategy—and it works."

The LorryMer experience offers a good example of the use of an integrated management system in practice to achieve strategic business objectives. Key elements of the LorryMer experience include the following:

- Alignment starts at the top—strategy drives intended business results and competitive advantage.
- To execute strategy, a company needs to consistently select, fund, and resource the programs that best align with the strategic objectives and contribute the highest business value.
- Effective program management practices deliver the intended strategic business results and create a competitive advantage for a company.

REFERENCES

1. Bowen, H.K, Clark, K.B, Holloway, C.A, Wheelwright, S.C. Development projects: The Engine of Renewal. *Harvard Business Review* 72(3), 1994, pp. 110–120.
2. Martinelli, Russ and Jim Waddell. "Aligning Program Management to Business Strategy". *Project Management World Today* (January-February 2005).

3. Pearce II, John A. and Richard B, Robinson Jr. *Strategic Management: Formulation, Implementation, and Control*, New York, NY: McGraw-Hill, 2000.

4. Gharajedaghi, Jamshid. *Systems Thinking: Managing Chaos and Complexity*. Woburn, MA: Butterworth-Heinmann Publishing, 1999.

5. King, William R. and Cleland, David L, *Strategic Planning and Policy*, New York, NY: Van Nostrand Reinhold, 1978.

6. Proctor & Gamble website, http://www.pg.com

7. Merck Website, http://www.merck.com

8. Pearce II, John A. and Richard B. Robinson. *Strategic Management: Formulation, Implementation, and Control*, New York, NY: McGraw-Hill, 2000: pp. 241–244.

9. Martinelli, Russ and Jim Waddell, "Aligning Program Management to Business Strategy", *Project Management World Today* (January-February 2005).

10. Cooper, Robert G, Edget, Scott J. and Kleinschmidt, Elko J, *Portfolio Management for New Products*. Cambridge, MA: Perseus Publishing, 2001: pp. 3.

11. Cooper, Robert G, Edget, Scott J. and Kleinschmidt, Elko J, *Portfolio Management for New Products*. Cambridge, MA: Perseus Publishing, 2001: pp. 87–88.

12. Cooper, Robert G, Edget, Scott J. and Kleinschmidt, Elko J, *Portfolio Management for New Products*. Cambridge, MA: Perseus Publishing, 2001: pp. 140–141.

13. Poli, M. and Shenhar, A.J, *Project Strategy: The Key to Project Success*. in Portland International Conference on Management of Engineering and Technology. Portland, OR: PICMET 2003.

14. Milosevic, D.Z, Inman, L. and Ozbay, A, "Impact of Project Management Standardization on Project Effectiveness", *Engineering Management Journal*, 2001. 13 (4): 9–16

15. Tipping, J.W, Zeffren, E. and Fusfeld, A.R. "Assessing the Value of Your Technology". *Research Technology Management*, 1995. 38(5): 22–39.

16. Major, J, Pellegrin, J.F. and Pittler, A.W, "Meeting the Software Challenge: Strategy for Competitive Success". *Research Technology Management*, 1998. 41(1): 48–56.

17. Nicholas, John, M, *Managing Business and Engineering Projects: Concepts and Implementation*. Upper Saddle River, NJ: Prentice-Hall, Inc., 1990.

18. Aronson, Z.H, et al. "Project Spirit—A Strategic Concept". *Portland International Conference on Management of Engineering and Technology*. Portland, OR: 2001.

19. Graham, Robert J. and Randall L. Englund. *Creating and Environment for Successful Projects*. San Francisco, CA: Jossey-Bass, 1997.
20. Shenhar, Aaron, *Strategic Project Management: The New Framework*. PICMET, 1999
21. Mintzberg, Henry, *The Rise and Fall of Strategic Planning*. New York, NY: The Free Press, 1994.

Chapter 4

Managing and Delivering
the Whole Product

In presenting the business case for program management in Chapter 2, we noted that consumer demand for customized solutions, increased integration of features and functions, and incorporation of the latest technologies are powering ever-increasing product, service, and infrastructure complexity. Many business leaders have come to the realization that their current methods of development are no longer sufficient to manage the higher levels of complexity and have turned to the program management model as the solution.

In this chapter, we describe how the program management discipline is deployed to manage increased complexity of design and process. In doing so, we utilize the concept of the *whole product* to demonstrate the value of program management in meeting ever-increasing customer expectations. First, we explain the whole product concept and how companies are utilizing it to win in business. Then we define the steps used to develop and deliver the whole product through the program management discipline.

The purpose of this chapter is to fully explore the following concepts of program management:

- The relationship between the whole product concept and program management
- How program management is used to manage complexity driven by customer wants and needs
- The role of program management in the development and delivery of the whole product

These concepts lead to an understanding of the value of program management as a critical business function within a company to effectively manage increasing complexity by developing and delivering complete solutions that fulfill customer expectations.

THE WHOLE PRODUCT CONCEPT

The concept of the whole product is not new. In marketing terms, the whole product is defined as the products and services that best meet the customer's wants and needs (see box titled, "The Concept of the Whole Product").[1]

The Concept of the Whole Product

The whole product concept is simple; there are two compelling value propositions for each company's products, services, or infrastructure capabilities, as follows: the expectation on the part of the customer that their wants and needs be met and the ability of the company to provide a product, service, or infrastructure capability that fulfills the wants and needs. Many times there is a gap between the two. To close the gap, the company must add an array of services and ancillary products to the original solution; therefore, creating the **whole product**. To understand this, let's examine the different forms of products.

- **Generic product** is what is shipped in the box and purchased.
- **Expected product** is the product that meets the *minimum* expectations of the customer.
- **Whole product** is the product providing the *maximum* chance to achieve the customer's buying objective.

In marketing battles, the generic product is the center of the battle for the early market. When the market shifts toward a mainstream market, more sophistication is needed to win, and the center of the market battle shifts to whole product solutions.

If we shift from a marketing focus to a product, service, or infrastructure development focus, the whole product can be defined as the integrated product solution that fulfills the customers' expectations.[2] It is important to note that the whole product can be a physical product, service, or infrastructure capability. If we purchase a personal computer, for example, we wouldn't consider it acceptable if we were delivered a box of circuit boards, a second box containing the enclosure and power supply, another box containing peripheral devices such as disk drives, network adapters, and sound cards, and finally an envelope containing the computer software applications and operating system on a series of compact disks. Rather, unless we're a computer hobbyist or a systems integrator, we *expect* to receive an integrated computer that we can unpack, set up on a computer station, plug in, and begin using. Thus, we want to experience the delivery of the whole product. For a good example of how market leaders practice the whole product concept, see box titled, "Delivering the Whole Product at Dell Computers."

Delivering the Whole Product at Dell Computers

No one understands the value of the whole product concept better than Dell Computers. Dell has built its business model on the whole product concept. It understands that customers have individual wants and needs when it comes to using a personal computer in their work or personal lives. This is the primary reason one will not find Dell computers in retail outlets. The user finds it only marginally valuable to purchase a basic, generic personal computer off the storeroom shelf. Instead, the user gains greater value by being able to customize his or her solution based on individual needs. Dell's online or phone ordering process provides that opportunity.

When one orders a Dell computer, he or she starts with a generic version of a computer model. From there, the customer can order specific hardware, software, support, and training options to create a personal version of the computer product that meets his or her individual usage needs. Dell, therefore, provides the whole product—the integrated product solution that completes the customer's expectations.

For companies that use the program management discipline to develop products, services, or infrastructure capabilities, the program manager is responsible for the delivery of the whole product to the marketplace or customer's environment. This means he or she must have a full understanding of what customers expect in the total solution. The project managers for the program are responsible for the development and delivery of the individual elements of the whole-product. Figure 4.1 graphically illustrates the whole-product concept in the development of a personal computer. In this simplified example, the whole product consists of the integration of six elements: the motherboard development, the memory circuit board development, the software development, the enclosure development, product manufacturing, and product testing. Other functions normally involved in the development of a personal

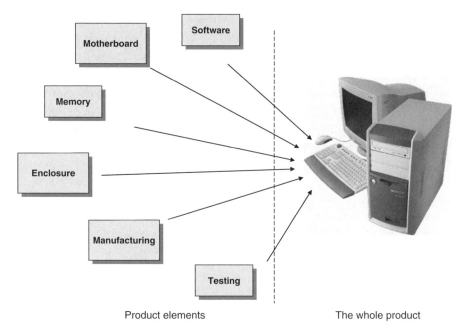

Figure 4.1   The whole-computer product.

computer—such as architecture, product marketing, quality, customer support, quality assurance, and finance—are purposely not shown for the sake of simplicity.

The development program is organized in a similar manner. Each of the six elements of the personal computer shown in Figure 4.1 is organized as a functional project, which is led by a project manager and consists of a team of functional specialists in the domain they represent (see Chapter 5). The job of each project team is simple: plan, develop, and deliver its respective element of the whole product to the other members of the program team.

The program manager, in turn, leads the development program and is responsible for integrating the elements into a whole product and delivering it to the customers and stakeholders. In this case, the whole product would be the personal computer consisting of the integration of the various circuit boards, the memory devices, the enclosure, the software stack, and the manufacturing and testing of the product. In addition to delivery of the product, the program manager is responsible for the achievement of the business objectives for which the product development effort was initiated, ensuring the whole product meets the expectations of the customers.

USING PROGRAM MANAGEMENT TO DELIVER THE WHOLE PRODUCT

As many companies have learned, meeting the customer's expectations becomes the means to achieve the strategic business objectives of the firm. If customers put a priority on receiving the whole product, the concept then needs to be part of a company's business strategy. Once it is part of the strategy, the program management discipline becomes responsible for developing and delivering the whole product to the customer. Managing the development and delivery of the whole product involves the following aspects:

- Understanding cross-project dependencies
- Determining the level of development complexity
- Selecting the management approach
- Planning the whole product program
- Executing the whole product program

Understanding Cross-Project Interdependencies

One should clearly understand the extent of the dependencies between the projects before choosing a management approach. In the development of a whole product solution, effective management of the interdependencies between the project teams is critical. The interdependencies between projects consist of a series of cross-project deliverables, coordinated tasks, cross-discipline decisions, and shared management of risks and problems that will be encountered on the product, service, or infrastructure development effort.

As the term implies, interdependent projects are those that have a dependence upon the delivery of an output from other projects. For development programs, project interdependencies normally equate to deliverables. A deliverable is a tangible output from one project team that is delivered to one or more of the other project teams. For example, a completed software module handed off from the software development project team to the system test team is a deliverable between two project teams involved on a program. Interdependency between the project teams is formed when a deliverable from one project team is needed to successfully complete the work of a second project team.

To illustrate how interdependencies between projects are a determining factor in program management, let's look at a simplified product

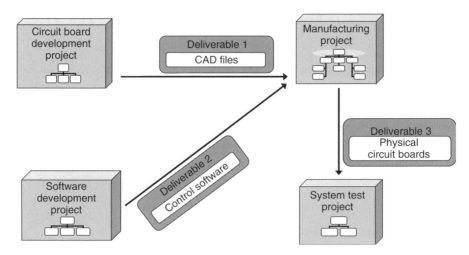

Figure 4.2   Cross-project deliverables.

development effort. Four of the projects represented on the product development effort are circuit board development, software development, preproduction manufacturing, and system test as shown in Figure 4.2.

The circuit board development project team is responsible for the design and development of the primary control board in the product, the software project team develops the software stack used for command and control of the product, the manufacturing project team builds the physical circuit boards, and the system test project team is responsible for all functionality testing.

Let's examine a few of the cross-project deliverables in Figure 4.2 that form the interdependencies between the projects. The three deliverables shown are: (1) the computer aided design (CAD) files delivered from the circuit board project team to the manufacturing project team, (2) the circuit board control software delivered by the software project team to the manufacturing project team, and (3) the physical circuit boards delivered from the manufacturing project team to the system test project team.

This series of deliverables creates a highly interdependent relationship between the projects. For the first deliverable, the manufacturing project team needs the CAD files from the circuit board project team for the production of the physical circuit boards. A hard interdependency is established between the circuit board development and manufacturing project teams.

The manufacturing team cannot build a circuit board that will be capable of powering and controlling the circuit boards without the command

and control software from the software project team. This, of course, would lead to failure of the system test project team to complete the test and validation of the product.

Finally, the system test project team will not be able to successfully complete, let alone start, the product functionality test cycle without the circuit boards delivered by the manufacturing team. The three deliverables described create a highly interdependent relationship between the four project teams. The program manager needs to manage these interdependencies as closely as the independent deliverables created by the project teams.

Determining the Level of Development Complexity

The level of development complexity is highly dependent upon the number of cross-project interdependencies and has a vital influence on businesses choosing either a project management-only approach or a program management approach. To illustrate, we will simply add two elements to the example above. First, we will assume that the product requires the development of application specific-integrated circuitry by an integrated circuit (IC) development project team. Second, we will assume that the product requires the development of an enclosure to power, cool, and contain the product, requiring the work of an enclosure development project team. Our expanded product development effort is shown in Figure 4.3, with each arrow representing a potential deliverable interdependency between the projects.

One can visualize that the complexity of the interdependency structure between the projects increases with the addition of just two product elements. The original three interdependencies from the previous scenario—CAD files, command and control software, and physical circuit boards—are still required, but many additional interdependencies are required as well. Examples of additional interdependencies due to the expanded development effort may include delivery of the following:

- IC characteristics to the circuit board team
- ICs to the manufacturing team
- Circuit board characteristics and dimensions to the enclosure team
- Power and cooling limitations of the enclosure to the circuit board team
- Power management software to the enclosure team

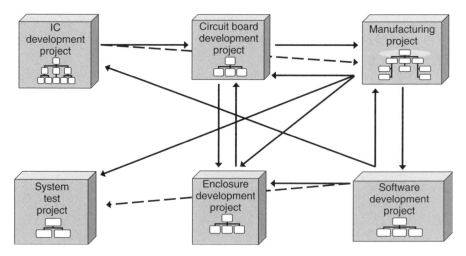

Figure 4.3   Increased complexity with increased interdependency.

In fact, a total of 30 potential interdependencies between the 6 projects are now possible (see box titled, "How Many Dependencies are There among Five Projects?").

To further add to the challenge, the interdependencies above may only represent those required for *one phase* of the development effort. The same number of deliverables, or more, may be required for each phase of the PLC. One can quickly see that as the magnitude of the development effort increases, the number of interdependencies between project teams becomes unmanageable without an effective approach to structure the development effort.

How Many Dependencies are There among Five Projects?

Some organizations use size as the key variable when determining complexity. The reasoning is that as the size increases, so does the number of management activities, resulting managerial deliverables in a development, and the number of dependencies among them. Such increased complexity, then, has its penalty—larger development efforts require more managerial work and deliverables to coordinate the increased number of interactions. This translates into more work for management.

Let's take the following two examples: a development effort with five projects and a development effort with 6 projects. If we assume that each project is dependent on all other projects, how many dependencies are there in each of the two developments? To calculate this, we will use a formula

$D_n = n(n - 1)$ in which D_n is the total number of dependencies and n is the number of projects (functional deliverables).

The development effort with five projects will have $D_n = 5(5 - 1) = 20$ dependencies, and the development effort with six projects will have $D_n = 6(6 - 1) = 30$ dependencies. It means that from five to six projects, the development complexity increases from 20 to 30 dependencies, which is a 50 percent increase! Complexity grows (if measured by the number of dependencies) by compounding, rather than linearly.[3] And, the difficulty of management compounds in a similar fashion.

Selecting the Management Approach

One could make the case that the project managers from the four projects shown in Figure 4.2 could adequately manage the development effort described on a cooperative basis without the need for a program manager. We do not disagree that this may be true. In fact, development efforts of this nature are managed every day by capable project managers. Many times, one of the project managers is assigned as the lead project manager of the project. The small team of managers is able to plan, execute, and deliver their respective elements of the solution, while cooperatively managing the interdependent deliverables between the projects. In effect, the management of the project is handed from one project team to another, as the development effort moves from design to development to manufacturing to system test. However, industry benchmarking, research, and personal experience shows us that this project management-only approach is sustainable to fairly low levels of project complexity.

Four serious problems are consistently encountered when complexity overrides the project management-only approach. First, as complexity rises, the project managers are required to spend the majority of their time planning, executing, and delivering their pieces of the development effort. This eventually results in a classic waterfall approach in which deliverables are heaved over the wall to the downstream team. Eventually the team comes to rely on the big bang event near the end of the development effort—when all elements of the product, service, or infrastructure capability are expected to integrate flawlessly. Unfortunately, this rarely happens, and the development effort can experience significant schedule and cost overruns due to rework and extended integration efforts. Schedule and cost overruns directly hit a company's bottom line.

A second problem is encountered when the schedule gets tight due to internal or external pressures. When schedule pressure exists, the natural tendency for project managers is to hunker down and execute their project deliverables, with little or no time allocated to managing the cross-project interdependencies or the fundamental business aspects driving the development effort. Business requirements such as design wins, ROI, and cost goals may be left unattended. In this scenario, the product, service, or infrastructure capability may achieve all technical features and functional requirements intended and may be delivered to the aggressive schedule target, but something in the external environment, such as an early release of a competing product, may cause the business aspect of the development effort to go awry. Therefore, while the project managers are focusing on the completion of their respective deliverables, no one is left to focus on the changing environment. The result is normally a product, service, or infrastructure solution that fails to meet customers' expectations. This negatively affects a company's bottom line.

Problem three is that someone needs to be responsible for delivering the whole product to the customers and stakeholders. As the scenarios in the first two problems begin to unfold, the project teams begin focusing on their respective elements of the solution exclusively; after all, they are functional specialists. Therefore, the technical aspects of each functional discipline take precedence over the integrated solution when, in fact, just the opposite has to occur to adequately meet customers' expectations, as the whole product concept teaches us. In the personal computer example, this project management-only approach will most likely create a good basic personal computer that the company can bring to the market. However, as discussed earlier, the basic computer is only one element of the whole product solution that the customer is expecting.

The last problem that can be encountered is that interdependencies between projects grow at a rapid rate. Eventually, the cooperative approach by the project managers to manage the mesh of interdependencies breaks down, and no one is left to manage the large number of interdependencies between the projects. This is a critical point. There are two vectors of complexity that need to be managed. The first vector is the definition of responsibilities and deliverables for each individual project manager and his or her project team. The second vector of complexity involves defining, sorting out, and managing the large number of interdependent tasks, responsibilities, and deliverables that require joint sharing of information, planning, and executing *between* the project managers and members of the various project teams. Both vectors must be considered when deciding to use either a project management-only or a program management approach to managing products, services, or infrastructure development efforts.

When the level of project complexity increases to the point in which it is no longer effective to utilize a project management-only approach (experience shows that this approach is only sustainable to fairly low levels of complexity) the most effective solution is to adopt the program management model for development.

Planning the Whole Product Program

Two activities that dominate program planning from the whole product perspective are defining the program strategy and developing the program plan. We define the program strategy as the approach, position, and guidance of what to do and how to do it to achieve the highest competitive advantage and the best value from the whole product. This is done by means of spelling out the four elements of the program strategy: the product, service, or infrastructure definition; the business case; the success criteria; and the integrated program plan. Each element is entirely focused on the whole product. For example, the whole product concept describes the feature set for the whole product, as well as other services and ancillary devices needed to maximize customer value. Details of developing a program strategy are covered in Chapter 3.

Developing the program plan requires a business's understanding of all the elements needed to provide the whole product solution, identifying the project deliverables and cross-project interdependencies and integrating multiple project plans into a master program plan that will meet customer expectations and achieve the intended business results.

When utilizing the program management model to plan the synchronized work of multiple interdependent projects to develop and deliver the whole product, it is helpful to view management responsibilities within the program in two dimensions: vertically and horizontally. This concept, as illustrated in Figure 4.4, is a core characteristic of the program management model. The figure shows a simple example of five functional elements that may be involved in a product development program—circuit board development, enclosure development, software

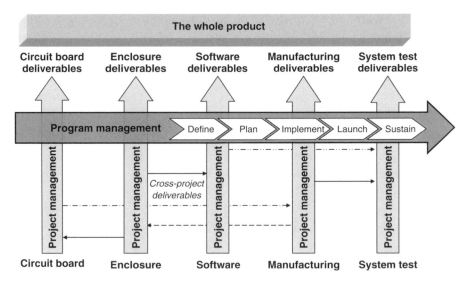

Figure 4.4   Horizontal and vertical dimensions of a program.

development, manufacturing, and system test. Once again, other functions such as system architecture, product marketing, quality, customer support, and finance are normally involved in a product development program but are not represented in the diagram for the sake of simplicity.

Both the vertical and horizontal elements involved in the management of a program are clearly evident in the figure. First, let's look at the vertical or project management element. The work of each functional organization involved in the development effort is structured as a project and is led by a functional project manager. These project teams of functional specialists are responsible for the development and delivery of their respective pieces of the product under development—each piece represents one element of the whole product. During program planning, each project team develops its respective plan, based upon interdependencies with the other project teams.

The work of the program manager cuts *across* the functional project teams; therefore, he or she manages the horizontal dimension of the program. To create an integrated product, service, or infrastructure solution, the program manager is responsible for ensuring the following three primary things: (1) the deliverables from the project teams form an integrated whole product solution, (2) the highly complex network of project interdependencies is synchronized and coordinated throughout the PLC, and (3) the program business case remains viable.

During program planning, the role of the program manager is the master integrator. As the project plans are being developed, the program manager and his or her core team ensure the plans integrate with one another to form the master program plan. Program planning is covered in detail in Chapter 6.

Executing the Whole Product Program

Program execution is about implementing the program plan and adapting to the operating conditions, as needed. We present a big picture view of program execution in relation to developing and delivering the whole product. Details of program execution are covered in Chapter 7.

The end result or output of the work accomplished by each project team is known as the project deliverables. As an example, the circuit board development project team is responsible for delivery of the various circuit boards that make up the whole product; the software project team is responsible for delivery of the software stack (application, drivers, and utilities) for the product; and so on for each of the functional project teams. Again, the project managers are managing vertically, or *within* their functional domain, to develop their respective element of the whole product and deliver it to the other members of the program team.

A distinguishing factor of the program management model from other forms of management is that the functional project teams cannot take a silo approach to developing and delivering their element of the whole product. Each project team within the program is highly dependent upon cross-project deliverables from the other project teams to be successful, as demonstrated earlier in this chapter. The cross-project deliverables are the second aspect of the horizontal dimension of a program. They form a highly interdependent relationship between the project teams that must be managed as closely as the effort to create the individual product deliverables. In Figure 4.4, the cross-project deliverables are represented by the horizontal lines between the projects. We call this cross-project interdependency the "space" between the project teams. Management of the space between the projects is the responsibility of the program manager and involves the identification and synchronized delivery of the project interdependencies. This is accomplished through interface defintion, cross-project coordination, communication, decision making, problem solving, and team-based risk management.

The third dimension of the program management model is time, specifically time to money or time to go live. Management of time is accomplished through the use of a PLC to synchronize the work of the cross-project, cross-discipline program team. The development of the interdependent cross-project deliverables must be synchronized in time to achieve the time to money goal. Thus, effectively managing the work output of multiple projects to achieve time to money is a competitive advantage provided by the program management discipline. The program manager utilizes the PLC to accomplish this synchronization in a concerted manner. Much like a conductor who ensures each instrument section within an orchestra is synchronized in time and measure until the concert is completed, the program manager utilizes the PLC to establish the cadence and synchronize the work of the project teams to develop the whole product.

The three dimensions of the program management model—the vertical management by the project managers, the horizontal cross-project management by the program manager, and the synchronization of work through time—are combined to create the integrated output of the whole product. Thus, program management enables competitive advantage through delivery of the whole product and accomplishing the business goals. It's about the business.

VIEWING THE WHOLE PRODUCT AS A SYSTEM

The final piece of the whole product concept involves viewing the development of the whole product from a company perspective. With such a perspective, the system itself is the whole product, which is defined earlier as the final product, service, or infrastructure capability under development. The system is comprised of subsystems or the elements delivered by each of the project teams. The subsystems are, in turn, comprised of components that are derived from technologies and human capabilities within each of the functional teams.

For those familiar with the concepts of systems development, the whole product approach to developing products, services, and infrastructure capabilities should seem very familiar. The whole product approach encompasses systems concepts with a slight twist of terminology. For example, in the systems approach, a system is decomposed into smaller modules called *subsystems* but called *elements* in the whole product

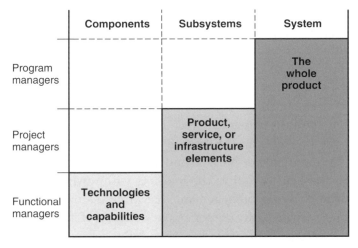

Figure 4.5   System element ownership.

approach. Figure 4.5 demonstrates the direct correlation between the whole product and a system.

It also shows the division of responsibilities for development of the whole product. Functional managers and their organizations are responsible for developing and delivering the technologies and human capabilities needed to create the elements of the whole product. The project managers are responsible for developing and delivering the functional elements of the whole product. Finally, the program manager is responsible for the integration and delivery of the whole product to the market or customer's environment.

Lastly, it is important to understand the environment within which the whole product exists. As shown in Figure 4.6, the whole product is a single system that is part of a greater business **ecosystem**, which is comprised of

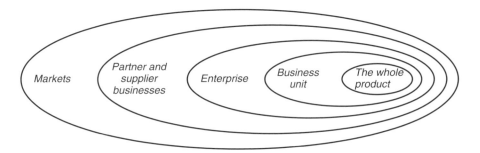

Figure 4.6   The product, service, or infrastructure ecosystem.

the **business unit** that is funding the development of the whole product, the **enterprise** that is comprised of multiple business units, the **partner and supplier businesses** that support the enterprise, and finally the **market** that consists of both customers and competitors within which the enterprise operates.

By viewing the business ecosystem in the model shown in Figure 4.6, one can see that the whole product has an enormous impact on the success of the business unit. Likewise, the impact on the business unit will affect the enterprise as a whole and have influence on the partner and supplier businesses, as well as the markets within which the enterprise operates. The program management discipline, as owner of the whole product, has responsibility for effective delivery of the whole product into the ecosystem. Thus, this discipline helps to execute the business strategy of the enterprise and positively affects partners and suppliers of the enterprise, as well as the markets that the enterprise serves.

SUMMARY

This chapter introduces the concept of the whole product, which we define as the integrated product, service, or infrastructure solution that fulfills the customer's expectation. The whole product concept demonstrates the true scope and nature of the program management discipline and explains how it can be utilized to manage ever-increasing levels of development complexity.

Managing the whole product involves multiple aspects. First, the whole product solution is broken into multiple elements that become the projects within the program. Then, understanding the project dependencies as the foundation of development complexity. Determining which management approach to employ is largely based upon the level of complexity. The project management-only approach to product, service, or infrastructure development is sustainable to lower levels of complexity. When complexity overrides the effectiveness of the project management-only approach, the program management model becomes a viable alternative. To deliver the whole product, program management involves the management of highly interdependent projects that must be collectively managed in a concerted manner to meet customer expectation and achieve improved business results. The program manager's role is to lead a cross-project team in defining, sorting out, and managing the large number of interdependent tasks, responsibilities, and deliverables that require joint sharing of information, planning, and execution between multiple project teams. The whole product can have a huge influence on the entire ecosystem within which a company operates. By owning the development and delivery of the whole product, the program management function owns achievement of a portion of the business objectives. Thus, it operates as a critical business function within the organization.

The Principles of Program Management

▼ The program manager is responsible for the definition, development, and delivery of the whole product, while the project managers are responsible for delivery of the individual elements of the whole product.

▼ Program managers manage horizontally or *across* the functional projects, while project managers manage vertically or *within* their functional domain.

▼ Managing the interdependencies between the project teams on the program is a critical role of the program manager.

▼ To enable the management of a higher level of development complexity, don't take a silo approach to managing and delivering elements of the whole product but rather an integrated approach.

REFERENCES

1. Moore, Geoffrey A. *Crossing the Chasm*. New York, NY: HarperCollins Publishing, 1991.
2. Martinelli, Russ and Jim Waddell. "Aligning Program Management to Business Strategy". *Project Management World Today* (January-February 2005).
3. Moore, James F. and Jonathan Reese. *Death of Competition: Leadership and Strategy in the Age of Business Ecosystems*. New York, NY: HarperCollins, 1996.

Part II

Managing the Program

While Part I dealt with the strategic purpose of program management, Part II delves into details of how to manage a single program from its inception to end of life. It consists of four chapters—The Program Team, Program Definition and Planning, Program Execution, and Program Processes.

Chapter 5 is about the program team. It emphasizes two critical factors that matter when structuring a program. First, the program team structure must support the highly integrated nature of a program. Second, the program team structure must enable the fundamental elements for team success—from coordination of its activities and interdependent deliverables to collective problem solving. A fundamental lack of understanding in the difference between a program team structure and a project team structure often stands in the way of successfully structuring a program. This is mainly because managers incorrectly use their experience in establishing project team structures for programs; there is also a wealth of available information on establishing project structures, which contributes to this problem. The most effective program structure we've experienced is the core team structure. With the PCT structure in place, an organization becomes well-positioned to achieve improved business results, both strategic and tactical in nature. Strategic benefits relate to establishing a framework to link business strategy and execution output. Tactical benefits include providing the foundation necessary to either break down product and process complexity or to fully identify and evaluate risk to the product, service, or infrastructure under development, as well as risk to the business.

Chapter 6 focuses on the define and plan phases of the PLC. The program definition process involves the steps necessary to integrate business strategy, customer and end-user research, and technology research into a viable product, service, or infrastructure concept. The viability of the concept is then validated through the creation and analysis of a comprehensive program business case. The program planning process spells out the steps necessary to channel the product, service, or infrastructure concept into a cross-project, multi-discipline implementation plan. When executed, the program implementation plan will become the means to achieve the business objectives intended. Lastly, this chapter provides a set of important behaviors that highly effective program managers exhibit during the program definition and planning phases.

The purpose of Chapter 7, Program Execution, is to focus on execution of the implementation plan to create the whole product, service, or infrastructure and prepare for the launch of the program output into the market or end-user's environment. It looks at the implement, launch, and sustain phases of the PLC. While the second phase, launch, has received some attention in literature, the sustain phase and process are seldom covered. We describe the steps necessary to keep the program business running as effectively and efficiently as possible after a product, service, or infrastructure capability is launched. Additionally, the steps to fully support the customers and end users ensure that the business objectives of the program are attained and that the program remains viable from a business perspective. The close out of the program, when it reaches end of life, is also covered in this chapter.

Chapter 8 describes the primary program processes that a program manager can utilize to make the operational aspects of a program—the PLC, schedule management, financial management, risk management, change management, and stakeholder management—more efficient, predictable, and repeatable. The PLC is a framework that can be used to synchronize the work, deliverables, and decisions on a program. The schedule management process is used to manage the horizontal collaboration between the project teams over time. The financial management process is used to manage the financial aspects of a program, such as cost estimation and control, financial feasibility assessments, and cash flow management. The risk management process can be used to identify and proactively deal with the various risk factors surrounding the program before they become problems that have to be reactively managed. The change management process is a program manager's greatest weapon to combat program scope creep. Finally, stakeholder management is a process with which a program manager can manage the communication, political, and conflict resolution aspects of a program.

Chapter 5

The Program Team

This chapter begins the discussion of *how* to successfully manage a program utilizing the program management model. The concepts we present are based upon best-known practices that we have encountered in our personal experience, research, and benchmarking studies with various companies representing multiple industries.

The discussion of how to manage a program must start with *how* to correctly structure a program. An effective program structure is the keystone to realizing the benefits of program management. Without careful consideration of how an organization's programs are structured, many or all of the benefits of program management will be unrealized.

Having said this, however, it is surprising how few companies have actually implemented an effective and consistent program structure. In practice, we see executives and middle managers taking two approaches that commonly lead to futile attempts to achieve an effective program management model. First, they experiment with multiple program structures, trying to find "the one" that will provide the best results, or they expend an inordinate amount of time, money, and resources trying to improve process and tools, without realizing and fixing the underlying structural problems.

The most common error businesses make in implementing the program management model is failing to understand the difference between a program team structure and a project team structure. Simply stated, projects, especially larger ones, tend to be vertically structured with multiple layers of organization. Programs, by comparison, need to be horizontally structured to create the cross-project, cross-discipline network necessary to promote effective coordination, communication, and decision making. [1]

In this chapter, we examine the role of the program team in achieving success, describe the PCT approach as the most effective way to structure a program, evaluate how this approach helps to achieve both the critical elements of team success and the benefits of the program management model, and finally, chronicle the evolution of a program team structure within an organization. The material in this chapter will help senior management and practicing program managers understand the following:

- The critical factors involved in program team success
- The advantages of the core team approach to structuring a program
- How the core team structure helps to achieve the benefits of the program management model

CRITICAL FACTORS FOR PROGRAM TEAM SUCCESS

One of the primary characteristics of a program is that it's a highly integrated endeavor in which the product, service, or infrastructure solution is more important than specific technological solutions. The program is synergistic in nature and is truly a case in which the sum of the parts is more valuable than any of the parts on its own. By way of example, a personal computer is of little value without the memory subsystem, the monitor, or the microprocessor. The value to the user is in the integrated product, not in the individual pieces.

Additionally, programs tend to be dynamic in nature with intense cross-discipline integration, in which the actions of one functional project team affects, supports, and reinforces the other project teams involved with the program. This high level of integration requires a program to be structured in a way that enables *all* functions to be part of the product, service, or infrastructure development process through all phases of the PLC. For programs to be successful, the program team organization must facilitate a high level of cross-discipline coordination and communication between the program manager, project teams, and stakeholders. Additionally, the program organization must enable the fundamental elements for team success.

Fundamental Elements of Program Team Success

Like all other teams, program teams have a set of fundamental elements that distinguish effective teams from ineffective ones. Program teams must be effective in the following areas for success:

- Team communication
- Cross-discipline and cross-project coordination
- Decision making
- Cross-discipline problem solving
- Team-based risk management

Team Communication

Highly effective teams communicate clearly, consistently, and frequently within the team structure, as well as with stakeholders outside of the team. [2] Effective communication permits the program team to make better decisions, evaluate information between the project teams to assess impacts on interdependencies, deal with interteam conflicts, and keep key stakeholders apprised of program status. The program team structure must enable open, nonhierarchical, and clear communication channels within the team. The program manager and the team leaders must be adept at both vertical and horizontal communication. This means being able to effectively communicate across a wide spectrum of groups and disciplines, including, in some cases, outside suppliers and partners of the firm (see the box titled, "The Russians Join Us Late at Night"). From a vertical perspective, the program team must be able to effectively communicate with senior management at the top of the organization down through the organization structure to the individual contributors on a program, such as a manufacturing operator on the production floor.

The Russians Join Us Late at Night

"Communication is the key," says Sri Rastogi. Rastogi is a project manager on a program that is geographically dispersed, with part of the team in Portland, OR; part in Houston, TX; and part in Moscow, Russia. This is how Rastogi views communication in geographically dispersed teams.

There is an eleven hour time difference between Portland and Russia, so finding a good time to communicate in person is tough. One of the things the Moscow team has done is shift their work day. They now come in about 10 or 11 o'clock in the morning, then go home anywhere between 8 and 10 o'clock in the evening. We now have overlap at the end of the day where we can usually find people in the office.

Instant messaging technologies have also helped a lot. I log in from home for an hour each night and turn on my instant messenger. If anyone in Russia needs to contact me during their morning, they can do so, and I'll respond immediately.

I don't know how it is for the rest of the company, but the fact that I make myself available at 11 o'clock at night on a daily basis during the development cycle is a necessity to help communication channels stay open on a geo-dispersed team. It's not rocket science, but it works!

Cross-discipline and Cross-project Coordination

An effective program team must be able to coordinate and integrate many complex activities and deliverables throughout the PLC; they quite often do so with a geographically dispersed team. Traditional hierarchical and functional team structures are inadequate to address the high level of cross-project coordination that needs to take place in a timely and cost-effective manner. [3] The program team structure must horizontally integrate the project teams to enable the program manager to effectively coordinate and channel the work of the teams toward a common output.

As an extreme example of the importance of good cross-project communication and coordination, nine months after being launched, the Mars Climate Orbiter was lost during its first pass around the red planet. The spacecraft became the victim of poor communication between project teams on the program. One team programmed navigational data to be sent to the Orbiter in English units (feet, inches, and pounds), while another team programmed the Orbiter to receive and interpret the data in metric units (meters, centimeters, and kilograms). The miscommunication resulted in the Orbiter entering the Martian atmosphere at 57 km instead of the intended 140–150 km, and the $125 million spacecraft was destroyed by heat caused by atmospheric friction. Among the primary contributing factors for the loss were the following, which reinforce the importance of good cross-project coordination and communication: [4]

- The operational navigational team was *not fully informed* about the details of the way that the Mars Climate Orbiter was pointed in space.
- The systems engineering function within the program that is supposed to track and double-check all *interconnected aspects* of the mission was not robust enough.
- Some *communications channels* among project engineering groups were *too informal*.

- The mission navigation team was oversubscribed and its *work did not receive peer review*.

Decision Making

Due to the complexity of most programs, program teams need to draw upon a wide array of information to make effective decisions. This means all disciplines and functional projects must be involved in generating information for and participate in the decision-making process. Because of the cross-discipline and cross-project nature of programs, information must be presented in a manner that all parties can understand and be able to utilize in the decision process. It is no longer the case that engineering utilizes technology-only data, or marketing utilizes customer-only data, or finance utilizes accounting-only data to drive program related decisions. Discipline-specific data and information must be combined with cross-discipline information to be effectively utilized for program-level decisions.

An effective program decision is described as one that has the following characteristics:

- Not reached by a single individual
- A sound solution to the problem
- Based upon input provided by each team member involved in the decision process
- Aligned with the program objectives and success criteria

Program teams are charged with making the following three types of decisions: (1) strategic decisions, (2) tactical decisions, and (3) operational decisions. Strategic decisions may include such things as which technologies to include in a design, what product cost is needed to maintain competitive leadership, and key partnerships or alliances in which to engage. Tactical program decisions may include which features to incorporate and exclude, the number of prototypes to build, and changes to the program schedule. Operational program decisions may include factory scheduling, which common cross-project processes and tools to employ, and how to communicate progress to program stakeholders.

The final key aspect of program decision making is that it must incorporate what we describe as both vertical and horizontal decisions. As stated earlier, program-level decisions involve cross-project and cross-discipline input and involvement. Additionally, effective program decision making involves the empowerment of the project teams to drive functional, project-specific decisions.

All vertical or project-specific decisions have the potential to affect other projects on the program. When this occurs, the decision must be ratified at the program level. Conversely, all program-level decisions will impact one or more project teams. Therefore, the decision must be evaluated across the projects before a final decision is implemented. This requires a high degree of communication between the program manager and the project managers on the program. Figure 5.1 graphically illustrates the difference between the vertical and horizontal decisions that occur on a program.

Let's look again at a personal computer development example to illustrate the interplay between the vertical and horizontal decisions involved on a program. Competitive intelligence has indicated that a competing product will be introduced within two months of the computer launch with a higher microprocessor-base frequency (frequency is a measure of how fast a computer will operate). A program-level decision is needed to decide if the base frequency of the computer should be increased to stave off the competitive threat. The microprocessor project team is charged with evaluating the technical aspects of increasing the base frequency and the impact to their element of the computer product. It is determined that the higher frequency is achievable but will require a larger physical size, higher power requirements, and a new frequency control algorithm. This information is then brought back to the program level

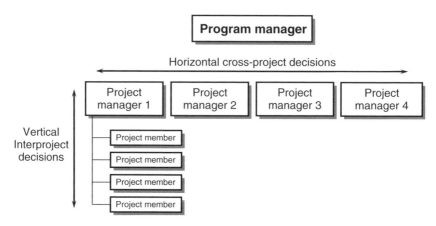

Figure 5.1 Horizontal and vertical program decisions.

for evaluation by the program manager and the project teams. The computer motherboard project team must evaluate the impact and feasibility of the increased physical dimensions of the microprocessor, the power supply team must evaluate the higher power requirements needed, the enclosure project team must evaluate additional cooling requirements due to the added power, and the software development project team must evaluate the impact and feasibility of developing new code for the frequency-control algorithm. When analysis by the project teams is complete, a program-level (horizontal) decision is made by the program manager with consultation input from the project managers to increase the base frequency of the microprocessor. This then will spawn a series of tactical discipline-specific project (vertical) decisions to implement the new course of action. The entire program team now shares responsibility for the results.

This example demonstrates how effective program decisions require well-coordinated and communicated interplay between horizontal and vertical decisions. For this process to be effective and efficient, two critical elements must be in place. First, the program team structure must support the high degree of communication and coordination between the project teams. Second, the program team must be empowered by the organization's management team to make the strategic, tactical, and operational decisions necessary.

By empowering the program team to make decisions, decision making becomes quicker and more effective. If a program team is required to rely on functional or line managers for decisions, these managers will quickly become a bottleneck in the program implementation, thus, reducing efficiency and effectiveness. [5] Program team empowerment means giving the program manager and his or her project managers the responsibility and authority to make decisions. In the book *The Power of Product Platforms*, Meyer and Lehnerd state that "there is no organizational sin more demoralizing to teams than lack of empowerment." [6] The program team must have confidence that their empowerment will not be undercut by the firm's management. If such action occurs, it must be immediately eradicated by the firm's senior management.

Just as the program manager must be empowered by senior management, the program team must also be empowered by the program manager to make key decisions. As a senior program manager from a major aerospace company told us, "Empowerment is something that builds over time through gaining the confidence of the leader." He also pointed out that there needs to be empowerment boundaries because, as he put it, "one big screwup on the part of someone wipes out ten attaboys."

Cross-discipline Problem Solving

Decision making is one of the final steps of effective problem solving. For a problem to be fully resolved, a series of decisions has to be made and the outcome communicated to all affected and interested stakeholders. Like decision making, there are many methods available for effective problem solving. The issue is not choosing the right method, as many methods can suffice, but rather what needs to be in place for a program team to resolve problems in a quick and effective manner to prevent roadblocks to progress.

Programs are complex, and problems encountered on programs can be equally complex. As a result, multiple points of view are needed to adequately resolve problems. Once again, a cross-discipline and cross-project point of view is needed. Because of the highly integrated nature of program teams, it is rare that a problem encountered on a program will affect a single project team. Most likely, multiple project teams will be affected; therefore, they need to be part of the problem solution.

The program team structure must involve the right people to resolve the problems that arise in a timely manner. Problem solutions must be developed with the big picture or program success in mind and must be supported by the entire program team to be effective. [7]

Team-based Risk Management

The final characteristic of an effective program team is cross-project or team-based risk management. As the term implies, team-based risk management involves a cooperative effort on the part of the program manager, the functional project managers, and program support personnel to identify and mitigate risks before they become problems. [8] Team-based risk management requires a systems perspective in which risks are viewed as potential problems for any of the project teams involved and for the success of the program as a whole.

The Software Engineering Institute (SEI) states that the team-based risk management approach is founded upon a shared vision for the program that is structured out of the individual views of the functional project teams. It is also anchored by open communication between all members of the program team. [9] Open communication is the vehicle by which the functional project-risk viewpoints are merged to form a shared-risk view for the product, service, or infrastructure capability under development.

Effective team-based risk management is built on nine core principles that are summarized in Table 5.1.

The relationship between functional project team risk management and program team risk management is illustrated in Figure 5.2. The program manager is responsible for evaluating risk of the whole product. Therefore, the program manager is responsible for any risk that can affect the success of the whole product, up to and including the business risk. In addition, he or she is responsible for the many risks involved

Table 5.1   Nine core principles of team-based risk management.

Principle	*Fundamentals of the Principle*
Shared vision	A shared vision for success is based upon commonality of purpose, shared ownership, and collective commitment
Forward-looking search for uncertainties	Thinking toward tomorrow, anticipating potential outcomes, identifying uncertainties, and managing program resources and activities while recognizing the uncertainties
Open communication	A free flow of information at and between all program levels through formal, informal, and impromptu communication processes
Value of individual perception	The individual voice that can bring unique knowledge and insight to the identification and management of risk
Systems or whole-product perspective	Product, service, or infrastructure development that must be viewed within the larger systems-level definition, design, and development
Integration within program management	Risk management must be an integral and vital part of program management practices
Proactive strategies	Strategies that involve planning and executing program activities based upon anticipating future events
Systematic and adaptable methodology	A systematic approach that is adaptable to the program's infrastructure and culture
Routine and continuous	A continuous vigilance characterized by routine risk identification and management activities throughout all phases of the PLC.

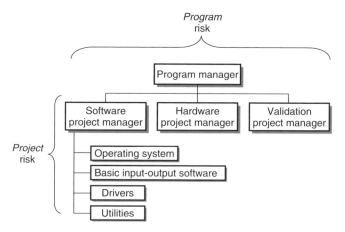

Figure 5.2   Program and project risk management relationship.

in the interdependencies between the projects. The project managers, by contrast, are responsible for managing the risk associated with their respective functions, which can affect the successful delivery of their element of the whole product.

Project-level risks that are identified as also being program-level risks become part of the program risk inventory, and the respective project manager assumes ownership for the resolution of the risk. We go into more detail on program risk management in Chapter 8.

Advantages of team-based risk management include the following. [10]

- *Multiple perspectives of a risk:* Brings functional project teams together in identifying and mitigating risks and opens doors to strategies that the entire program team can cooperatively implement; this is more effective than a single team working alone.
- *Broader base of expertise:* The combination of cross-project representation brings together a richer pool of experience and expertise in perceiving and dealing with risks.
- *Improved communication:* By openly sharing risks, the cross-project program team is able to draw on one anothers resources to more rapidly identify and respond to risks.
- *Shared vision:* Risk management efforts are focused on a broader vision of shared program success than project-specific risk management.
- *Better decision-making:* Team-based risk management gives decision makers a more global perspective of risk impact on the success of the program.

Through team-based risk management, the program team collectively possesses greater knowledge about risks to the program, thinks in a broader perspective, and manages the mitigation of risks more effectively than a single project team.

The foundation for effective team-based risk management, as well as for all the other elements of effective team performance described on page xxx, with the implementation of a solid program team structure. A program must be structured in a way that enables all functional project teams to be part of all aspects of the development process. The most effective program structure we have encountered and utilized is PCT structure, the specifics of which are covered in the following section.

STRUCTURING THE PROGRAM TEAM

To be successful, a program team must coordinate its activities and interdependent deliverables, effectively communicate what is being accomplished as well as roadblocks, and collectively solve problems and make decisions that support the program objectives. The full-program team consists of three primary entities, as illustrated in Figure 5.3—the program manager, the PCT, and the extended team. Although at least one consulting firm has claimed it developed the core team approach in the 1990s, [11] in reality its existence has been in place for several decades in companies specifically utilizing the program management model of development—particularly in the aerospace, defense, and automotive industries.

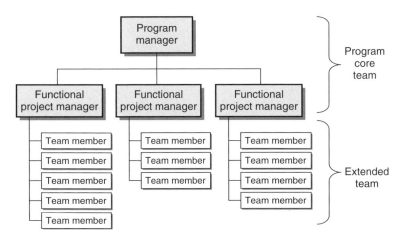

Figure 5.3   The program team structure.

The particular functions and support organizations that make up the program team are dependent upon the elements and scope of the product, service, or infrastructure under development. Program team membership may also vary as a program progresses through its life cycle. To remain effective, it is important that the PCT is limited to the key functional representatives within an organization. It is recommended that all programs within an organization be structured in a similar manner for consistency.

The Program Manager

The program manager is the leader of the program team and of the program in its entirety. In this capacity, the program manager is the champion for the product, service, or infrastructure capability under development and has primary influence over the work of the people involved with the program. In the core team structure, the core team members report to the program manager in an indirect relationship for the duration of their participation. The direct reporting and career development of the core team members still resides with their respective functional managers.

The program manager integrates the work of the functional project teams (see box titled, "My Job was to Integrate Two Cultures,") and as the leader of the PCT, he or she has the following duties:

- Establish the program objectives/priorities and set and negotiate the program critical success factors with senior management
- Define core team makeup and recruit core team members from the respective functional managers
- Exercise overall control of the program activities
- Motivate the team to make and hold commitments in support of the program objectives
- Direct the team for preparation of milestone and management reviews
- Establish consistent use of processes, methods, and tools for the program
- Establish the critical success criteria and measurement metrics that drive results
- Create a team environment of trust and respect
- Encourage managed risk taking

- Establish decision forums, decision methodology, and drive team decision making
- Assist the PCT project managers in resolving resource gaps within the projects and other issues that may impact their ability to achieve the objectives for the program
- Celebrate accomplishments and provide recognition as appropriate

It should be apparent that the program manager must earn the respect and right to carry out these duties based on prior experience, well-developed skills and capabilities, and respect and empowerment that is established over time. Having a qualified program manager who can fulfill these responsibilities is a prerequisite to an effective PCT structure. [12]

My Job was to Integrate Two Cultures

"All my professional life I have dealt with software development—banging out code," began Jerry Dorsey, now one of several project managers for the geographically, and culturally-dispersed Dacia program. "I have always managed local, co-located software development teams, so I was stunned when my boss summoned me and asked me to manage a project with an outsourced development team in Romania. I think I asked the same question three times—Romania?" Jerry had expressed interest with the program management office director to move into a program management role, and she worked with Jerry's manager to move him into a role that required integration skills. In this case, solving a cross-culture problem by integrating the Romanian team with the U.S. team.

Jerry continued, "At first I was shocked, since I didn't know the first thing about cross-cultural integration, but I began communicating with members of the Romanian team to learn about how they worked and what they valued." Jerry soon discovered that corporate culture and national culture often collide. "The Romanians were used to being tasked," said Jerry. "They had an attitude toward me that 'he's the boss,' and, therefore, I should have all the answers. The concept of brainstorming solutions, which is a common part of our company culture, was completely unknown to them. Because they lived under communism dictatorship for so long, they were used to people telling them what to do and just doing it."

Jerry continued, "They will also never say no. I could just give them more and more to do and they'd try to get it all completed. So, I had to learn how much I could actually give them by monitoring the progress of their deliverables. As long as they met their deliverables on time, I figured they weren't being overtasked."

The biggest lesson for Jerry, however, had little to do with managing the development of the software. As he explained, "Building strong personal relationships was the most critical element in integrating the Romanian team into our company and program culture. We were able to bring the key technical leaders from Romania to the United States early in the planning phase to meet and interface directly with their United States counterparts. There's no better way to build mutual trust! I also made a point to travel to Romania once every two to three months to get to know the Romanians and make myself directly accessible to them."

"At the end of the day," concluded Jerry, "this was a great experience for me personally and for my career. I got first-hand experience on what it means to be a program manager, and it's definitely an avenue I'd like to continue to pursue."

The Program Core Team

The PCT is the cross-discipline and cross-project leadership, and decision-making body of the program that is responsible for ensuring the program and business objectives, as well as customer satisfaction, are achieved. [13] The PCT consists of the functional project managers who represent their functions and provide leadership for the delivery of their function's element of the product, service, or infrastructure capability under development. Collectively, the PCT constitutes the management team of the program that works under the direction of the program manager. [14] The PCT must become a very tight and cohesive team that has a shared responsibility for the business success of the program. Each member of the core team must be committed to the success of the other members on the team. A primary responsibility of the program manager as leader of the PCT is to build a trusting, cohesive team and lead them to mutual success by way of program success.

The size of the core team is dependent upon the elements making up the item under development and complexity of the solution. This is defined by the program requirements, but typical PCT size varies between 4–12 members, including the program manager. To illustrate, let's look at an infrastructure development example in which a program is formed to convert a company's personal computing capability from a desktop to a laptop platform to increase worker productivity—the business objective. Figure 5.4 shows the primary interdependent projects that must be formed, planned, and executed to develop the whole product solution and achieve the business objective.

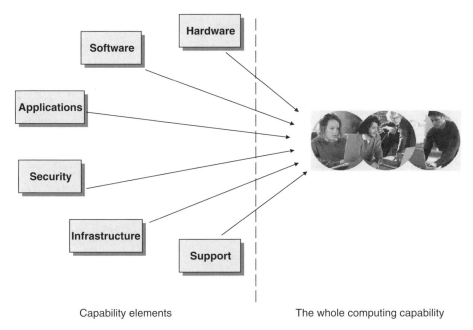

Capability elements   The whole computing capability

Figure 5.4   Infrastructure program elements.

The program consists of the following six primary elements that make up the whole solution: the hardware platform, the operating software, the application software, security, the infrastructure compatibility, and user support. Each element is organized as an interdependent project lead by a project manager who owns the delivery of their element of the computing capability to the rest of the program team. In the spirit of form following function, the PCT consists of six project managers and the program managers, as illustrated in Figure 5.5. Other support functions, such as quality and finance, may also be included in the core team.

Product development teams at Intel tend to be relatively large. Figure 5.6 shows a typical core team of an Intel program to develop a server

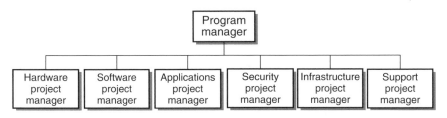

Figure 5.5   The infrastructure development PCT.

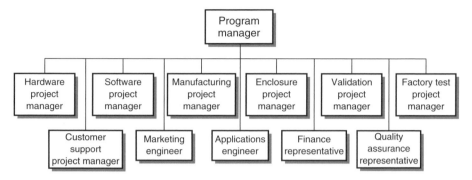

Figure 5.6   Intel server development PCT.

computer. The core team consists of 12 members, as follows:

- One program manager
- Seven project managers: hardware, software, manufacturing, enclosure, validation, factory test, and customer support
- Two functional individual contributors: marketing engineer and applications engineer
- Two support personnel: finance representative and quality assurance representative

This core team is fairly large and is driven by the functional elements needed for the successful development, delivery, and support of the Intel server product under development.

PCT members have to fulfill the following two types of responsibilities: Program team responsibilities and functional responsibilities. Therefore, it is said that core team members wear two hats—the program hat and the functional hat.

Program responsibilities of the core team members include the following:

- Deliver their element of the product, service, or infrastructure under development
- Define and execute the requirements of their functional organization
- Develop their project plan and work with the program manager to integrate their project plan into the master program plan
- Identify functional risks and manage the risks within their sphere of influence

- Manage the work commitments for their respective project team
- Work with other PCT members to deliver cross-project deliverables (see box titled, "An Engineer in Charge of Marketing")
- Assist in resolving problems
- Drive project-level decisions
- Participate in and provide functional consultation for program-level decisions
- Assist the program manager in balancing the program constraints
- Effectively execute established program processes, methods, and tools

Functional responsibilities of the core team members include ensuring the following:[15]

- Adequate expertise exists on the functional project team
- The functional project team is adequately staffed in numbers
- Representation of the functional perspective on the program
- Project specific objectives are met
- Functional issues impacting the team are raised proactively within the PCT

As the leader of the PCT, the program manager facilitates the communication and collaboration between the members of the core team. He or she also brokers any conflicts and issues that arise between core team members, the core team and the functional managers, and the core team and senior management. [16]

An Engineer in Charge of Marketing

This is a story about Chen Liu, a marketing project manager for a $1 billion company and current member of a product development PCT. "I've been with the company for 24 years now. The first 12 years I worked in the laboratories doing technical research, then 8 years as an engineering manager. I moved into marketing four years ago, which was a bit of a tough transition," explained Chen.

"As the marketing project manager on a PCT, I represent the broader marketing function. I coordinate the work of the various marketing functions so we are working in a cohesive manner for the program manager. There's the worldwide MarCom (marketing communications) team, the regional field marketing people, the regional marketing managers, and the application engineers. So my job is to create a marketing project plan, integrate that plan within the master program plan, and coordinate the work of the various marketing people."

When asked about how he works with the other members of the PCT, Chen was quick to explain.

"We sometimes joke about this in the marketing group, saying we're the center of all activity on a program. But, in reality, all functional teams probably feel the same way because our work is so intertwined. I interface a lot with finance concerning revenue projections and product sales pricing; I interface with manufacturing by providing forecasts that drive materials supply purchasing, as well as identifying the number of demonstration units the field will need; and I work with the customer service team to ensure our customers are fully supported. I interface a bit with the customer documentation people as a reviewer of the user's manual and a bit with the quality and validation teams to define quality goals based on customer expectations and various regulations. But the team I interact with the most is the engineering team to make decisions about features and functions, product cost, and ensuring the marketing and engineering requirements are aligned. This is where my engineering background is a real advantage."

Chen concluded by adding, "It's quite possible that I interface directly with every project team represented on the PCT."

Chen's story brings to life the realities of how interdependent the project managers and their teams become during the course of a program.

The Extended Team

The extended team consists of individual contributors that make up the functional project teams. They can be assigned to the program on either a full-time or part-time basis depending upon the work they are assigned. They are responsible for ensuring their respective project accomplishes all deliverables to the program within schedule, allocated budget, and with full functionality and performance.

The functional project managers lead the extended team and are the primary decision makers on each project team. It is common for many of the functional project teams to be organized in the same horizontal manner as the PCT.

For example, consider a software project manager who is a member of a PCT. In addition to representing the software function on the PCT, the software project manager also leads the software project team. The functional team leads on the project team may include the operating system team lead, firmware team lead, basic input-output software BIOS

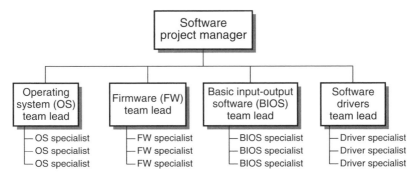

Figure 5.7   Software project team structure.

team lead, and the software drivers team lead, as shown in Figure 5.7. Each team lead will have a set of individual contributors reporting to him or her for the duration of the team's involvement on the program. The individual contributors are specialists in their respective functional areas or domains.

Management of the program occurs at multiple levels of the organization. Executive management sets the business strategy and objectives; the program manager and the PCT defines, plans, and implements the program objectives that help achieve the business objectives; the functional project teams plan and execute the specialized work to deliver their element of the product, service, or infrastructure capability; and the functional managers ensure the project teams are adequately staffed and that functional capability is sufficient to achieve the program goals.

The Role of the Functional Manager

With the program core team structure, the biggest fundamental change to roles and responsibilities within the organization is that of the functional manager. Under a functional or project structure, the functional manager has significant power and influence, but under the PCT structure, much of the power and influence for a specific program shifts to the program manager. Power and influence for the functional managers is diminished as they take on a program support role.

Under the PCT structure, the functional managers are freed from the daily execution details of a program and can focus on the capability growth of their function. They can now focus on hiring, training, and developing the functional specialists within their organization; maintain skills, best practices, and tools to sustain long-term functional expertise; make resource and work commitments in support of the program; and assign qualified functional project managers to represent their functions on programs. [17]

As the PCT becomes empowered to make program-level decisions and has control over the program resources, the functional managers can become uncomfortable with the new structure. This should always be viewed as a significant risk to implementation of a core team structure and needs to be resolved immediately by senior management.

The program manager must establish good working relationships with the functional managers supporting his or her program. Program managers need the help of functional managers to resolve cross-functional contentions, help make key trade-off decisions, keep programs adequately staffed with skilled resources, and provide technical review of cross-project work.

The Integrated Nature of the Core Team

The core team concept fully supports the highly integrated nature of programs through the horizontal structure that is the basis of the approach. Only two levels of management separate an individual contributor and the program manager or a core team member and executive management. The horizontal structure of the PCT ensures that responsibility is divided and shared among the functional project team managers and that no function has more influence than another. All work on the program team is jointly planned and executed by the members of the PCT and extended team. Tasks, deliverables, and timelines are highly integrated between the project teams, and the resulting integrated plans and schedules are overseen by the PCT and, ultimately, by the program manager.

Coordination and communication within the PCT occurs both horizontally and vertically. Figure 5.8 demonstrates the triangulation of collaboration and communication that takes place on a core team. Directions, decisions, and cross-team issue brokerage come from the program manager to the project managers; cross-project communication and work coordination occurs between the project managers; and status, decision consultation, and issue escalation come from the project managers to the program manager.

As demonstrated in Figure 5.9, when the third element of the program team, the extended team, is added the more traditional vertical or

Figure 5.8   Core team collaboration and communication triangulation.

hierarchical coordination and communication that is typical of a traditional functional project team structure becomes evident.

Decisions, direction, and program pass downs come from the functional project managers to the extended program team members. Work status, issues, and detailed risks come from the extended project team members to the respective functional project managers. The large majority of the cross-project coordination and communication occurs at the core team level, even though some detailed interaction will occur between individual contributors on the project teams.

The horizontal, cross-project structure that the core team approach provides is key to establishing the shared responsibility and understanding needed to achieve program success. In addition to enabling effective cross-discipline and cross-project coordination and communication, the

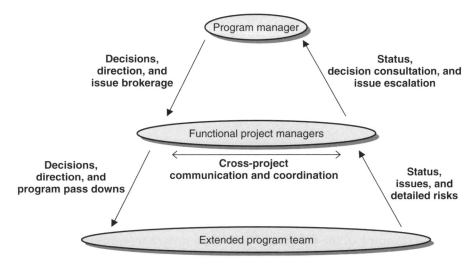

Figure 5.9   Vertical program coordination and communication.

PCT structure also enables the other elements needed for program team success; empowered decision making, cross-discipline problem solving, and team-based risk management.

With each function represented on the PCT by an empowered project manager, the team can make effective decisions based upon multiple points of view. With decision-making power contained within the PCT, decisions can be reached quicker than having to go outside of the team for functional evaluation and/or approval. The same can be said for effective problem solving. Problems can be evaluated and resolved more efficiently and effectively under the PCT structure due to the cross-discipline viewpoints and involvement embedded in the team structure. Under strong leadership from the program manager and the related tools available to them, core teams are effective in breaking through the problem solving and decision-making bottlenecks that occur with other forms of team structure.

As established earlier in this chapter, effective team-based risk management is critical to program success due to the high level of complexity and uncertainty involved. The core team structure establishes the shared vision for the program that is needed to adequately identify, assess, and resolve the key risks that threaten the success of the product, service, or infrastructure under development.

Hopefully, we've made the case that if an organization is utilizing a program management model or planning to implement the model, team structure in the form of a PCT is the keystone to successful execution of the organization's programs. The following section will describe a common evolution that organizations go through to develop an effective program structure. It should be noted beforehand that this type of evolution is not recommended. We recommend that an organization directly implement a PCT structure, even though it may be a large step from the current structure.

EVOLUTION OF A PROGRAM STRUCTURE

We continue to be surprised by how few companies have successfully implemented an effective program team structure. We believe that the primary reason for this low success rate is that there is a fundamental lack of understanding of the difference between a program team structure and a project team structure. Much of this may be the direct result of limited published documentation on the subject of program team structures.

Without a firm understanding of the differences, the natural tendency on the part of executive and middle management is to try to implement a project structure for their programs. This is mainly due to previous experience in establishing project team structures, as well as the wealth of literature available on the subject. Unfortunately, this approach will lead to a long and organizationally painful exercise, as described in the sections that follow.

Beginning with a Functional Team Structure

The functional team structure is probably the most common team structure utilized today. [18] In this approach depicted in Figure 5.10, each functional team sequentially works on their element of the program, then hands both the work output and program ownership to the next functional team in line.

For the example shown in Figure 5.10, once the architecture is completed, it is handed off to the hardware functional manager and team, along with ownership of the program. When the hardware design is complete, the design and program ownership is handed off to the software functional manager and team, and so on down the line to final integration and delivery.

This structure has many vertical layers within each of the functions involved in the development effort; therefore, strong functional expertise can be established. However, this structure is not effective for the program management model. First, it lacks the necessary horizontal structure needed for effective coordination and communication between the functional teams. There are too many barriers between the functional teams, which inhibit effective cross-team collaboration.

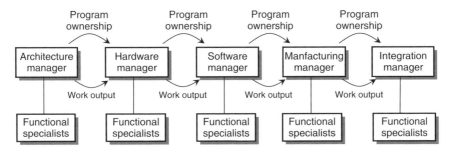

Figure 5.10 Functional team structure.

There is usually also a lack of shared responsibility and ownership of the product, service, or infrastructure under development. Under the functional structure, team work output and program ownership tends to be "thrown over the wall" when the development effort transitions from one team to another. When problems occur downstream, there many times is a lack of willingness on the part of upstream teams to reengage with the development effort, as they are typically already working on another program.

The silo structure of this approach also tends to be very slow and cumbersome due primarily to ineffective problem solving and decision making. When problems are encountered, the functional team that currently has ownership of the development effort becomes responsible for evaluation and resolution of the problem. Many times there is a lack of cross-discipline information needed to effectively define the right problem and establish the right solution. Instead, a lot of finger-pointing tends to take place, which can bring progress to a complete standstill. In addition, decisions are typically made outside of the program team under the functional structure and become the responsibility of the functional managers. With the large number of decisions necessary for a single development effort, a functional manager can quickly become a bottleneck to progress, especially if he or she is supporting multiple development programs.

The biggest drawback of the functional structure, however, is the lack of a shared vision of success among the functional teams, which is a critical element of the program management model. Each team is focused and rewarded on elements that are specific to that function and are not usually held accountable or rewarded for achievement of the overriding business results. In addition, the technical solution for each function tends to be more important than achievement of the business objectives. It is quite common under the functional structure to see a misalignment between functional goals and business goals.

Moving to a Multiple Project Team Structure

One of the major weaknesses of the functional structure for a program is the lack of a single person to manage a function's activities and deliverables. This responsibility falls on the shoulders of the functional manager, who is normally supporting multiple programs. The next logical

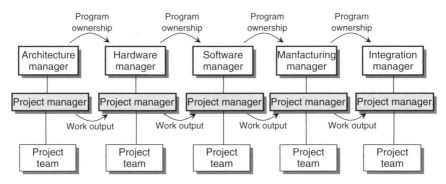

Figure 5.11   Multiple project team structure.

step in the evolution of a program structure is the assignment of project managers for each of the functional teams involved on the program. Thus, the multiple project team structure is established, as illustrated in Figure 5.11. The program now begins to look more like a collection of multiple related projects.

The primary advantages of this structure over the functional team structure is that true project management skills are introduced into the development effort, and the project manager for each function is focused on and responsible for his or her team's performance on a single program. These advantages benefit the functional managers of an organization most and actually do little to benefit the program.

The major drawbacks experienced in the functional team structure are still experienced in the multiple project team structure. The program is vertically structured, with little horizontal collaboration between the project teams. There is a lack of shared responsibility for the product, service, or infrastructure capability under development, with work output typically being sequentially handed from one team to the next. Good cross-discipline problem solving and decision making are not enabled and continue to be inhibitors to progress. Finally, the misalignment between project team success and program success is not resolved by this form of structure.

Evolving to a Coordinated Project Team Structure

Organizations normally come to realize that a silo organization, either with or without functional project managers, is not sufficient to manage the highly interdependent and complex nature of their programs. Astute

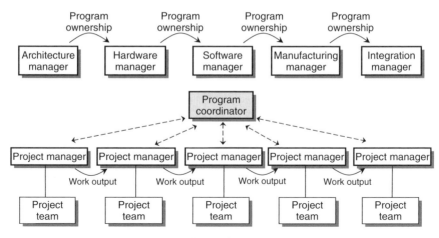

Figure 5.12   Coordinated-project team structure.

business managers realize that they need someone to facilitate cross-project coordination and communication to gain more efficiency in their development programs. The next step in the evolution of the program team structure is what we call the coordinated project team structure in which a program coordinator is used to oversee cross-project coordination and communication, as demonstrated in Figure 5.12.

Under the coordinated-project team structure, the program coordinator becomes responsible for initiating and facilitating the horizontal cross-project interplay that must be established for effective program execution. However, the program coordinator does not possess the organizational clout or requisite skills to effectively lead the functional project managers on the program. Additionally, the program coordinator has not been empowered by senior management for ownership and accountability relative to the results and success of the overall program. This person, therefore, tends to focus more on the tactical elements of program management such as coordinating cross-project schedules, facilitating communication between the project teams, and creating status reports for senior management.

With the coordinated-project team structure, some level of horizontal coordination and communication is established, with the barriers between the project teams softened but not eliminated. Senior management will now begin to sense the beginning of interdependent relationships between the functional project managers. However, without a true program manager leading the program, power and influence still reside within the functional organizations. [19] Therefore, most of the drawbacks with the functional team and multiproject team structures—such as lack of shared ownership for the product, service, or infrastructure under development, cross-project problem solving and decision making, and the misalignment between project team success and program success—still exist. Meanwhile, the program coordinator is tolerated but barely valued by the functional teams on the program.

Implementing the PCT Structure

The final phase of the program structure evolution occurs when senior management of an organization realizes that the functional silos need to be minimized to effectively execute its program. With that, a new horizontally focused program structure must replace the vertical structure legacy. In addition, an empowered leadership position is put in place to skillfully integrate the work of the functional project teams. Hence, the implementation of the PCT structure within the organization, as illustrated in Figure 5.13.

Under the PCT structure, program ownership belongs to the program manager, who is fully empowered by the senior management team of the enterprise. In this capacity, the program manager drives the cross-project and cross-discipline coordination and collaboration that is necessary to

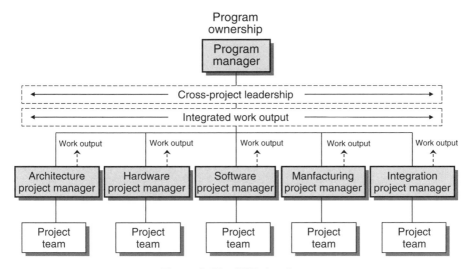

Figure 5.13   PCT structure.

eliminate the barriers brought upon by functional silos and, in the process, establish strong interdependent relationships between the project teams.

The PCT now possesses shared ownership of the product, service, or infrastructure that is under development. The core team structure is highly integrated, meaning there is joint consideration of trade offs, decisions, problem resolution, and risk identification and management.

It's important to reiterate that taking an evolutionary approach to implementing a PCT structure is highly inefficient (see box titled, "The Battle of Powerville," for an example of a failed transition of power). We strongly recommend avoiding this implementation approach. Be brave and take a revolutionary approach but effectively manage the change!

The Battle of Powerville

Rollercoaster History: Founded in the early 1900s, it took Miner Truck Company more than 60 years to reach the number one position in the mining vehicle industry—a position they would hold for nearly 30 years. However, in the early 1990s, things began to change. Annual sales began to drop, financials were in the red for several years, and the company dropped to number three in the industry. Additionally, it took the company 16 months to develop a new product, while competitors needed only 8 months. As a result, Miner Truck became a takeover target, and soon a Swedish multinational acquired it—ending the family-owned business of nearly 90 years.

Hopeful Future: Under new senior leadership from the Swedish parent company, things again began to look rosy during the next five years. Sales grew 200 percent, annual ROI increased to a respectable 14 percent, and the product development cycle time became shorter. For this to happen, some fundamental changes were made. The old functional structure was abandoned for a matrix structure and product development teams became cross-functional. The teams were led by a program manager and were empowered to make major program decisions. The original functional vice presidents (VPs) remained, but their decision-making power was transferred to the team. The first pilot program achieved the time-to-market goal of eight months, which was in line with the company's competitors.

Failed Future: The leaders of the functional groups, had the power to nominate the team members that would represent the functions on the development programs. Despite new team charters that empowered the team to handle decisions within their scope, the VPs saw this as an erosion of their power within the company. They, therefore, ordered their representatives on the team to not make any decisions without consulting them first. At first the team members resisted but later acquiesced over fear of negative effects to their careers. This turned the program team from a fully empowered PCT to a powerless coordinated project team, and the VPs into the primary decision makers for the program.

It took the senior management team of the Swedish parent nearly eight months to figure out why schedule performance on new product development programs was slipping dramatically. One month later, all the VPs and a program manager were let go.

Lessons learned from this story include the following:

- Major organization changes are not accomplished without shifts in the power structure.
- Shifts in the power structure can lead to power struggles on programs.
- Organizational interest is better served once the power struggles are resolved in alignment with the organizational goals.

ACHIEVING BUSINESS RESULTS

With the PCT structure in place, an organization becomes well positioned to achieve improved business results, such as rapid time-to-market performance, as illustrated in Figure 5.14.

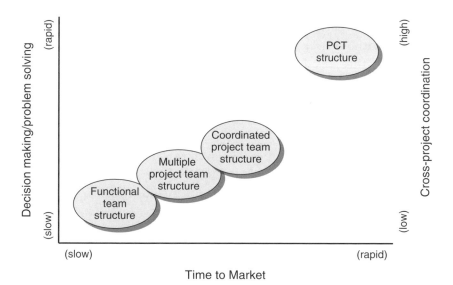

Figure 5.14   Higher performance with PCT structure.

The PCT structure provides the foundation necessary to begin breaking down product and process complexity; to establish consistent development processes and tools across the project teams; to fully identify and evaluate risk to the product, service, or infrastructure under development, as well as risk to the business to effectively manage resources across the program; and to establish the framework to link business strategy to execution output. This structure also effectively creates a firm foundation for realizing the business benefits that the program management model offers an organization.

SUMMARY

Two critical factors matter when structuring a program. First, the program team structure must support the highly integrated nature of a program. Second, the program team structure must enable the fundamental elements for team success. Common sense teaches us that to be successful, a program team must coordinate its activities and interdependent deliverables, effectively communicate what is being accomplished and roadblocks and collectively solve problems and make decisions that support the program objectives.

Surprisingly, there is a low success rate for establishing an effective program team structure. The primary reason for this low-implementation success rate may be that there is a fundamental lack of understanding of the difference between a program team structure and a project team structure. Without a firm understanding of the differences, the natural tendency on the part of executive and middle management is to try to implement a project structure for their programs. This is mainly due to their previous experience in establishing project team structures, as well as the wealth of information available on establishing project structures.

With the PCT structure in place, an organization becomes well positioned to achieve the business benefits that the program management model offers an organization. These benefits include rapid time-to-market performance; effective management of complexity; establishment of consistent development processes and tools across project teams; fully identifying and evaluating risk to the product, service, or infrastructure under development, as well as risk to the business; and effective management of resources across the program.

The Principles of Program Management

▼ Not understand the difference between a program team structure and a project team structure is the most common error businesses make in implementing the program management model.

▼ The program core team structure supports the highly integrated nature of a program and enables the fundamental elements for team success.

▼ The PCT is the cross-discipline, cross-project leadership, and decision-making body of the program.

REFERENCES

1. Martinelli, Russ and Waddell, J. "Demystifying Program Management: Linking Business Strategy to Product Development." *PDMA Visions Magazine*, (January 2004): pp 20–23

2. Eikenberry, Karen. *Elements of a High Performance Team*, Sideroads website: *http://www.sideroad.com/Team_Building/high-performance-teams.html*

3. Mohrman, Susan A., Cohen, Susan G., and Mohrman, Allan M. Jr. *Designing Team-Based Organizations: New Forms for Knowledge Work*. San Francisco, CA: Jossey-Bass Publishers 1995; pp. 5–6

4. NASA Website: http://nssdc.gsfc.nasa.gov/database/MasterCatalog?sc =1998-073A

5. McGrath, Michael E, Anthony, Michael T, and Shapiro, Amram R: *Product Development: Success Through Product and Cycle-time Excellence*. Stoneham, MA: Butterworth-Heinemann Publishers, 1992: pp. 87–90

6. Meyer, Marc H, and Lehnerd, Alvin P: *The Power of Product Platforms*. New York, N.Y: Free Press Publishers, 1997; (p 129)

7. Bianco-Mathis, Virginia and Fritzgerald, Bill, "*Cross-Functional Teams at AOL*", *OD Practitioner*, 34(2), 2002

8. Software Engineering Institute, "An Introduction to Team Risk Management". *SEI Special Report CMU / SEI-94-SR-1*, May 1994

9. ibid

10. Higuera, Ronald P, *Team Risk Management*, www.stsc.hill.af.mil/crosstalk/1995/TeamRisk.asp, 1995

11. McGrath, Michael E, Anthony, Michael T, and Shapiro, Amram R: *Product Development: Success Through Product and Cycle-time Excellence*. Stoneham, MA: Butterworth-Heinemann Publishers, 1992: pp. 62

12. Wheelwright, Steven C. and Clark, Kim B. *Revolutionizing Product Development*. New York, NY: Free Press Publishers, 1992; pp. 209–210

13. Martinelli, Russ and Waddell, Jim: "Program Management: Linking Business Strategy to Product and IT Development", *Project Management World Today*, (September-October 2003)

14. McGrath, Michael E., Anthony, Michael T and Shapiro, Amram R: *Product Development: Success Through Product and Cycle-time Excellence*. Stoneham, MA: Butterworth-Heinemann Publishers, 1992: pp. 206

15. McGrath, Michael E, Anthony, Michael T and Shapiro, Amram R: *Product Development: Success Through Product and Cycle-time Excellence*. Stoneham, MA: Butterworth-Heinemann Publishers, 1992; pp. 62

16. ibid, pp. 84

17. Martinelli, Russ, and Waddell Jim: "Program Management - Part II", *Management RoundTable Best Practices Report* (November 2003)

18. McGrath, Michael E, Anthony, Michael T, and Shapiro, Amram R: *Product Development: Success Through Product and Cycle-time Excellence*. Stoneham, MA: Butterworth-Heinemann Publishers 1992; pp. 70

19. Wheelwright, Steven C and Clark, Kim B: *Revolutionizing Product Development*. New York, NY: Free Press Publishers, 1992; pp. 193.

Chapter 6

Program Definition and Planning

The previous chapter began with the discussion of how to successfully implement a program utilizing the program management discipline and practices. This chapter continues the discussion with a focus on how to successfully define and plan a program. Program definition and planning are the first two phases of the PLC, as shown in Figure 6.1.

Simply stated, program definition involves the integration of business strategy with customer (or end user) and technology research to develop a product, service, or infrastructure concept. The feasibility of the concept is then tested through the process of developing and analyzing the program business case that, if successful, ensures the concept is viable from a business perspective. In effect, program definition aligns business objectives with strategy through the development of a new product, service, or infrastructure concept and then proves that the concept makes business sense and will serve as the means to achieve the intended objectives.

Program planning channels the work completed in program definition toward the creation of a cross-project, multidiscipline execution plan that will guide the work of the program team to turn the concept into a tangible product, service, or infrastructure capability.

The role of the program manager during program definition and planning is to facilitate and manage the horizontal collaboration between key representatives of the disciplines involved and to ensure that their work remains in synchronization to produce a well-defined concept, business case, and integrated execution plan. This chapter will put forth a process for program managers to channel research and strategic activities toward the creation of these key program deliverables.

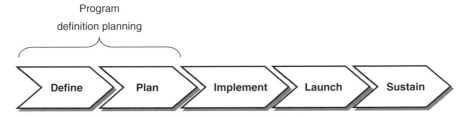

Figure 6.1   Program definition and planning and the PLC.

In doing so, this chapter will help program managers and others involved in programs understand how to do the following:

- Channel research and business strategy into a concept and business case
- Convert a concept and business case into a plan that aligns execution to business objectives and strategy
- Define and structure the projects within a program based upon the concept and requirements
- Integrate project plans into a cohesive program implementation plan

PROGRAM DEFINE PHASE

We described the closed-loop relationship between strategic management, program management, and project management in Chapter 3. Program definition is an extension of strategic management and is the critical element that binds all the functions together to effectively execute the business strategy. Inputs into the definition process include the results of the business strategic management efforts, the customer and/or end user research findings, and the results of technology research programs. Program definition involves integrating the goals, objectives, and ideas generated from these inputs to create potential product, service, or infrastructure concepts and then testing the feasibility of the concepts from a business perspective.

Concept Definition and Approval

Concept definition and approval is a critically important and difficult exercise due to the lack of incoming quantifiable data. As discussed in Chapter 2, this phase of a program is commonly referred to as the fuzzy front end because people are working with ambiguous information at this

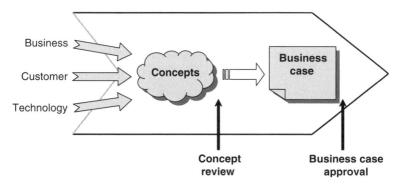

Figure 6.2 The program definition phase.

stage. Concept definition is as much of an art as it is a science because the people defining the concepts have to predict multiple scenarios that may play out anywhere from one to ten years in the future. As one product marketing research analyst from Intel said, "When I ask our customers what they envision a personal computer being able to do in the next five years, they look at me somewhat strangely and tell me that they're not sure. Right now they're just trying to focus on staying in business and staying relevant in their business."

Regardless of the well-documented challenges associated with the fuzzy front end, as we pointed out in Chapter 2, some of the greatest opportunities for business success lie in the work and the output of the program definition phase. The output of the synthesis of business, end user, and technology inputs to the program definition phase is a set of product, service, or infrastructure concepts (see Figure 6.2).

The role of the program manager during this initial stage of a program is to lead the small concept definition team, define a framework for collaboration, facilitate the cross-discipline team interaction, and focus the team's work on the business objectives desired.[1]

The initial program definition team is normally a small, highly experienced, cross-discipline team. Because a concept is an integration and synthesis of business, customer (or end user) and technology ideas, at a minimum, representatives of each of these disciplines must be part of the definition team. A small, but well represented, definition team may be composed of the following people:

- *Program manager*: Responsible for leading and facilitating the work of the program definition team in the creation of the product, service, or infrastructure concept and championing the intended business objective driving the need for the program

- *Customer and/or end user representative*: Responsible for assessing the market, representing customers' or end users' wants and needs
- *Engineering representative*: Responsible for assessing the early design considerations to determine if the new idea can be designed and built
- *Architect*: Responsible for technology research findings and ideas and for assessing the use of existing or new architectures and processes to create the product, service, or infrastructure capability
- *Financial representative*: Along with the program manager, responsible for assessing the financial feasibility of the concept

Concept-Definition Process

The following steps constitute a generic approach for a program manager to use in leading the program definition team through the development and approval of a product, service, or infrastructure capability concept. This process can be used as a foundation for creating a customized process by tailoring the steps for any organization.

Identify the Business, Customer, and Technology Objectives

The business, customer, and technology objectives that the program is intended to achieve must be clearly identified in this early stage and communicated throughout the life cycle. They become the cornerstone on which the product, service, or infrastructure concept and program are built, as well as the primary success criteria against which program execution and closure are measured. The objectives are the guiding principles that help the program manager and other key decision makers understand the scope of the program, make key trade-off decisions, and determine resource requirements. The objectives must be specific, measurable, realistic, and time related.[2]

The program manager will derive the initial business objectives from the output of the strategic management process and decompose them into more detailed requirements for the specific program being defined. Customer objectives can be derived from the information gained by performing customer and end user research. Technology objectives are the primary output of technology research efforts that invest in the development of new materials and technologies to be transferred to products, services, or infrastructure applications. The transfer of technology discussions occurs early in the strategic planning process and become more comprehensive as specific programs are defined.

The program manager uses the business, customer, and technology objectives to develop the initial critical success criteria for the program. The program strike zone is an exceptional tool for collecting and communicating the program success criteria (see Chapter 11); it can also be used by the program manager to keep the program definition team focused on achieving the intended objectives.

Develop the Product Concept

The heart of the definition phase is the product, service, or infrastructure concept. Technology research, customer research, and strategic planning efforts must converge to create a concept that will best achieve the business objectives of the program. Deriving a concept, or set of concepts, is an iterative process that involves the integration and synthesis of the ideas of the subject matter experts on the definition team. The program manager should lead the ideation process and continually focus the output toward the objectives identified. This involves the following: facilitating the cross-discipline collaboration, setting periodic concept reviews in front of primary program stakeholders to create tension in the system and expedite the concept definition, continually collecting and communicating any changes to the initial program critical success criteria, and testing the concept feasibility from a business perspective.

Conduct the Concept Approval Meeting

The final stage of the concept definition process is the presentation of the product, service, or infrastructure concept to the primary program stakeholders for approval. This is normally the first major decision checkpoint in the PLC. It is effective if the program manager facilitates the concept approval meeting and presents the initial program critical success criteria for review and approval by the stakeholders. The program manager can then turn the review over to the other members of the program definition team to present their aspects of the product, service, or infrastructure concept.

One of the following three decisions should be made during the concept approval meeting: (1) reject the concept(s) with the direction to redefine all or part of the concept(s), (2) reject the concept(s) and cancel the program, or (3) approve the concept(s) with direction to move into the back half of the definition phase to develop the business case.

BUSINESS CASE DEVELOPMENT AND APPROVAL

The program business case is a guiding document used by senior management to assess the feasibility of a program from multiple business perspectives; it is also used by the program team as its primary chartering document. It establishes the program vision by describing a business opportunity in terms of alignment to strategy, market, or customer needs; technology capability; and economic feasibility. It also provides a balanced view of the business opportunity versus business risk.

The program business case contains the most important information available at this early stage of a program. It will be used to determine if the business unit should invest financial, human, and capital resources to plan, develop, launch, and support a new product, service, or infrastructure capability.

The role of the program manager during this aspect of the program definition phase is to lead the definition team toward the creation of a successful business case and present the business case to the primary program stakeholders for approval (see Figure 6.2).

Program Business-Case Development Process

The steps in the following sections constitute a generic approach for a program manager to use in leading the program definition team through the development and approval of a product, service, or infrastructure business case. This process can be used as a foundation for creating a customized process by tailoring the steps for any organization. See Chapter 10 for a description of the elements commonly found in a program business case.

Define the High-level Requirements

Business and high-level technical requirements must be developed, documented, and delivered during the program definition phase. The program manager is responsible for collecting and documenting the high-level requirements during this phase of the program. The importance of diligently defining requirements cannot be overstated, as requirements are the basis for program planning and engineering activities. Poor requirements will result in poor planning that will result in poor execution and failure to meet the business objectives.

Once the concept is approved, gathering and documentation of requirements should begin. The program manager should work with the content experts who represent the various functional disciplines on the program definition team to collect the requirements. At the close of the definition phase, requirements must be documented in sufficient detail to describe the product, service, or infrastructure concept; key features and functions; end user needs; time-to-money window; and cost and financial targets at a minimum.

Analyze the Risk

Once the product concept is defined and high-level requirements documented, a thorough risk analysis can be conducted to properly estimate the probability of both technical and business success of the proposed program. This risk-based approach allows managers to make decisions based on greater knowledge of the product, service, or infrastructure concept under evaluation.

The program manager should lead the risk identification and assessment activities, with involvement of the program definition team. The objective of this exercise is to attempt to identify as many unknown events that may occur and potentially prevent the program from succeeding. With the risks identified, the team can assess the severity of each risk event, with respect to both the probability that the risk event will occur and the impact to the program if it does occur. It is important to note that this exercise happens very early in the program and little is actually known about the risk events. Therefore, quantifiable data to support the risk analysis will be difficult to develop. It is recommended that each risk be evaluated on a qualitative basis as high, medium, or low severity.

Assess Program Complexity and Strategic Alignment

The program manager and definition team should next evaluate the level of complexity and how well the program aligns with the business strategies of the organization. This information will be used to evaluate the program within the context of the entire portfolio of programs, once the program business case is approved. It is recommended that this assessment be completed as part of the program business case.

The program manager can evaluate the complexity of his or her program through the use of the complexity assessment tool described in detail in Chapter 10. Program complexity will help identify the categories of risk the program will be faced with and can be used to determine the amount of contingency reserve to build into the program budget and schedule—the more complex the program, the larger the reserve. Finally, the complexity assessment can give the program manager an indication of the level of skill and experience needed for the program team.

The program alignment matrix (see Chapter 10) can be used to establish the degree to which the program is aligned with the organization's business strategy. The alignment assessment aids the program manager to understand how well his or her program supports the strategic objectives of the firm. It also gives senior management another important dimension for evaluating the feasibility of a proposed program concept. With this in place, each program concept can be evaluated on the basis of cost, benefit, risk, and strategic importance.

Assess Feasibility

Multiple concepts may be developed during the definition phase. If this is the case, a feasibility assessment is required prior to the business case completion to select a single concept to present to the primary program stakeholders. With the concepts defined and a risk assessment complete, the program team can validate the concepts from business, customer, and technical perspectives. Each concept is evaluated on its own merit and then compared to the feasibility results of the other concepts. At the end of the evaluation, a single concept should be selected and the business case completed.

Conduct the Business-Case Approval Meeting

The final step in the program definition phase is to present the program business case to the primary program stakeholders for approval. This is the first phase/gate decision checkpoint in the PLC. The program manager typically presents the program business case to the executive decision-making body in charge of the business's portfolio of programs.

One of the following three decisions should be made during the business-case approval meeting: (1) reject the business case with direction to repeat the appropriate elements of the program definition phase, (2) reject the business case and cancel the program, or (3) approve the business case with direction to move into the planning phase of the PLC.

Successful completion of the program definition phase should culminate in the commitment of funds and resources to develop a detailed implementation plan and the addition of the program to the business's portfolio and the program road map.

PROGRAM PLANNING

"Planning is everything." That's how a program manager from Ford Motor Company describes the importance of program planning. It seems to be innately attractive to many individuals to just jump in, start taking action, and doing things. Action-oriented individuals many times have a difficult time with planning activities. How many times have you heard it said, "Let's just do it; we know what needs to be done and how to do it."? Many believe that spending time in meetings doing planning is a waste of time. However, as we have observed on several occasions, these same individuals eventually learn to appreciate the value of planning before execution begins in earnest once cost overruns, schedule slips, and program cancellations begin to occur. Most of us have learned through experience that early and effective planning does prevent poor performance and problems such as late time to money, missed commitments to customers, and quality issues.

In the program planning phase, all work is focused on developing an integrated master program plan. Program plan development involves integration of information from all functional elements represented in the definition and scope of the program. The primary objectives of the program planning phase are threefold, as follows: (1) develop and understand the scope of the program, (2) develop the integrated cross-project implementation plan, and (3) submit and review the implementation plan with executive decision makers to gain approval to enter into the program implementation phase of the PLC. After the program plan is developed and approved, it becomes the road map to program success. Figure 6.3 illustrates the sequencing of the primary steps in developing a program implementation plan.

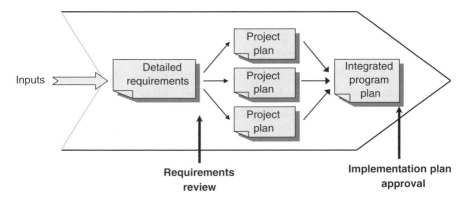

Figure 6.3   The program planning phase.

The Program Planning Process

The steps in the following sections constitute a generic approach for a program manager to use in leading the PCT through the development and approval of a detailed program implementation plan. This process is based upon the planning models used by multiple companies that we deem highly effective in program planning. The process described can be used as a foundation for creating a customized process by tailoring the steps for any organization, keeping in mind that program planning can be situational in nature (see the box titled, "Different Programs, Different Planning Situations").

Different Programs, Different Planning Situations

We certainly subscribe to good program planning. The question always arises, however, how much detail is needed in a program plan? The answer to this question usually lies in the type of program to be planned.

Using a simplified 2×2 matrix of program types (see Figure 6.4), a program manager can determine the amount of detail needed in his or her program plan.[3] The program type is based upon the following two variables: technology novelty and program complexity.

A routine program is one having a low level of technology (less than half of the technologies are new) and low complexity (few cross-project interdependencies). An administrative program is low in technology but high in complexity (many cross-project interdependencies). A technical program is high in new technology use (more than 50 percent of the technologies are new) and low in complexity. A unique program is high in both technology content and the number of cross-project interdependencies.

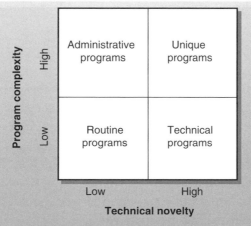

Figure 6.4   Generic program types.

The amount of detail needed in a program plan depends upon the type of program, referred to as situational planning. In general, the higher the program complexity, the greater detail required in the program plan. The greater the technology content, the greater the uncertainty or technical risk of the program outcome. This calls for a plan with fewer details but more schedule and budget buffers included in the plan.

Collect Inputs

The first step in developing a program plan is to collect as much information about the program as possible. The program manager should have the information from the program concept and business case at his or her disposal, but if this is not the case, this information should be the first to be collected.

A clear understanding of the strategic business objectives that drive the need for the program is critically important to ensure the program implementation plan is in alignment with company strategy; therefore, any output information from the strategic management process should be collected. This includes all information about the program that is contained in the portfolio of programs and the program road map.

The high-level requirements in support of the program business case also need to be collected, as they become the foundation for developing the detailed requirements. The high-level requirements should provide information on customer and end user needs, primary features and functions, technology strategies and targets, and business requirements such as cost and schedule targets.

Form the PCT

Program planning is a team effort involving the program manager and his or her PCT members. Following the approval of the business case, and in conjunction with collecting program inputs, the program manager will set out to form the full PCT. As described in Chapter 5, program team structure is a case of form following function. In other words, the make up of the PCT members is driven by the functions that need to be involved in the program to provide the whole solution, as described by the program concept and business case.

The role of each functional project manager is to develop the detailed plan within his or her domain. The role of the program manager is to integrate the functional project plans into one cohesive program plan. Depending on the size of the program, planning can require a large commitment of time and energy from the PCT.

A key responsibility for the program manager is to set the vision for the team. The program vision grounds the team on where the business is going, what it takes to get it there, and how this particular program fits into that picture. [4] The program manager should ensure that the vision and requirements for the program are sufficiently understood to the point that the members of the team internalize the program objectives as their own and are as motivated as the program manager to achieve the program's success.

Document Detailed Requirements

Creation of a good program plan is contingent upon how well the detailed requirements are defined and documented. If gaps exist in the detailed requirements, gaps will exist in the program plan. These gaps normally *will* get filled but later in the program in the form of scope changes, which become more expensive as time progresses due to increased amounts of rework. Likewise, low quality or nonspecific requirements *will* have to be redefined when someone tries to interpret them and perform real work against them. The later the reconciliation occurs, the more expensive it becomes.

The detailed requirements begin with the high-level business, technical, and customer requirements delivered at the end of the definition phase of the program. Each member of the PCT is responsible for ensuring the requirements needed to perform their specialized functional work are included and adequately stated in the detailed requirements documentation. The detailed requirements must be clear, concise, and thorough, as they are the basis on which the functional project teams will create their work breakdown structures and functional plans.

Once completed, the detailed requirements constitute the program scope and should be managed as a living document (see the box titled, "The Situational Approach to Scope Management"). From the requirements, a program-level work breakdown structure (WBS) can be developed if the program manager feels it will be helpful and useful to the PCT. It is recommended that the program manager initiates a comprehensive stakeholder and peer review of the requirements to check that they are complete with respect to the information known about the program at the given time and that the quality level is sufficient.

It needs to be understood that more will be learned about the program as time progresses, and there may be many changes in requirements during the remainder of the planning phase and the implementation phase that must be captured and documented in the detailed program requirements.

The Situational Approach to Scope Management

It is no secret that scope needs to be defined in a timely manner. We have witnessed some programs that were fully scoped early in the PLC, and others that weren't fully scoped until later in the PLC—with both approaches being praised as sufficient and timely. So why the variation in the scope timeline between the programs?

The answer again is that there are varying types of programs, with varying requirements for scope timeliness. The 2×2 matrix shown in Figure 6.4 identifies the four program types—routine, administrative, technical, and unique. Detailed program scope can be frozen at varying points in the PLC, depending upon the program type (see Figure 6.5).

Routine programs have low levels of technical uncertainty and cross-project interdependencies; therefore, program scope can usually be frozen quite early in the planning phase of the PLC. Administrative programs require more cross-project collaboration, which means that program scope freeze takes longer than for routine programs. However, scope freeze can be achieved by the midpoint of the planning phase.

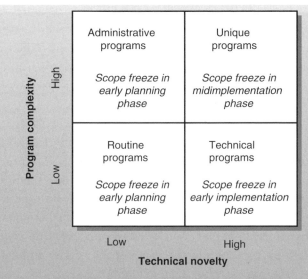

Figure 6.5   Scope freeze by program type.

By contrast, technical and unique programs carry a high level of technology risk that usually delays the completion of the detailed scope definition. For technical programs, therefore, program scope freeze usually occurs early in the implementation phase, even later for unique programs due to the large number of cross-project interdependencies involved.

This process is referred to as situational scope management in which program scope and scope freeze is dependent upon the type of program defined.

Determine Project Scope

Once the program-level requirements and scope definition are completed, the requirements must then be dissected by each of the project teams on the program to create their respective project-level WBS and project plan. The WBS is a deliverable-oriented grouping of all functional elements that organizes and defines the total work scope of each project.

The WBS is a hierarchical representation of the project that facilitates the scheduling, budgeting, resource allocation, and control activities for the functional project manager and the program manager. It is a key element of the planning process utilized to transform requirements into functional project elements (tasks) and deliverables. When managers create a project-level WBS, they must understand the project deliverables within the program and envision each functional project as a hierarchy of goals, objectives, and tasks that must be accomplished to successfully begin and end a project.

Plan the Projects

With each project WBS complete, each project team can begin developing their respective initial project plan. We use the term *initial project plan* intentionally, as a final project plan cannot be developed until the integrated program plan is created. At that point, the initial project plan is updated to reflect the changes that were made during the integration process.

The program manager should set and communicate a common format and content expectation for each of the project plans. Primary elements that should be included in the project plans include the following:

- Project description and objectives
- Project deliverables
- Critical success criteria
- Project schedule
- Project budget
- Team structure
- Resource profile and staffing plans
- Project risk analysis
- Tracking and controlling methods
- Change management methods
- Team communication strategy

To keep the work of the various project teams synchronized, the program manager should set clear milestone dates for project plan maturity and completion. He or she should also set up periodic project plan reviews with each of the project managers to ensure planning work is progressing and to assist them in eradicating any barriers to successful completion. As changes to the program-level requirements and scope occur, the program manager should quickly communicate the changes to the entire PCT, so the changes can be incorporated in the project plans.

Create the Integrated Master Program Plan

Once the initial project plans are completed, the program manager can lead the PCT through the creation of an integrated master program plan. The program plan incorporates the work of all project teams, the interdependencies between the teams, and all other functional representatives that are part of the program team.

The program plan will vary by program type and situation (see box titled, "Situational Program Planning") but should include the details necessary to execute program implementation, launch, transfer and/or sustaining, and closure.

Situational Program Planning

Program planning approaches, as well as the details contained in each program plan element, vary by program type. This is the basis of situational program planning. Figure 6.6 summarizes this variation according to the four program types previously identified—routine, administrative, technical, and unique.

Routine programs are low in technical novelty and complexity and are, therefore, quite stable in nature. This normally results in clear business goals and a well-defined scope, requiring a simple WBS with few levels of detail. Program schedules can be created using simple Gantt charts, which informally recognize the cross-project interdependencies without showing them in detail. With precise business goals, scope, and a simple schedule, the team can develop a precise bottom-up cost estimate and risk response plan.

Administrative programs are high in complexity and low in new technology introduction. Business goals and scope are normally well-defined, stable, and detailed. Added complexity requires mapping of the many cross-project interdependencies, but the technical stability allows for standard scheduling techniques. The same added complexity generally means larger program size, with higher financial exposure, justifying the need for detailed bottom-up cost estimates reconciled with financial targets contained in the program business case. Risk is primarily related to the increased number of interdependencies between the project teams; therefore, risk management should be tiered, focusing on both program-level and project-level risks.

Technical programs have a high number of new technologies that often requires time to fully define the business goals, scope, and WBS for the program. Still, the goals, scope, and WBS are simple due to the low level of complexity. For the same reasons, the schedule becomes fluid and readily reshaped to account for goal and scope changes. The rolling wave approach (RWA), or similar approach, can be used. This means that only the schedule for the following 60–90 days are planned in detail, while the remainder of the program schedule is represented by milestones. Similarly, cost estimates are fluid as well. A detailed cost estimate for the next 60–90 days can be detailed, while cost estimates for the remainder of the program are at the summary

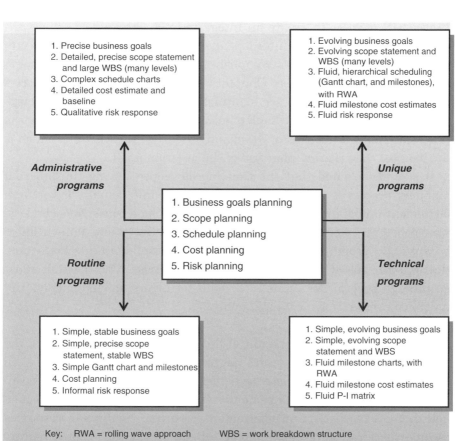

Key: RWA = rolling wave approach   WBS = work breakdown structure

Figure 6.6   Situational program planning overview.

or rough order of magnitude level. The increased technical novelty results in increased technical risk and the need for more rigorous risk management implementation and tools.

For unique programs, business goals, detailed scope definition, and WBS development takes time to evolve as a result of many new technologies and cross-project interdependencies. The evolving nature of scope leads to the need for fluid schedules. Program mapping and rolling wave scheduling processes can be used to contend with the fluidity. Similarly, cost estimates for milestones are more detailed in the near-term and more summary level for the longer term. Combined novel technologies and complexity push risks to the extreme, making it the single most challenging element to manage. In response, a rigorous risk management plan is needed, as well as a combination of tools such as the P-I matrix (see Chapter 10) and Monte Carlo Analysis (MCA).

Program Schedule: To create an integrated plan, the work of the various project teams and the interdependencies between them first have to be comprehended. This is accomplished through the program mapping and scheduling processes (see Chapters 8 and 11). The interdependencies between the project teams are identified and planned through the mapping process, which when complete also yields an initial timeline for the program. This is the basis for developing the program schedule, which is the heart of the program plan and of the program itself.

It should be noted that the timeline developed during the program mapping process is a low-confidence-level schedule, usually lower than 50 percent probability of success (same as flipping a coin). Nevertheless, we consistently hear war stories of missed delivery and launch dates because the program was planned on a low-confidence-level schedule that was the result of a program mapping exercise. The schedule must go through several iterations to account for errors, misjudgments, and unknown events (risks) that *will* occur.

Resource Planning: Once the initial program schedule is developed, the functional project managers and the program manager can identify the resources needed to complete each task within the required time constraints. It is important to remember that resource estimates include both human resources and nonhuman resources. Nonhuman resources include supplies, materials, and capital equipment that people require to complete their work. The result of resource planning is the program resource profile, which details the resource demand for the program.

The program manager is responsible for working with the functional managers of the organization to make sure each functional project is adequately staffed. If resource capacity is insufficient to meet demand, the program scope and/or timeline must be reconciled to balance capacity and demand—the triple constraints apply to a program as well. In some cases, time estimates need to be adjusted based upon the skill and experience levels of the people assigned to the various tasks.

Detailed Program Budget: The program budget represents the organization's financial investment in the program. Much like the development of the program schedule, each project team is responsible for the development of their respective project budgets. The program manager works with the PCT to integrate the project budgets into an overall program budget. Budgeting is addressed in the planning phase of the program management model, but is a key element of each phase of the life cycle. A rough order of magnitude estimates are established in the definition phase to determine financial feasibility of the program, the detailed budget is established in the planning phase, and budget expenditures are monitored and controlled during the implementation, launch, and sustain phases.

The Program Strike Zone: The initial program critical success criteria that were established in the definition phase of the PLC can now be completed based upon the program planning work to date. A tool called the program strike zone (see Chapter 11) is very effective for identifying the success criteria against which the program will be measured for completion and for establishing the boundaries within which the PCT can manage the program and make decisions without direct involvement of senior management. The program strike zone needs to directly reflect the program objectives established in the business case, as well as key elements of the program plan such as completion date, product cost, budget, and quality targets. As part of the plan approval process, the program manager and senior management negotiate and agree on the success criteria and what actions will be taken when success criteria thresholds are compromised.

Risk Assessment: Once the primary program elements are developed, the PCT should perform a detailed program-level risk identification and assessment. Each functional project manager should come into the program planning process with a set of project-specific risks that his or her team identified and assessed. The program manager is then responsible for collecting all risks from the functional teams and leading the PCT through a risk analysis exercise to determine which of the risks are program-level risks and which are project-only risks (see Chapter 8). The program risks are then categorized and prioritized by potential impact to the program. As discussed in Chapter 11, the P-I Matrix tool can be used to identify program risks, assess the probability of occurrence and potential impact, and provide a representation of risk severity to facilitate effective program decision making.

It is recommended that only the highest-level risk events be formally included in the program plan. However, all risk events need to be tracked and managed through the program risk management process for the duration of the program. During planning, the risk information should also be utilized to estimate schedule and budget buffers that are used to protect the program success criteria in the event any of the risks come to fruition.

Update the project plans

With the primary elements of the integrated program plan now completed, each project plan needs to be updated to reflect any changes that occurred during the integration process. This step ensures that the project plans stay in synchronization with the master program plan.

Validate the Program Business Case

Once the detailed program plan elements are completed, the program manager and key PCT members need to validate that the program business case is still viable. At a minimum, this involves reevaluating the program financials, alignment index, and benefit versus risk analysis, plus verifying that the results still support the strategic business results of the program.

This seems like an obvious step in the planning process, but in practice it is a step that is often overlooked for a number of reasons. Once detailed planning and execution activities begin in earnest, it becomes difficult to view the program from business and strategy perspectives. It is crucial that the business case is periodically reevaluated to keep the program viable from a business perspective. This is especially true at major decision checkpoints of the program.

Conduct Implementation Plan Approval Meeting

The final step in the program planning phase is to present the implementation plan to the primary stakeholders for approval. The implementation plan approval is the next major phase/gate decision checkpoint in the PLC. The program manager typically presents the program implementation plan to the executive decision-making body, with assistance from key members of the PCT. Additionally, the updated program business case should be reviewed to ensure that the program remains viable from a business perspective.

One of three decisions should be made during the implementation plan approval meeting: (1) reject the implementation plan with direction to repeat the appropriate elements of the program planning or definition phase (2) reject the implementation plan and cancel the program, or (3) approve the implementation plan with direction to move into the implementation phase of the PLC.

Successful completion of the program planning phase should culminate in the full commitment of funds and resources to develop the product, service, or infrastructure capability on the part of executive and functional managers. Communication to customers and end users concerning the details of the program can now begin to take place.

PROGRAM DEFINITION AND PLANNING PROCESS OVERVIEW

Table 6.1 summarizes the steps involved in defining and planning a product, service, or infrastructure program.

IMPORTANT BEHAVIORS

There are a set of important behaviors that we have seen demonstrated on successful programs by both program managers and functional project managers, specifically during the definition and planning phases of a program.

Table 6.1   Program definition and planning steps.

Program Definition Phase	Program Planning Phase
• Form concept definition team	• Collect inputs
• Identify business, customer, and technology objectives	• Form the PCT
• Develop the product concept	• Document the detailed requirements
• Conduct the concept approval meeting	• Determine project scope
• Define the high-level requirements	• Plan the projects
• Analyze the risk	• Create the integrated master program plan
• Assess program complexity and strategic alignment	• Update the project plans
• Assess feasibility	• Validate the program business case
• Conduct the business case approval meeting	• Conduct the implementation plan approval meeting

Willingness to take the lead: The program manager needs to take a leadership role during both the program definition phase and program planning phase. The early program team is primarily composed of specialists from multiple functional disciplines that normally are not effective in leading a cross-discipline team. When the program manager fulfills this role, all viewpoints are considered and the work is focused toward achievement of the program vision and business objectives intended.

Empower the team: The program manager must be willing to share the empowerment granted to him or her with the functional project managers on the program team. Likewise, the project managers must share their empowerment with their respective team members. Full-team empowerment is a powerful tool in quick and effective decision making in which people closest to an issue are able to evaluate the situation and decide the proper course of action.

Demonstrate integrity and trustworthiness: The program manager and project managers should set the example in the areas of working with a high degree of integrity and build the trust of team members and program stakeholders. As a leader, integrity and trust are the foundational elements of credibility.

Take ownership: The program manager needs to display total ownership for the management and outcome of the program. He or she needs to be ready to take total responsibility for his or her own actions and the actions of the team. Additionally, the program manager should champion the program inside and outside the organization.

Exhibit fairness: The program manager needs to exhibit fairness and balance between cross-discipline representatives on the program team. This means being able to set aside any historical bias toward or against a particular discipline that he or she may have had in the past. Exhibiting fairness also means setting the expectation that all others on the program team behave in the same manner. Failure to do so can foster unhealthy team dysfunction.

Be organized and disciplined: Program managers and project managers alike need to be good at organizing and managing the details on a consistent and continual basis. Programs are complex undertakings, and without a high degree of organization on the part of the team leaders, program details can quickly become unmanageable.

Communicate effectively: The program manager is the voice of the program during the definition and planning phases. He or she must be willing and able to communicate in all directions and at all levels within the organization. It also means being good at both verbal and written communication methods, being consistent in delivering messages, and being able to tailor the communication for the level and interest of the recipient.

SUMMARY

The program definition process involves the steps necessary to integrate business strategy, customer and end-user research, and technology research into a viable product, service, or infrastructure concept. Then, test the viability of the concept through the creation and analysis of a comprehensive program business case.

The program planning process involves the steps necessary to channel the product, service, or infrastructure concept into a cross-project, multidiscipline implementation plan. When executed, the program plan will guide the work of the cross-project implementation team to turn the concept into a tangible product, service, or infrastructure capability that will become the means to achieve the business objectives intended.

A set of important behaviors that highly effective program managers exhibit during the program definition and planning phases ensure that the work is accomplished efficiently and effectively.

The Principles of Program Management

▼ Program definition and planning are no more and no less than preparing for the future.
▼ Program definition is an extension of strategic management and is the critical element that binds all the functions together to effectively execute the business strategy.
▼ A PLC is a framework to guide all program planning activities.

▼ The intent of the planning phase is to develop the program scope, integrate cross-project implementation plans into the master program plan, and submit the plan to gain approval by the senior sponsor(s).

▼ Program planning techniques vary in accordance with the program situation.

REFERENCES

1. Martinelli, Russ. "Taming the Fuzzy Front End", *Project Management World Today* (July-August 2003).
2. Doran, George T. "There's a S.M.A.R.T. Way to Write Management Goals and Objectives." *Management Review* (November 1981): pp. 35-36.
3. Shenhar, Aaron J. "One Size Does Not Fit all Projects: Exploring Classical Contingency Domains." *Management Science*, 47(3), 2001: pp. 394-414.
4. Wheelwright, Steven C. and Kim B. Clark. *Revolutionizing Product Development*. New York, NY: Free Press Publishers, 1992.

Chapter 7

Program Execution

The previous chapter continued the discussion of how to successfully manage a program utilizing the program management discipline and practices by focusing on program definition and planning. This chapter will complete the discussion by describing how a program is successfully executed.

The role of the program manager during program execution is to manage the horizontal collaboration between the project teams and to ensure that the work of the multiple project teams remains in synchronization, with cross-project interdependencies and deliverables executed as planned. The project managers and other members of the PCT manage the work of their respective functional teams and ensure that the deliverables, milestones, and success criteria are achieved.

In practice it is common for a program to transition from a program manager who specializes in definition, planning, development and launch, to a program manager who specializes in maintenance of business and sustaining activities between the launch and sustain phases of the PLC. The role of the program manager does not change, only the person leading the program team. However, we are beginning to see more companies that are assigning a single individual throughout the life of a program—from concept to end of life. We mention this not to advocate one approach over another but to point out that there is not a single "best approach" to managing a program.

This chapter will put forth a process for program managers to use in program execution. It will explain how to create the intended product, service, or infrastructure capability; launch the solution into the market or customer environment; and sustain the program until end of life.

In doing so, this chapter will also help program managers in the following ways:

- Understand the process for implementing the program plan
- Describe the primary steps involved in launching a product, service, or infrastructure capability
- Discuss key elements in sustaining a program to end of life
- Explain behaviors of successful program managers during program execution

PROGRAM EXECUTION AND THE PLC

Program execution is focused on carrying out the work documented in the master program plan and the functional project plans; it encompasses the implementation, launch, and sustain phases of the PLC, as shown in Figure 7.1.

The primary objectives of the implementation phase are to (1) successfully accomplish all of the design and development work necessary to complete the elements of the product, service, or infrastructure capability under development; (2) integrate the functional elements to create the whole product, service, or infrastructure; and (3) prepare for the launch of the program output into the market or customer's environment.

With the elements of the product, service, or infrastructure capability developed, integrated, and the solution approved for release to the customer or end user, the program enters into the launch phase. The objectives of the launch phase are to (1) methodically and successfully transition the product, service, or infrastructure capability from the development environment (sometimes referred to as the preproduction environment) to the customer environment; (2) ensure that customers are

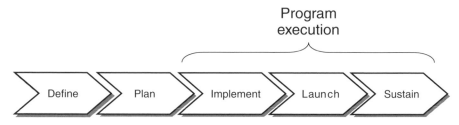

Figure 7.1   Program execution and the PLC.

fully supported; and (3) ensure all production and sustaining systems are in place and effectively operating.

Once the necessary start-up activities to turn the product, service, or infrastructure capability into a sustainable and on-going entity are completed, the program enters into the sustain phase of the PLC. The sustain phase of the PLC is the final phase of the program and traditionally signifies the transition from development to support and maintenance of business. The primary objectives of the sustain phase are to (1) keep the program business running as effectively and efficiently as possible, (2) fully support the customers and end users, (3) ensure that the business objectives of the program are attained and that the program remains viable from a business perspective, and (4) efficiently close out the program when it reaches end of life.

PROGRAM IMPLEMENTATION PHASE

The majority of the program team anticipates the program implementation phase of a program. It is when an intangible concept becomes a real, usable, tangible asset for realization of the business objectives that the program is meant to achieve. It is also the part of the program in which the quality of the program plan and the ability of the program manager to lead will be put to the test.

The job of managing a program becomes much more challenging during the implementation phase due to a couple of natural factors. First, the size of the program team and the number of cross-project interdependencies to track and manage grows rapidly between planning and implementation. Figure 7.2 demonstrates this phenomenon. The program profile shows that many elements of the program, such as staff size, budgeted dollars spent, and number of interdependencies, are at their highest levels and peak during the implementation phase of a program. These factors need to be closely managed both vertically within the projects by the project managers and horizontally across the projects by the program manager to ensure that the program stays in alignment with the business objectives. The challenge for the program manger is to remain at what we call the 10,000-foot level of a program and focus on managing the horizontal cross-project collaboration. This means that the program manager has to empower the project managers to manage the details of the five-foot level within their functional discipline. The program manager cannot get

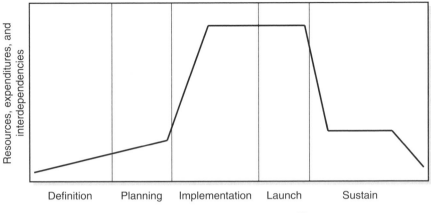

Figure 7.2 Typical program profile.

sucked down to ground level for very long, or the horizontal collaboration begins to break apart.

The second factor that complicates the management of the implementation phase is that all program stakeholders suddenly become aware of a program's existence, seemingly overnight. It is as though many stakeholders don't pay attention to a program until it becomes *real* in their eyes, meaning that until a plan has been approved and resources are working to implement it, a program doesn't really exist. When this happens, the program manager has to spend more time away from the program team to manage the expectations and inquiries of the stakeholders—plus, manage the politics now surrounding the program.

In the implementation phase, work of the program team is focused on creating the product, service, or infrastructure capability that was conceived and planned; then, preparing for its introduction into the market or customer environment. Figure 7.3 illustrates the sequencing of work associated with program implementation.

The Program Implementation Process

The following steps constitute a general approach for a program manager to use in leading the PCT through the implementation phase of a program. This process is based upon the models used by multiple companies that we deem highly effective in program execution. The general process described can be used as a foundation for creating a customized process by tailoring the steps for any organization.

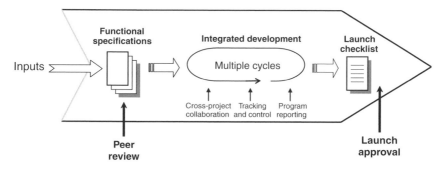

Figure 7.3   The program implementation phase.

Collect Inputs

Like the other phases of the PLC, thus far, the first step in the program implementation phase is to collect as much information coming out of the planning activities as possible. Foremost, the program manager should have the program implementation plan at his or her disposal and should have intimate knowledge of its details.

The detailed requirements in support of the program implementation plan also need to be collected, as they become the foundation for developing the detailed specifications. The detailed requirements are used to create a design, which is used to create a set of deliverables that meet the needs of the end user.[1] The detailed requirements should provide information on customer and end user needs, primary features and functions, technology strategy and targets, and business requirements such as cost and schedule targets.

The program success criteria are the final critical input needed to begin the implementation work. At this point in the program, the entire set of success criteria should be established. They are the set of business, technological, and customer targets that the program plan is aimed toward achieving. These become the boundary conditions within which the program manager and his or her program team can work without senior management intervention. All members of the extended program team should have access to and be familiar with the program success criteria.

Staff the Extended Program Team

Upon approval of the baseline plan, and in conjunction with input collection, one of the first activities that takes place during program implementation is the selection and assignment of all human and nonhuman resources to the program. The program manager is responsible for ensuring the PCT is fully staffed, while the functional project managers are responsible for staffing their respective project teams. This is not to say that the program manager does not assist the project managers in obtaining the qualified personnel, if necessary.

The program manager is not responsible for selecting the resources for the entire program. Rather, he or she ensures each project manager has the ability to staff their project team with the number of resources required, as well as fulfill critical skill requirements. Where gaps exist in these areas, the program manager will work with functional management, executive management, and the human resources department to fill resource gaps. If critical gaps cannot be filled, changes to the program plan may be required. In early PCT meetings, during the implementation phase, the program manager should ask for project team staffing reports on a weekly basis to gain visibility into how well the resource ramp is progressing. Once the teams are staffed to the level detailed in the program implementation plan, the program manager should only need to review staffing levels if there is a change in personnel or a change in requirements and scope.

Document Design Specifications

Design specifications are discipline specific and are derived from the detailed requirements of the program. They drive the project team execution and provide technical data concerning functional characteristics, interfaces, integration requirements, operational characteristics, as well as the technological and design constraints of the various subsystems and components.[2] The specifications are the guiding documents for each of the specialized project teams.

The program manager is not directly involved in the creation and documentation of the design specifications. His or her responsibility is to monitor the documentation of the design specifications across the project teams by requiring that the project managers report on specification status during the PCT meetings. This is especially crucial in the early part of the implementation phase until the specifications are fully documented. A program manager for DaimlerChrysler Motor Company shared with us that she will occasionally attend specification peer review sessions to send the message to the execution teams that this activity is important to her and to create a sense of urgency within the program team. She also admits that she always comes away with a better understanding of the program she is leading.

Manage Cross-Project Collaboration

Managing cross-functional collaboration is not new to most experienced project managers. But at the program level, program managers not only have to manage cross-functional collaboration but also cross-project collaboration across the various disciplines needed to create the whole product, service, or infrastructure capability under development. This brings us back to the horizontal nature of programs and the need for program managers to be able to manage in the horizontal dimension (see Chapters 1 and 4).

Cross-project interdependencies: Managing the horizontal or cross-project collaboration during the implementation phase of a program first involves managing the highly complex network of project interdependencies in the form of deliverables. The program manager is responsible for ensuring these cross-project interdependencies remain synchronized and coordinated. The program map, which was created and matured during the planning phase of the program, is one of the most helpful tools in managing the implementation of the deliverables. If constructed properly, the program map shows each cross-project deliverable, which project team is responsible for delivering it, when it is to be delivered, and to which project team(s) it is to be delivered. The program manager can use the program map in the PCT meetings as a rolling wave planning tool, focusing the core team on cross-project deliverables due within a four week rolling window, for example. This practice will facilitate the cross-project discussions on status, risks, and issues that need to take place between the project teams.

Risk management: As discussed in the last chapter, team-based risk management is a powerful cross-project collaboration practice. Through team-based risk management, the PCT and extended program team have a greater knowledge of the risks to the program. This helps them think of risk in a broader perspective, in terms of the impact of risk outside of their functional domain and how cross-project risks must be managed collaboratively to be successful.

The program manager is the risk management advocate on the program and is responsible for evaluating, managing, and communicating the overall risk of the program. The PCT is responsible for identifying and tracking the key program-level risks across the functional elements of the program, and the functional project managers are responsible for identification, assessment, and management of all program and project risk events specific to their functions. The project managers are also responsible for bringing new risk events identified within their function to the program level for cross-functional evaluation and inclusion in the program risk portfolio.

Effective risk management must be practiced throughout the PLC as discussed previously. During the implementation phase, risk management becomes tactical in nature. Trigger events and hinge factors for each of the risks identified in earlier phases need to be monitored and response plans acted upon as necessary. In addition, new risk events should be identified and analyzed throughout execution of the program. It is recommended that the program manager schedule time in the core team meetings on a periodic basis to review risk management activities and progress.

By utilizing the program risk management process similar to the one described in Chapter 8, along with tools such as a risk scorecard and P-I matrix (see Chapter 11), the program manager can effectively manage the overall risk of the program throughout the implementation phase.[3] Program risk must trend downward over time, as cumulative risk is an indicator of program health. As the program approaches the end of the implementation phase, program risk will be a deciding factor on whether to transition from the program implementation phase to the program launch phase of the PLC.

Change Management: Change management is a critical process necessary to control the scope of a program. Historically, uncontrolled scope creep has been a primary cause of a program team's failure to meet its intended goals and success criteria.[4] During program implementation, rapid change is common. Shortly after a baseline plan is approved, changes begin to occur. It is important that a program manager establishes a robust change management process (see Chapter 8) and cross-project change control board (CCB) to evaluate change benefit versus cost to the program. Change management normally begins as early as the definition phase of a program and coincides with documentation and ratification of program requirements. Therefore, the primary factor that the program manager needs to deal with when entering the implementation phase is to ensure the right representatives are a part of and are actively participating on the CCB.

The program manager needs to be the advocate for change management on the program and be persistent and diligent in enforcing consistent behavior across the program. He or she is responsible for implementing the CCB and chairing the CCB meetings, which may be part of the PCT team meetings or a separate meeting. This usually is determined by the number of changes introduced to the program. Many program managers have a fast-track process that they implement via e-mail or collaboration software to handle changes needing a quick decision. The fast-track process, however, needs to be used as the exception instead of the norm to remain effective.

A simple change management process, such as the one described in Chapter 8, a change request form, [5] and a change control log are all that is needed to successfully manage change to the program baseline plan. The program manager should focus on these important elements for successfully managing program change: (1) All proposed changes should come into the process and be reviewed by the CCB, (2) changes should be reviewed repeatedly and consistently, (3) the CCB should have the right membership and active participation on the part of the participants, (4) the CCB should be empowered to approve changes within the program success criteria boundaries, and (5) an established escalation process should be in place to disposition those changes that are outside of the CCB's empowerment boundaries.

Managing cross-project interdependencies, managing program risk, and managing change are three significant program management processes that facilitate the cross-project collaboration efforts involved in the implementation phase of a program. The key factor in all of these processes is getting the project teams to consistently communicate and actively work together to create the integrated solution.

Track and Control Program Progress

Program implementation never occurs exactly as planned, creating a need for effective tactical program tracking and control practices on the part of the program manager and PCT members. A program manager can neither eliminate changes in the environment or customer needs nor avoid all errors made during program definition and planning, but he or she can use good tracking and control techniques to minimize the impact of these factors on the program objectives. Tracking progress of work to the program implementation plan consumes a large portion of a program manager's time and mind share during the implementation phase.

One of the most powerful techniques for minimizing negative impact on a program is proactive program management, which is described in the five boxes throughout this chapter, beginning with "The Five Practices of Proactive Program Management."

The Five Practices of Proactive Program Management

This is a true story from one of the authors who taught seminars for PMI International. A few days before the seminar, the author got a call from a CIO of a large company who was very straightforward: "I have eight employees who I am debating whether to send to the seminar and would like to know how you teach proactive management practices." After the CIO and the author agreed that being proactive means taking the initiative by anticipating and acting, rather than reacting to program events, the CIO went on to explain why it is so important in his program management practices.

Specifically, the CIO firmly believes that this practice is among the most important in program management and sees it less as an attitude and more as a teachable practice. The CIO went on to say, "Based on this interview and how well you explain how you teach proactive management practices, I will decide if it is of value for my employees to attend your training."

After swallowing the lump in his throat, the author told the CIO that he teaches "being proactive" as one of the key practices of program management, intermingled with five other practices, which are divided in three groups:

Direct Mechanisms

- Process
- Tools
- Metrics

Physical Infrastructure

- Organization

Cultural Infrastructure

- Culture

The author emphasized that it was not that these practices are developed from scratch and only for the sake of program managers being proactive. Quite to the contrary, these practices already exist for enabling a myriad of other practices and need minor adjustments to serve the proactive behavior. The CIO responded, "Let's see if being proactive per your teachings is good theory, or good practice. If it is good practice, my employees will attend your training, otherwise not."

The explanation of each of the five practices associated with proactive management practices, as the author described them to the CIO, follow later in the chapter.

Whenever possible, the control of the program should be administered within the team. This is a function of the empowerment from senior management to the program manager. Trust and credibility will be further enhanced in the eyes of management when they observe program managers and teams properly monitoring progress and performance and taking the appropriate corrective action when the need arises.

Program tracking and control consists of the following elements:

- Determining what elements of the program to track
- Deciding what metrics and tools to utilize
- Managing program deviations
- Reporting progress and problems
- Requesting assistance

What to track: Deciding on what elements of the program to monitor is the foundation of good management tracking and control practices (see the box titled, "The First Practice of Proactive Management: Program Management Process"). This is an area in which a program manager can get buried in the details of the program, if not careful. The program manager should focus on monitoring program-level elements such as major program milestones, overall program budget, cross-project risks, and changes to the program success criteria. He or she should then delegate the detailed tracking and control to the project managers. Project managers will track the progress of the critical elements associated with their functional specialty, with guidance from the program manager on standard elements that he or she wants tracked on all projects.

The First Practice of Proactive Management: Program Management Process
This is the most important of the three direct mechanisms—the process, tools, and metrics. The following example demonstrates the utility of proactive management through process.

We use the schedule control process to show how proactive management should be practiced, using a model referred to as proactive cycle of program control (PCPC). It is performed through the following five steps:

Step 1: What is the variance between the baseline and the actual program schedule?
Step 2: What are the issues causing the variance?
Step 3: What is the trend or the preliminary predicted completion date, if the current performance is continued?
Step 4: What new risks may pop up in the future, and how could they change the preliminary predicted completion date?
Step 5: What actions should be taken to deliver on the baseline, rather than on the predicted completion dates?

The purpose of the first step is to establish the variance—that is, whether the program or a specific activity is ahead or behind the time plan. Understanding what issues cause the variance is the second step. Given variance, productivity, and current issues, the third step strives to identify where the program schedule may end in the future. But because the future is unknown, it is better to explore what new risks may pop up in the future and further endanger the schedule completion. This is the focus of step four. Finally, step 5 nails down the most important aspect—act, act, act! Or, in other words, *how* to act to crash all variances, disturbances, and derailments and come out on top, reaching the schedule goal.

The major point about proactive management is that the line between the program past and future is between steps 2 and 3. This means that steps 1 and 2 describe the program history, the past, and there is no way to change it. History is less important than the future on which you can act and shape. Steps 3 and 4 (the trend) lead to step 5 (the future). The power is in step 5, the future.

Every review of schedule progress, no matter what tool is used, should be driven by these five steps of schedule-oriented PCPC. Do it every time, whether you review the schedule as an activity, milestone, or entire program. If you do it this way, you will reach the heart of schedule control—being proactive and always looking into the future.

Metrics and tools: Selection of the program elements to be monitored will influence the metrics used to measure progress, and the tools used to collect the measurements. The same process applies to the elements that will be monitored on each of the projects. The program manager, however, needs to drive standardization of metrics and tools across the project teams (see boxes titled, "The Second Practice of Proactive Management: Program Management Tools," and "The Third Practice of Proactive Management: Program Management Metrics").

The program critical success factors are the fundamental metrics of the program and are fully defined at the completion of the planning phase. During program implementation, the program manager should monitor the progress of the program toward achievement of the success factors, as they represent the targets for successful program completion. [5] The program strike zone is an excellent tool to identify the critical success factors of a program and to help the organization track progress toward achievement of the key business results desired (see Chapter 11). In our experience, this is one of the most powerful tools available to program managers for tracking and communicating the progress of a program to both senior management and the full-program team.

The program dashboard is a tool that highlights and briefly describes the status of a program (see Chapter 11) based upon the primary metrics selected. It provides specific status information relative to progress of a program toward achievement of major business goals. The report is color coded to signify the status. Green indicates that the program is on track and progressing well. Yellow indicates a key goal of the program is in jeopardy of being missed and represents a "heads-up" to management. Red indicates that a key goal has been compromised and immediate management action is required to recover the situation. In the case of yellow and red status, the report also indicates the actions that the program team is taking to resolve the issue(s). It is expected that management action and priority will intensify as programs are turned yellow and red.

The Second Practice of Proactive Management: Program Management Tools

This is the second practice of proactive management. But these are not software tools, though they can be easily computerized. Rather they are enabling devices, systematic procedures or techniques, that are used is to reach an objective. In our case, the goal is to help a program manager behave proactively. To demonstrate, we will describe an example tool for proactive management.

Calling any tool proactive is a misnomer. No tool is really proactive—people are. But we nevertheless term them proactive, if they can help a person be proactive. The example tool of our choice is the program dashboard (detailed in Chapter 11). The dashboard is a tool that highlights and briefly describes the status of the program by reporting on progress toward achievement of the major business goals of the program.

The secret is to design the tool to operate per the five steps of the PCPC: (1) Variance, (2) cause, (3) current trend, (4) future trend, and (5) act. So, its internal mechanism is completely ready to deliver proactive behavior of one type, reporting performance. For example, the schedule elements of the dashboard provides: (1) schedule variance (target versus actual performance); (2) cause (written explanation attached to the dashboard); (3) future trend (the dashboard predicts future state of the program, based upon past and current performance); and (4) act (the dashboard states actions required to overcome issues and reverse a negative trend). Hence, the dashboard delivers value by focusing on the future, which is the schedule trend. The most valuable program information is that which enables the program manager to act and shape the future as early as possible.

Managing variances: The purpose of program metrics and tools is to monitor the progress of the program team toward implementation of the program as laid out in the master program plan. If effective, the metrics and tools will give the program manager an early indication that deviations in team performance compared to the program plan have occurred. However, it is not enough to simply detect a variance between actual performance and planned performance. The important aspects of managing the deviation is understanding what it means, what caused the deviation, then determining what to do to correct it. [6] The program manager may need to adjust project team activities or resources to bring performance back into alignment with the plan. He or she may also need to adjust the implementation plan to compensate for estimation errors or uncontrolled changes.

The Third Practice of Proactive Management: Program Management Metrics

The purpose of metrics in proactive management is to warn us of early signs or trouble looming over the program horizon so that the program manager can proactively put the program back on the planned course. Similar to tools, we will first describe a few metrics and then discuss their proactive value.

The first metric is predicted PI (Profitability Index, predicted each month for a program). PI is an independent metric that serves as an early warning signal that the target PI is endangered and may need an urgent action to change it. The second metric, known as estimate at completion (EAC) predicts the final program cost. It is part of multiple metrics of earned value analysis (often used by NASA product programs and DOE software programs).

The value of these metrics is that they combine steps of the PCPC—current trend and future trend. In a company in which being proactive is a must, the predicted PI and EAC may make or break their programs. Like proactive tools, both predicted PI and EAC try to illuminate the future in terms of program profitability and cost trend and predict what program profitability and cost will be at program completion.

Reporting progress and problems: Consistent communication of program progress is a critical element of cross-project collaboration on the PCT. Any significant deviations or changes to the program must be communicated to the project teams by the program manager. In like fashion, any changes that have occurred on any of the projects must be communicated to the program manager and other project teams by the project managers. Late or ineffective communication can quickly result in rework, delays, and added cost during the implementation phase of a program. To ensure open and honest communication between program team members, the program manager needs to create an environment of trust, in which the messenger of bad news will not be punished.

Requesting assistance: The final step in tracking and controlling program progress is recognizing when to ask for assistance and making the request. The request may be for schedule relief, additional budget, and additional long-term or short-term resources. At times a program manager can be hesitant to request help outside of the program team, which is a behavior that has doomed more than one program.

Report Program Progress

The program manager spends a lot of his or her time reporting the status of the program to various stakeholders. This reporting takes the form of both formal and informal communications, from hallway conversations to formal program reviews and decision-checkpoint meetings. Regardless of whether the status report is formal or informal in nature, the message should remain consistent.

So how does the program manger know he or she is reporting a consistent, comprehensive, and accurate message about a program? The key is in consistent, comprehensive, and accurate collection of status from the project teams and the remainder of the PCT members.

PCT reviews: The PCT should review detailed status for each of the projects on a regular basis. For most programs, the primary agenda for each weekly PCT meeting is project status. Reporting project status facilitates the flow of information within the program, as depicted in Figure 5.9 in Chapter 5.

The program manager needs to be specific on what elements of the projects he or she wants reported, what metrics to use, and what level of detail to include to gain consistency in reporting across the project teams. In practice, this will take some work and time in coaching the core team members, especially with a new team.

An effective tool for establishing cross-project reporting consistency and concise communication of project status during the internal program status reviews is the project indicator. An indicator is a brief, one to two page presentation slide set that shows pertinent project status (see Figure 7.4).

By working with the project teams in developing a consistent format for the indicators, and a consistent set of metrics to report, the program manager will receive a comprehensive and concise report of project status that he or she can use to develop a formal or informal program report.

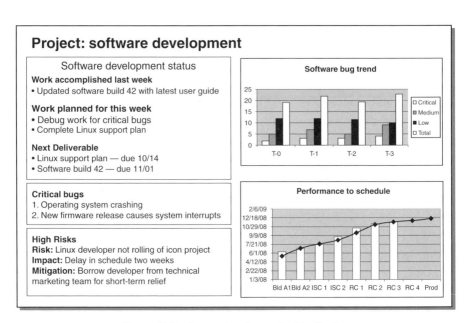

Figure 7.4   Example of a project indicator.

Program reviews: The program review is an organizational meeting in which each program currently funded within the organization will be reviewed by senior management. Each program manager is required to present the status for the program he or she manages. Even though the program review is a status meeting, it can also be utilized as a decision-making forum for factors that affect a program's key objectives.

The chair of the program review meeting is the executive with financial responsibility for the organization in which the program is funded and executed. He or she is typically a vice president or general manager of development and engineering, business unit manager or, in larger organizations, the director of program management. Required participants in the program review meeting include the business unit manager's staff as well as other invited executives. Attendance by the business unit manager and his or her staff is mandatory. The program review is normally held on a monthly basis. Details associated with preparing for and conducting a program review are covered in Chapter 11.

For consistency in reporting format and message, it is most effective to have an established program status indicator template for use by all program managers. Much like the project indicator used in the PCT meetings, the program status indicator is a brief one to three page presentation slide set that shows pertinent program status (see Figure 7.5).

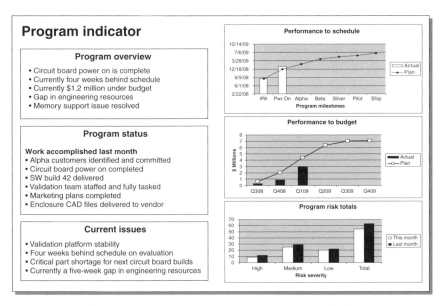

Figure 7.5   Example of a program status indicator

Much of the information contained in the indicator is derived from the most recent project indicators and presented in a summary.

In addition to the program status indicators, the program strike zone and program dashboard tools (see Chapter 11) are normally used in the program review to communicate status toward achievement of the program success factors and the business goals. The objective of the program manager in the review is to communicate both operational and strategic status of the program at the executive level and to use the forum to gain executive help when needed.

Complete the Program Launch Preparation Checklist

As program implementation nears completion, the program manager and core team must begin preparation for program launch. For the launch phase to be executed successfully, preparation for launch must begin during the program implementation phase. This is an example of the overlap between PLC phases.

The program launch plan is part of the overall program plan and is completed during the planning phase. In preparation for the launch approval decision-checkpoint meeting at the end of the implementation phase, the program manager will lead the team through preparation activities, including the completion of the launch-preparation checklist.

The launch preparation checklist is a summary-level listing of the primary launch preparation activities, with a complete or incomplete indication shown for each activity. Figure 7.6 demonstrates this type of checklist, which lists a set of primary launch preparation activities and completion status. Keep in mind that each program is unique and, therefore, each will have its own set of activities, as defined in the program plan. The checklist will be a primary item for evaluation during the launch approval meeting.

Conduct Launch Approval Meeting

The final step in the program implementation phase is to present the status of implementation and launch preparation to the primary program stakeholders for review. Program launch approval is the next major phase/gate decision checkpoint in the PLC. The program manager typically presents the status to the executive decision-making body in the form of a launch proposal in the approval meeting, with assistance from

Figure 7.6   Example of a launch preparation checklist

key members of the PCT. As in every program decision-checkpoint meeting, the business case should be updated and reviewed to ensure that the program is still viable from a business perspective prior to moving to the launch phase.

One of three decisions should be made during the launch approval meeting: (1) reject the launch proposal with direction to repeat the appropriate elements of the program implementation phase, (2) reject the launch proposal and cancel the program, or (3) approve the launch proposal with direction to move into the launch phase of the PLC.

Successful completion of the program implementation phase should culminate in the full commitment of funds and resources to launch the product, service, or infrastructure capability into the market or customer environment and to begin sustaining activities (see the box titled, "The Fourth Practice of Proactive Management: The Program Management Organization").

The Fourth Practice of Proactive Management: The Program Management Organization
The organizational elements that enable proactive management are decision making and accountability. First, decision making means the ability of the organization to speedily design, collect, and analyze data to make timely and proactive decisions. Second, accountability means knowing who makes the

decision, who deploys it, and who is held responsible for the decision. Both organizational features can enhance proactive management of a program.

For example, let's assume that a program deliverable includes multiple work packages owned by different functional project teams on the program and that program status shows schedule-trend problems. If a company has a nimble organization that has no difficulty handling critical situations that are organizationally complex, they would detect them and act proactively, using decision making and accountability. First, decision making would mean a quick decision by the program manager and the PCT. Second, roles in the decision deployment would be clear and swiftly fulfilled by the program manager and relevant PCT members.

PROGRAM LAUNCH PHASE

The objective of the program launch phase is to transition the product, service, or infrastructure capability into its intended state of ongoing use. This may be represented by a launch of a new product into production and delivered to customers, the launch of a new information system into the operational environment, or the launch of a new set of services for use by the intended end users.

A successful launch is recognition by senior management that the necessary start-up activities to turn the product, service, or infrastructure into a sustainable and on-going entity have been achieved. It also signifies the transition of the program from solution design and development to solution sustaining and maintenance of business. Figure 7.7 illustrates the sequence of work associated with a program launch.

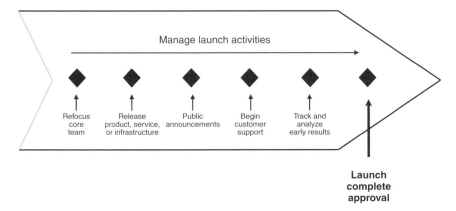

Figure 7.7   The program launch phase

Program Launch Process

The following steps constitute a general approach for a program manager to use in leading the PCT through the launch phase of a program. This process is based upon the models used by multiple companies that we deem highly effective in launching products, services, and infrastructure capabilities. The general process described can be used as a foundation for creating a customized process by tailoring the steps for any organization.

Refocus the PCT

The program launch phase signifies the first time the extended program team begins to decline in numbers and the PCT focus shifts from development to delivery activities. A refocus of the team is needed to redirect the work of the program team toward the specific activities of a product, service, or infrastructure launch. Most of the definition, design, and test teams that have been part of the program begin to exit, and teams such as customer support, product support, sales and advertising, and supply chain begin to increase in number and have a more prominent role in management and execution of the program.

The primary responsibility of the program manager is to facilitate the transition of team structure and makeup as quickly and efficiently as possible. This involves integrating new members into the appropriate discussion, decision, and work forums, opening and facilitating new lines of communication and performing team building activities to sustain the identity and motivation of the team. Part of this is communicating the program vision once again and setting direction on the business and program goals.

Release the Product, Service, or Infrastructure Capability

This step involves formally releasing the output of the program into the operational environment of the business. For a product development program this means beginning to produce the product in saleable quantities. Production processes, supply-line processes, and distribution channels are all turned from development to regular production status. Additionally, the order and fulfillment process begins to take formal orders and fulfills those orders to customers.

For a service or infrastructure development program this means formally moving the capabilities from the test environment to the operational environment and integrating the capabilities with all other systems within the organization. At this stage, end users are able to exercise the capabilities to their full extent.

Release Public Announcements and Begin Marketing Campaigns

This represents the formal declaration to the public and general customer base by the company that a new product, service, or infrastructure capability has been released and is ready for sale or use. It is generally accompanied by a publicity blitz at trade shows, in newspapers and trade periodical articles, and other forms of getting the word out regarding general availability.

Begin Customer and Product Support

As customers begin to buy and use the product, service, or infrastructure capability, the customer support team begins to respond to requests for assistance and information. All procedures developed by the team now become operational, and customer support metrics are collected.

Likewise, technical support of the product, service, or infrastructure capability begins as well. This may include such things as performing maintenance procedures, fault isolation, testing and repairs of failed units, or technical operational assistance at the customer site. All performance metrics for products, services, or infrastructure capabilities are also collected.

Track and Analyze Early Results

The tracking and analysis of the early customer and product or capability support metrics are critical in effective management of the launch phase. With any new product or capability release, defects or other problems may show up once the customer begins using it. Besides ensuring the customer's problem is resolved as quickly as possible, it is also important to understand the cause of the defect or problem and fix it so it doesn't continue to reoccur. This field data is also of great value to the architects that may be defining the next generation of the product, service, or infrastructure capability.

Conduct Launch Completion Approval Meeting

The final step in the program launch phase is to present the launch status to the primary program stakeholders for review in the launch completion approval meeting. Program launch completion approval is the next major phase/gate decision checkpoint in the PLC. The program manager typically presents the program launch status to the executive decision-making body in the form of a launch report, with assistance from key members of the PCT.

One of two decisions should be made during the launch completion approval meeting. Management should either reject the results of the launch report, with direction to continue the launch process, or approve the launch report, with direction to move into the sustain phase of the PLC.

Successful completion of the program launch phase should culminate in the full commitment of funds and resources to turn the product, service, or infrastructure capability into a sustainable and on-going entity within the market or customer environment.

PROGRAM SUSTAIN PHASE

When a program transitions from the launch phase to the sustain phase, it transitions from definition, development, and delivery to support and maintenance of business activities. It is important to realize that this new set of activities is just as crucial for achieving the business objectives than those of program definition, planning, and implementation.

The sustain phase of a program is critical for several reasons. Most importantly, it is the phase in which the business objectives of a program are achieved (or not, as the case may be). The sustain phase is also when improvements in the product, service, or infrastructure capability, or the processes supporting them, can be made to increase the profitability of the program. There is also a lot of direct interaction with the company's customer base during this time, more so than any other phase of the PLC. Therefore, customer support activities during this phase are important and will greatly influence the customer's perception of the company. Finally, the sustain phase is the part of the program where the majority of the key learnings about the business, products, or capabilities and the customer can be collected and fed back into the organization for improvement purposes.

Unfortunately, modern literature and training seem to ignore the critical nature of this phase. In most cases, any information about the sustain phase only encompasses the events surrounding program or project closure and auditing, ignoring the other activities. In the following

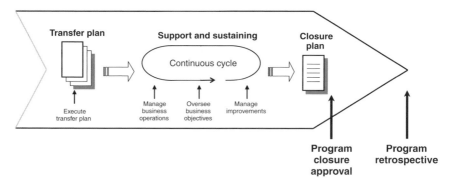

Figure 7.8   The program sustain phase.

section we provide much detail about this missing information and focus on the primary responsibilities of the program manager and the program team, as illustrated in Figure 7.8.

Transfer of Ownership

It is quite common for a program to experience a transition of program managers during the early stages of the sustain phase. This is due to a couple of primary reasons. First, some companies have a set of program managers who specialize in development and a second set who specialize in support and maintenance of business. Second, some companies transfer ownership and management of a program from one organization to another when it transitions into the sustain phase—normally from a research and development organization to an operations organization. However, even though the program manager may change, the program management function does not.

In conjunction with the transfer of ownership of a program from one manager to another, the program team will likely transition too. Program support and maintenance of business requires a different set of specialists and project teams. Most significantly, nearly all the teams associated with developing and launching the product, service, or infrastructure capability will exit the program, and the customer and product support, sales and marketing, supply chain management, and product improvement specialists will take a more prominent role in managing and executing the program.

During this time, the program manager should lead the program team through implementation of the program transfer plan. The transfer plan is part of the overall integrated program plan that was presented in Chapter 6. Implementation of this plan will drive the transfer of functional responsibilities and tasks to the sustaining team.

Manage the Business Operations

A significant portion of a program manager's work during the sustain phase is keeping the business surrounding the program operating efficiently. This involves ensuring the cross-project program team continues to provide excellent customer support; provides product support in the form of maintenance, repairs, and replacements; and manages the supply chain and distribution channels cost-effectively.

Oversee the Business Objective Achievement

The responsibility for managing the business for the program and functioning as the GM's proxy continues through the sustain phase. This, in fact, remains one of the program manager's primary responsibilities as described in Chapter 12.

During this phase, the business objectives from which the program was spawned should be realized. For example, the ROI predicted in the business case should be achieved, the cost savings targets should be met, market segment share gains should be realized, or a new market or market segment should be opened to the company. Tracking and overseeing the strategic goals and tasks of the program is called strategic program control.

Achievement of the business objectives is one of the most rewarding aspects of managing a program during the sustain phase of the PLC. Continued focus on the program vision and business objectives remain the means to ultimate program success.

Manage Improvements

Part of the overall program plan and business strategy may be to make improvements to the product, service, or infrastructure capability during the sustain phase of the program. These may be minor or major in scope depending upon the goals of the improvement exercise. Likewise, improvement in maintenance of business or support processes and procedures may also be undertaken during this phase.

For example, Intel will normally make both product and process improvements during the course of the sustain phase of a product development program. Improvement to the product in the form of the addition of new features or new technologies are performed to increase the life of the product in the market, to increase or maintain the average sales price (and gross margin), or to increase sales volume. When there are decreases in the product bill of material costs through supplier negotiation and economies of scale purchases, the cost of goods sold decreases and gross margin increases. Improvements in the manufacturing processes and procedures also decreases the cost of each product sold.

Like all other aspects of a program, managing improvements during the sustain phase requires effective cross-project collaboration between specialists by the program manager.

Close Out the Program

Program closure begins with program planning. The closure plan should be developed as part of the integrated program plan during the planning phase and executed during the sustain phase. When this occurs, the program manager leads the team through all activities necessary for discontinuing the program.

Exact elements of the closure plan vary from company to company, as well as from program to program. But, in general, closure implementation by the program team encompasses the following activities:

Conduct the closure approval meeting: The closure approval meeting is the final decision-checkpoint meeting of the PLC. The objective of the meeting is to review the closure plan and formally approve the discontinuation of the program and of the sale or use of the product, service, or infrastructure capability. The senior executive team of the business is the decision body.

Manage product, service, or infrastructure phase out: Upon approval to close the program, the activities necessary to formally discontinue the sale and/or support of the product, service, or infrastructure capability begin. This includes notification to the customers, vendors, and all support personnel; plan for final buys and production runs; disposition of inventory; finalization of program documentation; and closure of all accounts payable and receivable.

Phase out may also encompass the timing and functional overlap with a replacement product, service, or infrastructure capability. In this case, the program manager will work closely with the program manager of the replacement program to ensure a smooth transition from the old system to the new.

Conduct the Program Retrospective

A program retrospective should be conducted to assess, document, and discuss the successes and recommended improvements for future programs. Collecting and acting upon program key learnings is a necessary step in becoming a learning organization and, in doing so, strengthening the foundation for the next generation of programs. [7] When collecting data, include inputs from the PCT, extended team, customers, sponsors, and anyone else affected by the program. Successful elements of the program should be documented, as well as any elements that require improvement.

The role of the PCT is to evaluate, prioritize, and select the key learnings from the information collected. The functional project managers are responsible for gathering input from their functional teams. The program manager is responsible for documenting and presenting the key learnings to the organization.

Findings should be communicated to key stakeholders in the organization, including the program team, other program managers, functional managers, and executive management. The information should also be used to build a key learnings report and identify process improvement initiatives, as appropriate. Senior managers should understand and convey that the exercise is directed toward continual improvement of the organization, rather than letting it come across as finger pointing or assigning blame. A manager does not want to limit or minimize the open flow of needed information.

We present the program retrospective at the end of the program, but best-practice companies typically perform a program retrospective at the end of each phase of the program. The benefit of this approach is that the information is fresh, more comprehensive, more focused on a single phase, and learnings are fed back into the organization more quickly.

PROGRAM EXECUTION PROCESS SUMMARY

Table 7.1 summarizes the steps involved in executing a product, service, or infrastructure program.

Table 7.1 Program execution steps.

Program implementation phase	Program launch phase	Program sustain phase
• Collect inputs • Staff the extended program team • Document design specifications • Manage cross-project collaboration • Track and control program progress • Report program progress • Complete launch-preparation checklist • Conduct launch approval meeting	• Refocus the PCT • Release the product, service, or infrastructure capability • Release public announcements • Begin marketing campaign • Begin customer and product support • Track and analyze early results • Conduct the launch completion approval meeting	• Transfer of ownership • Manage business operations • Oversee business objective achievement • Manage improvements • Close out the program • Conduct the program retrospective

IMPORTANT PROGRAM MANAGER BEHAVIORS

There are a set of important behaviors that we have seen demonstrated on successful programs by both program managers and functional project managers during the execution phases of a program.

Persistence: One of the most important behaviors that a program manager needs during program execution is persistence. During the life of a program, the numbers and wide variety of problems and barriers that arise are enormous. It can take a tremendous amount of drive and persistence to achieve program success. The job is hard work and not for the faint at heart.

Innovative: Due to the number and complexity of problems and barriers mentioned above, many times it will take a great deal of innovation and ingenuity to find winning, acceptable, and reasonable solutions. The program manager must create solutions in a cost-efficient and timely manner and still successfully achieve the objectives of the program.

Firm but flexible: We have made the point throughout the book that the program manager is considerably more than a facilitator for a program. Successful program managers are champions for their programs and are strong leaders that need to be capable of quickly rewarding the team when it is doing well but also possessing the toughness to manage through the hard times. This includes being comfortable with pulling team members aside privately and providing constructive feedback when required.

Networking: The program manager needs to be astute at broad-based communication and networking and also be aware of what's happening across the organization as it directly or indirectly affects a program. He or she should be communicating regularly with senior management and the major stakeholders of the program to ensure that the base of support is there and is continuing.

Program culture also plays a role in effective program execution, as described in the box titled, "The Fifth Practice of Proactive Management: Program Management Culture."

The Fifth Practice of Proactive Management: Program Management Culture

A company with a strong performance culture usually features a select few corporate values like "we value customers first." It also identifies and defines lower-level cultural values that we call on-the-job behaviors. These are specific behaviors that are taught to each employee and are expected to be applied while an employee performs his or her work. When proactive behavior is required, it becomes the norm, which is followed and later evaluated through the reward system. When embedded into the reward system, proactive management practices provide adequate incentives to program managers to exploit the behavior.

The Rest of the Story...

Having heard the five practices of proactive management, the CIO agreed to send his people to the author's training seminar to learn about the practices in detail. We summarize below how the five practices enhance proactive management of programs.

Direct mechanisms to adjust the following

- The *process* to include proactive management practices
- *Tools* to be compatible with proactive management practices

- *Metrics* to be compatible with proactive management practices and tools

Physical infrastructure

- Adapt the program *organization* to ensure it provides speedy decision making and clear responsibilities

Cultural infrastructure

- Modify program *culture* by embedding proactive behavior into management systems and on-the-job behavior.

SUMMARY

Program execution is focused on implementing the work established in the program plan and the functional project plans and encompasses the implementation, launch, and sustain phases of the PLC.

The program implementation process involves the steps necessary to successfully accomplish all of the design and development work necessary to complete the elements of the product, service, or infrastructure capability under development; integrate the functional elements to create the whole product, service, or infrastructure; and prepare for the launch of the program output into the market or customer's environment.

The program launch process involves the steps necessary to methodically and successfully transition the product, service, or infrastructure capability from the development environment to the customer environment, ensuring that customers are fully supported and all production and sustaining systems are in place and effectively operating.

The program sustain process involves keeping the program business running as effectively and efficiently as possible, fully supporting the customers and end users, ensuring that the business objectives of the program are attained and that the program remains viable from a business perspective, and efficiently closing out the program when it reaches end of life.

A set of important behaviors that highly effective program managers demonstrate during program execution ensure that the work is accomplished efficiently and effectively.

The Principles of Program Management

▼ Program execution includes the implementation, launch, and sustain tasks
▼ Strategic program control is accomplished by tracking and overseeing strategic program goals and tasks
▼ Tactical program control is accomplished by tracking and controlling tactical program tasks
▼ The proactive cycle of program control controls future program trends

REFERENCES

1. Frame, J.Davidson. *The New Project Management*. San Francisco, CA: Jossey-Bass, 1994.
2. Keogh, Jim, Avraham Shtub, Jonathan F. Bard, Shlomo Globerson. *Project Planning and Implementation*. Needham Heights, MA.: Pearson Custom Publishing, 2000.
3. Smith, Preston G. and Guy M. Merritt. *Proactive Risk Management*. New York, NY: Productivity Press, 2002.
4. Archibald, Russell D. *Managing High Technology Programs and Projects*. Hoboken, NJ: John Wiley & Sons, 2003.
5. Milosevic, Dragan Z. *Project Management Toolbox*. Hoboken, NJ: John Wiley & Sons, 2003.
6. Lewis, James P. *Fundamentals of Project Management*. New York, NY: AMACOM publishers, 1997.
7. Wheelwright, Steven C. and Kim B. Clark. *Revolutionizing Product Development*. New York, NY: Free Press Publishers, 1992.

Chapter 8

Program Processes

The previous three chapters described how to successfully manage a program utilizing the program management discipline and practices. This chapter concludes the discussion by describing the primary program management processes that program managers use on a daily or frequent basis when managing a single program. Effective use of program management processes help to make the operational aspects of a program more efficient, predictable, and repeatable. Most importantly, program-level processes help to ensure that the work being performed by the multiple project teams within a program is being managed consistently. This means consistency in scheduling techniques, risk management techniques, and so on.

The primary program processes discussed in this chapter are not meant to be implemented as stand-alone processes, but rather as a suite of processes. The primary program processes include the following:

- The program life cycle
- Schedule management
- Financial management
- Risk management
- Change management
- Stakeholder management

We first present a detailed description of the program life cycle (PLC) as a guiding framework for program definition, planning, and execution. Then, we present the key processes to support definition, planning, and execution of a program.

The purpose of this chapter is to help practicing and prospective program and project managers, as well as other program players, understand the following:

- The key program management processes needed to manage a program
- The steps involved in each process
- How the processes can be implemented on a program

Effective implementation and execution of the primary program processes guarantee the highest probability of attainment of the intended business results. It should be noted that the formality in implementation of the processes is dependent upon the size and complexity of a program, as well as the experience level of the program manager and team. In general, for larger and more complex programs and less experienced personnel, more formality and rigor in process usage may be required.

THE PROGRAM LIFE CYCLE

The PLC is a series of phases and gates that a program progresses through from concept to end of life. The PLC is meant to be used as a framework for decision making and project team synchronization, which guides the work, deliverables, and decisions on the part of the program manager, program team, functional managers, and senior executives.

There are many forms of the PLC, most of which are specific to a particular type of development program or industry. Some life cycles are phase based and others are nonphase based, such as the spiral life cycle commonly used in software development (see box titled, "Falling into the Spiral").[1] Regardless of the type, it is most effective if a company tailors its life cycle to support its development methods and processes. We utilize a common five-phase PLC to describe how a program is effectively managed. The five phases are define, plan, implement, launch, and sustain (see Figure 8.1).

As stated earlier, the PLC is not really a process that can be segmented into a series of steps. Rather, it should be used as a framework to guide the activities of the program team toward a series of critical decision checkpoints. For many program managers, the PLC serves as the cornerstone of the program upon which the other program processes, methods, measures, and tools are built.

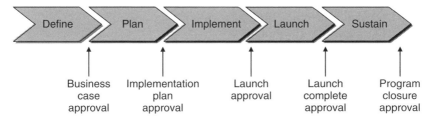

Figure 8.1   The Program Life Cycle.

Falling into the Spiral

Although the phase-based PLC seems to be slow because it is built on consecutive, linear phases, it allows for supreme program speed (overlapping the phases) and dealing with major risks through risk reviews at gate decision-checkpoints (see Chapter 11, the Program Review). However, some use the spiral model of PLC to manage programs in highly uncertain software development and information systems programs (see Figure 8.2). Its basic

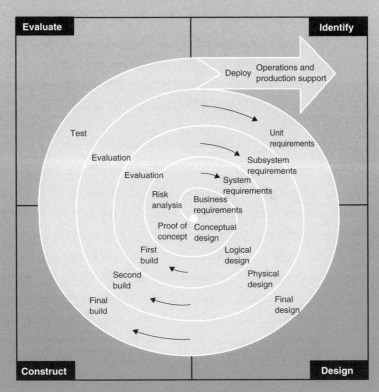

Figure 8.2   Spiral model of PLC.

benefit is the ability to help reduce program risk by responding to the changes in a program faster than other models.[2] It also has phases—identify, design, construct, evaluate—in which the same activities get rooted to incrementally reduce program risk. The spiral model gradually brings a sense of maturity and stability to a program through a series of iterative design, develop, and test cycles throughout the spiral phases.

PLC Elements

The following section describes the primary objectives and work associated with the five phases of the PLC shown in Figure 8.1.

The Define Phase

The work involved in the program define phase is normally twofold. It develops a product, service, or infrastructure concept based upon the culmination of strategic planning results, customer or end-user research, and technology research; and it also develops a business case to verify the feasibility of the concept from a business perspective. To successfully transition from the define phase to the planning phase, the program business case that supports the concept must be reviewed and approved by the senior management team sponsoring the program.

The Planning Phase

The program planning phase involves development of an integrated cross-project program implementation plan. The high-level (summary) and low-level (detailed) requirements are first defined to accurately scope the size and complexity of the program. Once the scope is determined, the program requirements are decomposed into a set of project requirements and detailed specifications, which are used to develop each project plan. The project plans are then integrated into a cohesive program plan with cross-project interdependencies, deliverables, and program milestones clearly identified. Finally, the business case is updated and analyzed based upon the detailed planning results to ensure the program is still viable. To successfully transition from the planning phase, the program implementation plan must be reviewed and approved.

The Implementation Phase

With a detailed plan in place and approved by the primary program stakeholders, program implementation work can begin in earnest. Each project team performs the work detailed in their project plan, while the program manager and PCT coordinates, monitors, and controls the work from a holistic, or systems perspective. The end result of program implementation is a tangible product, service, or infrastructure capability that represents the original concept and is ready for launch into the market or customer environment. To successfully transition from the implementation phase, the program plan must be fully implemented, all requirements proven, and the launch preparation checklist reviewed and approved.

The Launch Phase

Once the program requirements and launch criteria are met, the program is ready for primary stakeholder approval to transition into the launch phase of the PLC. All plans for executing the product, service, or infrastructure launch are executed in the launch phase. Supply chain, factory ramp (if a product), marketing and promotion, and early customer support preparation are the most common activities associated with this phase. When the launch criteria are met and the product, service, or infrastructure capability becomes a stable and sustainable entity, the program is ready for transition out of the launch phase and into the final phase of the PLC.

The Sustain Phase

Once the product, service, or infrastructure capability is launched and ramped, the program moves into the sustain phase. At this point, the program traditionally transitions from development to maintenance of business activities to keep the program business running as effectively and efficiently as possible. Many times, improvements in supply chain, production, or unit cost are made during this phase to increase (or maintain) gross margin or cost savings.

The second key aspect of the sustain phase is customer support. Customer support may include activities such as telephone support and troubleshooting, user training, set up or repair services, and collection of customer usage information to feed into the define phase of future programs. The sustain phase, and the program as a whole, concludes with the formal product, service, or infrastructure end-of-life decision and activities.

Special Considerations

PLC Phase Overlap: The work involved during the PLC phases is not strictly sequential, even though the description in the last section suggests that. Rather, in most cases, the work overlaps between phases to decrease the overall development cycle time through the PLC, which is the rationale behind concurrent engineering. For example, program planning activities usually begin prior to the define phase formally ending and the program business case being approved. Likewise, early program implementation will begin before the full-program plan is developed and approved. The same scenario exists through all phases of the life cycle and is well covered in literature.[3]

The overlap of work becomes *at risk* work, which means if the exit criteria for the preceding phase is not met and the program is cancelled or significantly redefined, the overlapping work may be of no value. Companies that focus on decreasing development cycle time become well versed in balancing the amount of at risk work they commission against the amount of cycle time reduction they are targeting.

Program versus Project Life Cycle: So what is the difference between a program life cycle and a project life cycle? With respect to form and function, the answer is nothing. The difference lies in *how* the program life cycle is applied.

First, on a program, the program life cycle becomes the de facto life cycle for both the program work and the work of each project team. For the work of multiple project teams to stay in synchronization, one life cycle model must be used by the entire program team. Development methods may differ between project teams, but the methods must fit within the overall framework of the program life cycle.

Second, a project life cycle normally focuses on the planning and implementation phases, whereas the program life cycle focuses on end-to-end development—from concept to end of life. This is one of the most fundamental and important differences between the program management model and a project management model. The up front conceptual and feasibility work is *part* of a program, not separate from it as is usually the case on a project. Likewise, back-end sustaining and support activities are also part of a program, not a separate operational responsibility.

SCHEDULE MANAGEMENT

By definition, the program schedule is the heart of the program plan and of the program itself. It becomes the foundation on which resource estimation and allocation is based, detailed budgetary information is developed, execution of work elements is monitored, and timeline success is measured. The program schedule and the PLC are the primary tools to facilitate and manage the horizontal collaboration between the projects.

Program schedule management differs from project schedule management in one key aspect. Detailed schedule management is the focus at the project level, and summary, or integrated schedule management, is the focus at the program level. This requires a modular approach where the schedule is disaggregated and partitioned according to the projects. Schedule details for each module are worked out by the project managers and project teams and then integrated by the program manager and PCT to gain the full perspective of the program. This approach helps to prevent the program manager from being bogged down in scheduling details and is an excellent example of so-called hierarchical scheduling.[4]

The program manager should keep in mind that the most detailed schedule is not necessarily the best schedule. Too much detail can divert attention to one aspect of a program—the schedule—and away from the other critical aspects of managing a program and leading a team. If the program manager focuses on the summary-level schedule and lets the project managers focus on the detailed schedule for their respective projects, a good balance for effective schedule management is achieved.

Process Elements

The major steps involved in developing and managing the program schedule are shown in the process flow diagram in Figure 8.3. This process is generic in nature but is based upon specific processes of companies that we consider exceptional in process definition and development. One can see from the process flow that creating a program schedule is not a simple task, as eight of the ten steps involve creating and negotiating a final

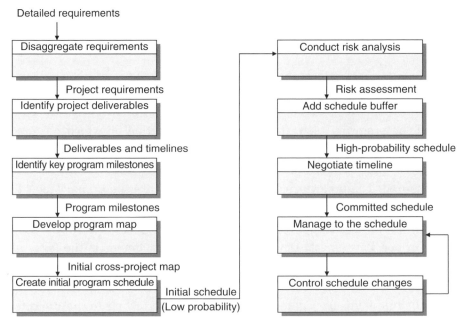

Figure 8.3   Schedule management process flow.

program schedule. The other two steps involve managing and controlling the schedule.

Disaggregate Requirements

The first step in the schedule management process is to review the detailed requirements and partition them by project team. The project specific requirements can then be used to develop the project-specific work breakdown structures WBS. The WBS helps bring to light the deliverables required from each project team and the work required to generate each deliverable.

Identify Project Deliverables

With the project-specific WBS generated from the detailed requirements, each project team can now identify each deliverable for which it is accountable. Once the deliverables are identified, the project teams can then determine the work required to generate the deliverables based upon an assumed set of resources. The final aspect of this step is to generate a timeline to complete each deliverable.

Identify Key Program Milestones

In conjunction with the work from the previous step, the program manager can identify each of the primary program milestones. At a minimum, these include the major decision checkpoints for the program and should also include other significant milestones, such as system power on, release dates, and validation start and end dates. Once these are identified, the program manager should provide estimated dates for the milestones based upon the timeline knowledge contained in the program business case. The program milestones and dates provide excellent guidance to the project teams in the development of the project-specific schedules.

Develop the Program Map

The program manager can now lead the PCT and other key stakeholders in the development of the initial program map (see the process described in Chapter 11). This is the first step in creating an integrated program schedule. The program map shows the critical deliverables for each project team throughout the PLC. More importantly, for the program manager, the program map shows the cross-project interdependencies that exist for the program.

Program mapping is an iterative process that may take several days and multiple iterations to complete. The primary program milestones generated in the previous step become the guiding schedule constraints for the mapping process and can be the source of multiple iterations, if the initial mapping results do not support the timeline targets presented in the program business case.

Create the Initial Schedule

Once the program map is created, it can be used to generate the initial program timeline. The mapping process, in fact, results in an estimated timeline based upon the detailed requirements. We do not advocate a particular scheduling method or tool—individual companies and program teams should determine that.

Whatever the method or tool used, it is critical to realize that the schedule generated in this step will be aggressive and carries a low probability of success. We have worked with many program and project teams that have committed to the schedule created from the program map, and nearly all of the teams failed to hit their committed delivery or launch date.

Why? Because the teams failed to realize that they were dealing with best-case estimates from the program team. Estimation errors are guaranteed, unforeseen issues will arise, sustained-work productivity is less than 75 percent, as well as unaccounted for resource allocation and skills gaps. To increase the probability of schedule success, the next three steps in the process are crucial!

Conduct Risk Analysis

Because programs operate in a less than perfect world, a comprehensive program risk analysis needs to be performed. This exercise, commonly performed in a workshop format, will flush out many of the potential roadblocks to success. Impact assessments should be stated in terms of impact to the program schedule. Because the program schedule is being generated with this process, only risks that have a potential impact on the overall program schedule need be considered.

The end result of the risk analysis is a short list of the risk events (10–15) that can have the most significant impact on the program schedule.

Add Schedule Buffer

Based upon the risk analysis generated in the previous step, the program manager can estimate the buffer amount he or she needs to include to increase the probability of schedule success. There are multiple ways to estimate the buffer amount. Three of the most common and effective approaches are to (1) utilize historical schedule variance data from previous programs, (2) employ critical chain scheduling tools,[5] and (3) use simple statistical analysis techniques (see the box titled, "Calculating Schedule Buffer").

The job of the program manager is to determine the amount of schedule risk he or she is willing to assume, calculate the amount of schedule buffer needed, and then add it to the master program schedule. It is recommended that some of the buffer be added to the end of the schedule to account for any unforeseen problems that will arise and the remainder of the buffer should be added around the trigger events for the risks identified. The program manager now has a high-probability schedule that he or she can put in front of senior management for negotiation and approval.

Calculating Schedule Buffer

One method of estimating schedule buffer that we have found accurate and useful is the simple statistical analysis technique known as three-point estimation. The following formula, for a weighted average-per beta distribution, can be used to calculate expected program duration:[6]

$$\text{Duration} = \frac{o + 4m + p}{6}$$ Where: "o" is the optimistic duration

"m" is the most likely duration
"p" is the pessimistic duration

The initial schedule generated from the program mapping exercise is commonly used to determine the optimistic duration. For the pessimistic duration, the team can use the risk analysis data that has the most significant schedule impact and add it to the optimistic duration. Historical variance averages for similar programs can be added to the optimistic duration to yield the most likely duration.

For example, suppose we have a program that has a duration of 16 months as determined by program mapping—this is the optimistic time. Historical data shows similar programs have been late by an average of 10 percent. Ten percent of 16 months (1.6 months) is added to the optimistic duration to give a most likely duration of 17.6 months. The risk analysis shows a maximum schedule impact of three months for one of the high risk events identified. This is added to the optimistic duration to give a pessimistic duration of 19 months. The times and estimated schedule duration are the following:

Optimistic time = 16 months
Most likely time = 17.6 months
Pessimistic time = 19 months

$$\text{Estimated time (app.)} = \frac{o + 4m + p}{6}$$

$$= \frac{16 + 4(17.6) + 19}{6} = 17.5 \text{ months}$$

The good news is that the calculation above gives a solid statistical estimate for schedule buffer of 1.5 months. The bad news is that it will only raise the probability of schedule success to about 68 percent. Most program managers are not comfortable dealing with a 32 percent chance of failure. To increase the probability of success, a standard deviation can be added to the expected program duration calculation. Where one standard deviation is defined as follows:

$$S = \frac{p - o}{6}$$ For the example above, the standard deviation (app.) is :

$$S = \frac{19 - 16}{6} = 0.5 \text{ months}$$

By adding one standard deviation (half of one month) to the expected time calculation above, the probability of schedule success increases to approximately 80 percent, two standard deviations added (one month) equates to 95 percent probability, and three standard deviations (one and one half months) give the program manager a 99 percent probability of schedule success. We recommend not going above a 95-percent confidence level.

If two standard deviations (one month) are added to the estimated schedule above, the total duration of the program is 18.5 months. This calculation tells the program manager that by adding 2.5 months of schedule buffer, the probability of schedule success is now at 95 percent. By adding a little over 15 percent buffer to the best-case schedule, the probability of success has nearly doubled—this demonstrates why schedule buffers are a program manager's best friend!

Negotiate Timeline

Now the fun begins! Chances are that the high-probability schedule generated in the previous step does not align to the timeline originally requested by senior management. The program manager may now have to negotiate the amount of schedule buffer that can be included in the program schedule. However, the program manager has two significant weapons in his or her negotiation toolbox—a comprehensive risk analysis showing schedule risks and a buffer estimate based upon historical data or statistical analysis. The program manager can now negotiate from a position of power using *real data*. This approach forces the negotiation to be based on the data surrounding the program, not on emotions.

Manage to the Schedule

The program team now has a schedule with which to complete planning activities and to manage the remainder of the program. Managing to the schedule involves tracking actual progress against the schedule and comparing it to planned progress. This should be performed in a two-tiered approach. The project managers should manage the schedule details within their projects, and the program manager should manage the cross-project schedule at the summary level.

Control the Schedule

When a variance between planned versus actual schedule performance is detected, or when issues and changes are encountered, any change to the master program schedule must be controlled. Standard schedule-control techniques such as management of slack time, resource outsourcing, and scope reduction can be employed to bring the program schedule back on track.[7]

Program Implementation

Exactly how the schedule management process is implemented on a program is dependent upon factors such as the size and complexity of a program, as well as the experience level of the program manager and project managers. The process should be tailored for the specifics of an organization and program team. See the box titled, "Implementation of the Schedule Management Process," for a starting point about how to tailor the process.

Implementation of the Schedule Management Process

Schedule estimation activities during the define phase of a program focus on the development of a high-level development timeline for use in the program business case. Because this is a gross estimation, not a detailed estimation, there is schedule risk introduced into the business case—especially for financial estimates such as time to breakeven. A sensitivity analysis based on time is a valuable inclusion to the business case.

Schedule estimation becomes very detailed during the program planning phase. The team must attempt to estimate as accurately as possible the duration of the development effort. To increase the probability of schedule success, buffers should also be included in the detailed schedule.

Schedule management also varies as a program progresses through the PLC phases. Standard scheduling methods and tools are normally useless when applied during the definition and early planning phases of a program. Work in this part of the program is highly iterative, which does not lend itself to standard scheduling techniques. Additionally, the individuals involved in the early phases of a program do not usually take kindly to micromanagement. One of the quickest ways to get uninvited to the definition team meetings is to show up with a detailed schedule.

The most effective way to manage the timeline during the definition phase is commit to firm concept and business-case approval meeting dates and then schedule periodic stakeholder meetings leading up to the approval meetings to rapidly close on the deliverables. The same method is also effective during the planning phase of a program.

During the program execution phases, standard scheduling methods and tools can be applied to manage the schedule. The program manager must be willing and capable of diving into the schedule details when a project team encounters schedule problems but only on a temporary basis. When the situation is resolved, the program manager must once again allow the project manager to manage the details of his or her project.

FINANCIAL MANAGEMENT

Managing the program finances is a key responsibility of the program manager for the business aspects of a program (see Chapter 12). Program management is a business function, so the program manager needs to be well versed and capable in this area. This involves setting and achieving the financial targets, cost estimation and control, financial feasibility analysis, cash flow management, and budgetary management. The financial management process provides a framework for managing the financial aspects of a program.

Process Elements

The major elements involved in managing the financial aspects of a program are shown in the process flow diagram in Figure 8.4. This is a generic process flow that can be tailored as needed to fit the needs of any organization.

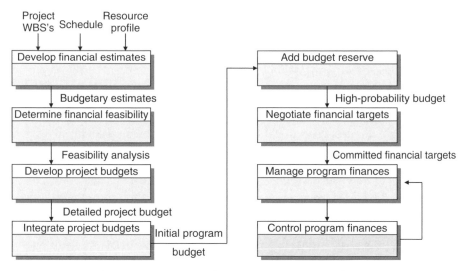

Figure 8.4   Financial management process flow.

Develop Financial Estimates

The first step in the financial management process is to develop budgetary financial estimates in support of the organization's planning activities. This may include program budget estimates for use in the program business case; ROI estimates; financial break-even estimates; and estimated average selling prices and/or product, service, or infrastructure costs. This activity requires a strong partnership with a financial representative for the program who understands the intricacies of the financial department of the company.

Determine Financial Feasibility

As part of the program business case, the program manager is responsible for a robust financial feasibility assessment. The feasibility assessment tests the viability of the program by evaluating the financial estimates developed in the previous step to determine if the program is a viable investment option for the business. As explained in Chapter 10, this analysis should answer the following questions:

- How much will the program cost to implement?
- How much will this program contribute to the company bottom line?
- Is the program worth investing in?

Develop Project Budgets

Once the financial feasibility of a program is determined and approved, the next step is to engage in detailed cost estimation. Like schedule management, the detailed cost estimation activities occur at the project level and are "rolled up" and managed at a summary level by the program manager. Each project manager uses traditional project-cost estimation techniques and information contained in his or her project WBS, detailed schedule, and staffing plan to estimate the resource costs for the project. In addition, any fixed costs such as equipment and subcontractor support should be included in each detailed project budget.

Integrate the Project Budgets

With the detailed project budgets complete, the program manager can integrate them into a program budget that represents the total investment in the program. Any costs outside of the project budgets, like the cost of the program manager's salary, are added to the program budget in this step. The program budget should include all variable costs, fixed costs, and overhead charges that the program will incur. The budget should encompass the cost to develop the product, service, or infrastructure capability and also the cost of the launch and sustain phases of the PLC.

Add Budget Reserve

Using the risk assessment data generated during the scheduling process, the program manager can estimate the amount of buffer he or she needs to include to increase the probability of program financial success. By taking an average financial exposure for the top ten risks for example, the program manager can determine the amount of budget risk he or she needs to add. Another commonly used technique is to use a standard percentage of the overall budget; a good rule of thumb is five to ten percent. Once estimated, the program manager should add the buffer to the budget to develop a high-probability program budget that he or she can put in front of senior management for negotiation and approval.

Negotiate Final Targets

In this step, the program manager negotiates the financial elements of the program with a senior manager or the customer. Financial targets such as the ROI, profit margin, manufacturing costs, and time to breakeven can all become negotiable items as long as they still comply with the business objectives of a program. In addition, the program budget itself, especially the budget buffer, is commonly negotiated. It helps tremendously if the program manager comes to the negotiating table armed with the data supporting the financial elements of the program to focus the negotiation on the data, not the overall targets. At the end of the negotiation, the program manager has a set of program financials that he or she is accountable for achieving.

Manage Program Finances

The program team now has the financial information needed to manage the remainder of the program. Managing program finances involves tracking actual progress against budget and comparing it to planned progress. This should be performed in a two-tiered approach. The project managers should manage the financial details within their projects, and the program manager should manage the finances at the summary level.

Control Program Finances

When a variance between planned versus actual financial performance is detected, or when issues and changes are encountered, the variance must be effectively controlled to ensure the financial viability of the program. The specific control techniques that are employed will depend upon the type of financial variance encountered. For example, if the bill of material cost for a product exceeds the targeted amount, the team can look for lower cost part substitutes or recommend removal of one or more features to eliminate material cost. Managed consumption of the budget buffer is also a viable option in budgetary variances.

The program manager also has to monitor and control the cash flow on the program, especially in smaller or less cash-rich companies (see Chapter 12). Once a program enters the implementation phase of the PLC, it can become a large financial drain on the company. Effective cash flow management is necessary for the program finances to remain solvent.

Program Implementation

Implementation of the financial management process can be tailored for the specific needs of a business and particular characteristics of a program. See the box titled, "Implementation of the Financial Management Process," for a starting point about how to tailor a specific process.

Implementation of the Financial Management Process

Financial management in the define phase of a program is focused on the development of the financial aspects of the business case. Initial financial targets, the estimated program budget, cost of goods sold, average selling prices, and financial risk are all elements that may be included in the business case. In addition, the business case normally includes a detailed financial-sensitivity analysis.

Financial estimation becomes very detailed in the program planning phase. The program team must try to estimate the program budget and all other financial aspects of the program as accurately as possible. Then, negotiate with their senior managers or customers for a win-win financial plan for the program.

During the remaining phases of the PLC, the focus shifts to monitoring and controlling the program financials as it is executed. Good financial metrics and data collection systems are needed to give the program manager periodic visibility into all current financial aspects of the program. To illustrate the importance of financial visibility, Jeff Singleton, a former Boeing program manager, explained that on a program consisting of 100 people (large by some standards, low by others), his program expenditures exceeded $120,000 per day to pay for the work of his program team. That's $600,000 per week!

This points out the importance of monitoring and controlling the program expenditures and changes to the program budget when the program is in execution. It is also a testament for having a dedicated finance representative on the PCT.

To ensure successful implementation of the financial aspects of a program, special consideration should be paid to the financial competency of the program manager and utilizing an effective budget estimation process.

Financial Competency: As a business leader within a company, the program manager must possess strong financial competency (see Chapter 13). In the role of the GM proxy, the business unit manager hands the financial responsibility of a program to the program manager. This means that he or she needs to have the financial acumen to carry out this responsibility. Short of getting an MBA in finance, the program manager should look for education and training opportunities to bolster his or her ability to understand, manage, and speak to the financial aspects of managing a program.

Budgeting Methods: We have witnessed multiple methods used to develop detailed budget estimates, most of which have been successful. The three most common are bottom-up estimation, top-down estimation, and iterative estimation. Any of the three will work, as long as both the program manager and his or her senior management can come to an agreement on the financial targets at the end of the day. From the program manager's perspective, it is our experience that the iterative approach to budget estimation yields the best opportunity for deriving a budget that has the maximum buy-in from both the program and senior management teams.

RISK MANAGEMENT

Developing new products, services, or infrastructures is risky business by nature, especially if a company wants to achieve or maintain competitive leadership. Technology risk alone is a constant, and risk taking can provide a means to gaining competitive advantage. However, risk taking does not mean taking chances. It involves understanding the risk/reward ratio, then managing the risks, or uncertainties, that are involved in each development effort. Failure to do so can lead to substantial loss for the enterprise, including the possibility that the program will fail to achieve the business results intended.[8]

By understanding and bounding the uncertainties on a program, the program manager is able to manage in a proactive manner. Without good risk management practices, the program manager will be forced into crisis management activities as problem after problem presents itself. The program manager is then forced to constantly react to the problem of the day (or hour). Risk management is a preventive process that allows the program manager to identify potential problems *before* they occur and put corrective action in place to avoid or lessen the impact of the risk. Ultimately, this behavior allows the program team to accelerate through the product development pipeline at a much greater rate.

Understanding the level of risk associated with a program is crucial to the program manager for several reasons. First, by knowing the level of risk associated with a program, a program manager will have an understanding of the amount of schedule and budget reserve needed to protect the program from uncertainty. Second, risk management is a focusing mechanism for the program manager. It provides guidance as to where critical program resources are needed—the highest risk events require adequate resources to avoid or mitigate them. Lastly, good risk management practices enable informed risk-based decision making. Having knowledge of the potential downside or risk of a particular decision, as well as the facts driving the need for the decision, improves the decision process by allowing the program manager and team to weigh potential alternatives, or trade offs, to maximize the reward/risk ratio.

Process Elements

The major components of the risk management process are depicted in the generic process flow diagram in Figure 8.5. Similar steps are described in Chapter 11, in which the program risk tool and the P-I Matrix are presented in more detail.

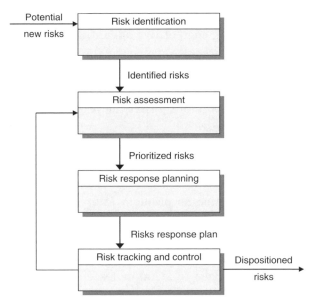

Figure 8.5   Risk management process flow.

Risk Identification

The first step in the risk management process is to identify all of the events that could possibly affect the success of the program. Although risk identification is the first step in the risk management process, it is not a one-time event. Risk identification is an iterative process that occurs throughout the PLC. Some program teams first begin by identifying the categories of risk, such as technology risk, market risk, business risk, and human risk, and then use brainstorming and other problem identifying techniques to identify all potential risk events within each category.

The key element of this step is to attempt to identify *all* potential risks. Do not make judgment at this step on whether a risk is of real concern, rather, that is the next step in the process. When risk identification is done well, it can be overwhelming, especially early in a program in which the number of uncertainties is at the highest. Remember that the goal of risk identification is to flush out as many potential risks as possible to get them on the table for discussion.[9]

Risk Assessment

Not all risk events require action. The risk assessment step is needed to sift through all of the risk events identified in the previous step to identify those that pose the most serious threat to the success of the program. The result is a prioritized "short" list of project and program risks that the team can then manage.

Scenario analysis is one of the most common methods for assessing risk events.[10] Scenario analysis involves analyzing each risk event in terms of the outcome, weighing the severity of the impact of the outcome, evaluating the probability if risk event will happen, understanding when the risk event may occur, and determining whether a risk event is isolated to a single project or will affect multiple projects on the program.

An effective technique to assess the impact of a risk event is to write an "if/then" statement. The if is the identified risk event, and the then is the impact of the event. An example if/then statement follows: *If* the technical support engineers are not assigned to the program by July 1, *then* the program launch date will be delayed two weeks. The *impact* is a delay in the program launch, and the *severity* of the impact is two weeks.

We recommend that a program team begin with a qualitative approach to risk assessment, at least for the first iteration of analysis. By this we mean assessing whether the severity of impact and the probability of occurrence is high, medium, or low for each of the risks identified. This gross analysis will accomplish two important things. First, it will quickly prioritize the risks so the highest risks can be identified for immediate action. Second, it gives the program manager and team an understanding of the overall risk level of the program. The P-I Matrix, which is detailed in Chapter 11, is an excellent tool for this type of risk assessment.

If additional analysis is needed, more quantitative methods of risk assessment can be utilized on the next iteration. For the more sophisticated programs of medium and large size, we recommend the Monte Carlo analysis technique for quantitative analysis. Monte Carlo analysis uses the identified risks to calculate various program compilation datas based on the probability of occurance values for the risks.[11] The important thing to consider is to ensure that the value of the quantitative information gained is greater than the cost of obtaining the information.

The final aspect of risk analysis involves segmenting the risks into program risks and project risks. Program-level risks are those that have cross-project implications and may affect the success criteria of the program. We'll explain this in more detail in the next section.

Risk Response Planning

At this point in the process, the PCT has a relatively short list of program-level risks identified, assessed, and prioritized. The next step in the process is to decide which risk events need action plans to either eliminate the possibility of their occurrence or mitigate the impact they will have on the program, if they indeed do come to fruition. The most common risk response options available to a PCT are risk acceptance, avoidance, mitigation, and transference.

Risk acceptance: This involves accepting the impact of a risk event when it occurs but fully understanding the severity of the impact prior to its occurrence. Risk acceptance is not a common response option in low-risk tolerant companies and industries, but it is relatively common in high-risk tolerant companies and industries.

No risk mitigation or avoidance plans are needed for risk acceptance, and no changes to the program plan are required. However, a contingency plan is needed to identify alternative courses of action if the specific risk event occurs.

Risk avoidance: Risk avoidance involves changing the program plan to avoid a potential cause of an identified risk event, therefore, removing the probability that the particular risk event will occur. For example, a program team may recommend removal of a particular new technology because the risk of the technology is too high and may adversely affect the program's success. Risk avoidance requires a risk-response plan, as well as appropriate changes to the overall program plan and relevant project plans.

Risk mitigation: Two basic strategies for mitigating a risk event are reducing the probability that the risk event will occur and reducing the impact that the risk event will have on the program if it occurs. Risk mitigation is a powerful technique because it targets the root cause of the risk event.[12] It requires a risk-response plan and may also require changes to the program plan. Risk-mitigation plans need to detail specific actions; a particular owner for the response activity; the identification of trigger events such as a date, milestone, or action to trigger deployment of the response plan; and estimated impact to the program if it is deployed.

Risk transference: If a program team recognizes that it lacks the expertise or the ability to respond to a risk event internally, they may choose to transfer the risk response to another party. This, however, neither removes the risk from the program, nor does it remove the responsibility for the risk from the program manager. It simply transfers the response to the risk. Warranties and insurance are a form of risk transference, as is outsourcing of knowledge work.

Risk Tracking and Control

The last step in the risk management process involves monitoring the trigger events associated with the program risks, identifying new risks, and executing risk response plans or contingency plans when risk events occur. The program manager and the PCT need to track and control the program risks with the same diligence that they track and control the program schedule and finances. They also need to keep an open mind that new risk events will come to light throughout the duration of the PLC that will need to be assessed and potentially responded to.

Program Implementation

Risk management should not be viewed as a one-time event. It's not uncommon practice for a program team to present the risks associated with the program at each phase decision-checkpoint meeting, then set the information aside until the next gate review. Practicing risk identification without practicing risk management is a fatal flaw. Risk management must be practiced consistently during all phases of the PLC. Figure 8.6 illustrates the iterative nature of risk management.

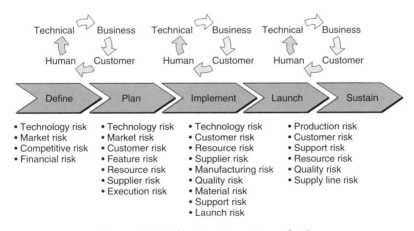

Figure 8.6   The iterative nature of risk.

Implementation of the risk management process can be tailored for the specific needs of a business and particular characteristics of a program. For a starting point from which to tailor a specific process on a program, see the box titled, "Implementation of the Risk Management Process."

Implementation of the Risk Management Process

During program definition, the definition team should be focused on the risks supporting the two key decision points during this phase—concept approval and business case approval. The risks associated with the concept approval are normally technological in nature but can also include financial and competitive risks of significance. In support of the business case approval decision, the program team should do thorough identification and analysis of technology, business, and market risks. Risk analysis at this point should focus on identifying the top ten risk events to be used as part of the business case approval decision criteria.

During program planning, the PCT must perform detailed risk identification and assessment both at the project level and the program level. The relationship between project team risk management and program team risk management is illustrated in Figure 8.7.

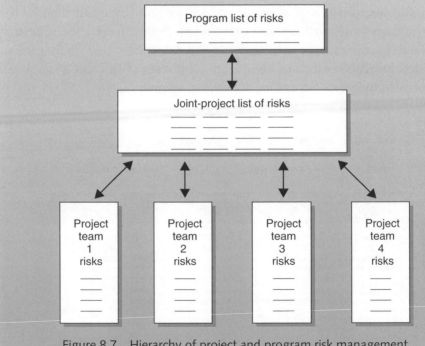

Figure 8.7   Hierarchy of project and program risk management.

Risk identifier	Risk description	Risk ownership				
		Project 1	Project 2	Project 3	Project 4	Program
1	Risk 1 description	✓				x
2	Risk 2 description				✓	
3	Risk 3 description				✓	
4	Risk 4 description			✓		x
5	Risk 5 description				✓	x
6	Risk 6 description		✓			x
7	Risk 7 description		✓			
8	Risk 8 description		✓			
9	Risk 9 description			✓		
10	Risk 10 description				✓	x

Figure 8.8   Sample of joint-project risk list.

Program risk list				
Risk identifier	Risk description	Risk level	Response	Owner
1	Risk 1 description	High	Avoid	Project team 1
4	Risk 4 description	High	Mitigate	Project team 3
5	Risk 5 description	Medium	Mitigate	Project team 4
6	Risk 6 description	Low	Accept	Project team 2
10	Risk 10 description	Medium	Transfer	Project team 4

Figure 8.9   Sample of a program risk list.

Each functional project team creates a discipline-specific list of risks to be reviewed at the program level on a continuing basis. The program manager leads the joint review of all risks to identify those that have cross-project implications and those may affect the success criteria of the program. The result of the team review is a joint-risk list (see Figure 8.8), which identifies each risk as other a project-level risk only or as both a project- *and* program-level risk.

From the information contained in the joint-project risk list, the program team generates and modifies the program risk list (see Figure 8.9).

During the execution phases of the PLC, risk management activities focus on tracking and control of the risk events at both the program and project level. Additionally, new risk events need to be added to the risk lists and evaluated by the project teams and elevated to the program level when appropriate. It's best that the PCT and project teams evaluate risk every two to four weeks to ensure current risks are being worked and to identify new risks. The earlier a risk is identified, the more time the team has to respond to it.

To ensure successful implementation of risk management practices on a program, a safe environment should be created for the team to discuss potential problems, an understanding should be establish that risk management is not an exact science, and the risk-tolerance levels of program stakeholders should be considered.

Don't shoot the messenger: For the risk management process to be effective, the program manager needs to establish an environment in which the team members feel comfortable raising potential problems and concerns. If the program environment is such that bad news and mistakes are punished rather than tolerated, team members will be reluctant to speak freely. This will choke the risk management process.

Risk management is not an exact science: When we say risk management is not an exact science, we're referring to the fact that the program manager is always working with subjective risk information. After all, we're trying to predict what *may* happen in the future, not analyzing what *did* happen in the past. Therefore, two things come into play. First, it's not possible to identify and manage every risk. There will always be problems that have to be contended with (problems are risks that weren't identified or managed). Second, don't get lost in trying to *quantify* every risk. This can be contentious because so many people have created quantification models and algorithms to help with managing risk. These models and algorithms are great tools for people in some industries, but in the high-technology environment, there is such a high number of risk events that a program manager could spend all of his or her time running the probability projections and end up neglecting the other aspects of managing the program.

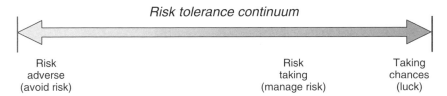

Figure 8.10   The risk tolerance continuum.

Understanding organizational risk tolerance: Risk tolerance is the level of risk an organization is willing to accept to achieve its business objectives. Every individual and every organization has a different level of risk tolerance, with corporate culture and values being a primary driver behind acceptable tolerance levels. Figure 8.10 demonstrates risk tolerance as a continuum, from complete risk avoidance on one end to total disregard of the consequences of risk on the other.

For example, companies in the aerospace and defense industries have a low tolerance for risk due to customer requirements and product use conditions. Products in these industries must have low-failure rates; therefore, companies developing the products adopt a low-risk tolerance approach. As a result, products have redundancies built in and many risk avoidance steps are included in the program plan. By contrast, in the commercial high-technology industry, risk taking is necessary to attain and maintain a competitive advantage through the use of rapidly changing technologies. As a result, products have few built-in redundancies, and program plans must incorporate schedule and budget buffers to allow for risk management actions.

So how does this affect the program manager? First, the program manager must understand the individual risk tolerance of key individuals involved with the program; the sponsor, functional managers, project managers, and the extended program team. Second, understanding tolerance levels, provides excellent guidance for how much risk a program manager can assume, and what response actions (avoidance, acceptance, mitigation, transference) are accepted as preferred options within the organization. Unfortunately, there is no mathematical formula, no statistical calculation, and no model to help manage the various levels of risk tolerance the program manager will encounter—that's why we refer to this as the "art" of risk management. Experience, understanding the corporate culture, and developing personal interaction with the people on and surrounding the program are the best methods for learning to manage organizational risk tolerance.

CHANGE MANAGEMENT

Change management is a critical process necessary to control the scope of a program. Historically, uncontrolled scope creep has been a primary cause of a program team's failure to meet its intended goals and success criteria, in particular, schedule and budget targets. Change can come from all directions—customers, sponsors and other key stakeholders, program team members, and errors introduced in the definition and planning phases. Unfortunately, coping with managing the changes introduced to a program presents a difficult challenge for most program managers. The good news is that establishing a change management process can be relatively easy.

Most humans, even those populating a program team, are not very fond of change, especially if it affects them personally. As a result, they tend to resist it.[13] To combat this natural phenomenon, the program manager can foster a prochange mindset within the program team by viewing change as an opportunity to improve the status quo. This means that change may improve the product, service, or infrastructure capability that the program team is developing, which, in turn, may improve the business results of the enterprise. At the same time, the program manager must then become the change management advocate on the program.

The program manager is responsible for implementing the CCB and chairing the CCB meetings. He or she must be the advocate for change management on the program and be persistent and diligent in enforcing consistent behavior among the players involved.

Process Elements

The major components of a generic change management process are depicted in the process flow diagram in Figure 8.11.

Submit Change Proposal

The first step in the change management process is for the person requesting the change to submit a written change request proposal. This proposal

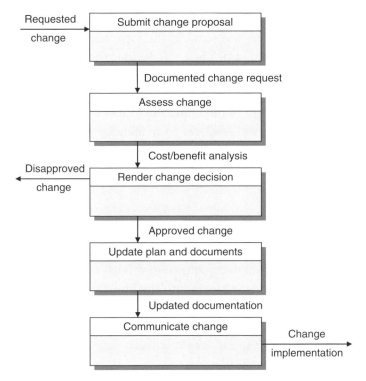

Figure 8.11   Change management process flow.

should include a detailed description of the change requested, a justification for the change, and the benefits to the program for implementation of the change. The justification and benefits are critical information that the program manager should require for any change request. Cursory statements such as "the customer has requested the change" is not adequate justification and should be rejected by the CCB.

Assess Change

The CCB will review the change request proposal, make a decision, and forward the proposal to the program team for cost and implementation options estimation, if approved. The CCB may also deny the requested change. This first decision point is critical. Change request assessment takes resource time away from members of the program team that would ordinarily be spent developing their program deliverables. By filtering poorly justified or nonvalue add changes, the CCB prevents wasted effort on the part of program resources.

Once the change is assessed by the appropriate members of the program team, the change benefits and costs in the form of schedule, budget, and resource impact are evaluated. The CCB decides the fate of the requested change by voting to approve, disapprove, or kick the change back for further evaluation. If the impact of the change causes one or more of the critical success factors to push outside of its boundary condition, the change decision has to be elevated to a member of the organization's senior management team. If approved, the change goes to the next step in the change management process and the affected critical success factors must be updated.

Update Plan and Documentation

For the change to be implemented by the entire program team, it has to be included in the program documentation driving the product, service, or infrastructure design and development. The impact of the change also has to be fully comprehended in an update to the program plan and respective project plans. This includes new deliverables and tasks, changes to the schedule, and changes to the program budget.

Communicate Change

The last step in the change management process is to broadly communicate the change and the impact of the change to all affected and interested program stakeholders. Communication should include the change description, impact to the program, and the benefit of implementing the change. This follows our philosophy of no surprises to management (and customers).

Program Implementation

Implementation of a change management process needs to be tailored for the needs of an organization and particular characteristics of a program. For a starting point for tailoring a specific process, see the box titled, "Implementation of the Change Management Process."

Implementation of the Change Management Process

Change management should begin on a program as soon as the requirements are developed. This can be as early as the definition phase if it makes sense to manage the high-level technology and business requirements driving the business-case development.

Change management is a must when the detailed requirements are documented in the early stages of the program planning phase. If the requirements continue to change without being adequately managed and documented, the results of the program implementation plan will not accurately reflect the program requirements. This situation results in a reconciliation between the requirements and plan later in the program, when it is much more costly and painful. Better to manage the changes to the program requirements *before* the program plan is fully developed, communicated, and committed.

During program execution, changes have to be tightly managed because even the smallest changes can have significant impact due to the potential high cost of rework that may be needed. Late changes also have the risk of introducing performance errors that may result in failures after program launch.

There is a point on every program where change, even the smallest change, cannot be permitted to achieve a successful program launch. Each program manager and CCB needs to determine exactly when this point occurs on a particular program, but, in general, it is prior to final evaluation and testing.

Once the program output is launched and the program enters the sustain phase, changes can again be introduced and managed as part of system upgrades and derivative deliverables. As the program nears its end of life, all change proposals must again come to an end.

To ensure successful implementation of a change management process on a program, special consideration should be given to establishing a fast track process for time-sensitive decisions and defining change management protocols to drive desired behaviors.

Fast-track change requests: All change requests must be reviewed and evaluated in the shortest time possible to prevent unneeded overhead burden on the program. In addition, a fast-track change approval process can be put in place to greatly speed up the change management process in the case of critical change decisions. This is normally instituted through an e-mail system that requires minimal CCB involvement and limited time to review, analyze, and recommend a change decision. The fast-track change approval practice needs to be maintained as the exception, rather than the norm, to remain effective.

Change management protocols: Change management protocols define how changes are to be addressed once the program is underway and should be clearly communicated by the program manager[14]. The purpose of the protocol is to ensure the change management process functions as efficiently as possible. The easy part of establishing an effective change management process on a program is defining the process; the hard part is changing the behavior of the people expected to follow the process. Change management protocols are meant to help with the behavioral challenges that will be encountered. Example protocols include the following:

- All proposed changes must be made in writing and include justification for the change and benefit to the program and business.
- The change management process must not be sidestepped by escalation to a senior manager.
- All CCB members are required to attend all change management meetings or send an empowered designate.
- Any change impacting the critical success factor boundaries must be elevated to a designated senior manager for approval.

STAKEHOLDER MANAGEMENT

A stakeholder is commonly defined as anyone who has a vested interest in the outcome of a program. More importantly, for a program manager, a stakeholder is anyone who can influence, either positively or negatively, the outcome of a program. Stakeholder management is a process with which a program manager can increase his or her acumen in managing the political, communication, and conflict resolution aspects of his or her program to ensure a positive outcome.[15]

The program manager who practices effective stakeholder management generates a greater probability of success for his or her program. Stakeholders come to the program with a variation of expectations, demands, personal goals, agendas, and priorities that many times are in conflict with one another.

Effective stakeholder management practices help the program manager identify and separate the highly influential stakeholders from the rest. They then plan and execute a strategy to rationalize and resolve competing agendas between influential stakeholders by striking a balance between their expectations and the realities of the program.[16] This is accomplished by building personal relationships with the program's primary stakeholders and managing the relationships to gain positive support.

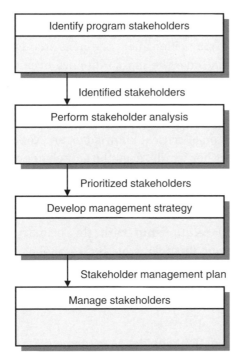

Figure 8.12   Stakeholder management process flow.

Process Elements

The program manager needs to acquire political skills and knowledge to balance the stakeholder's expectations and develop constructive relationships with the most influential stakeholders. The stakeholder management process illustrated in Figure 8.12 can be used to attain the political acumen and knowledge.

Identify Program Stakeholders

The first step in the stakeholder management process is to identify all the program stakeholders. The list of stakeholders identified in this first step should be comprehensive, much like an initial risk list. The stakeholder list should include both internal stakeholders (senior managers, functional managers, finance/accountants, program team members) and external stakeholders (customers, users, suppliers). Tools such as a stakeholder map are effective in helping a program manager identify the stakeholders.

Perform Stakeholder Analysis

Once all stakeholders are identified, the next step is to analyze the list to identify those who have the most influence over the program. These stakeholders are referred to as the primary stakeholders. With them identified, the analysis next turns to determining what kind of influence each stakeholder has on the program, such as decision power, control of resources or possession of critical knowledge, and their level of allegiance for the program. In other words, does the stakeholder prefer the program to succeed, to fail, or is he or she neutral?

The analysis information is critical to the program manager. He or she does not have time to manage all program stakeholders. To effectively build an influential relationship with the primary stakeholders, the program manager must flush out who those stakeholders are, understand what level and type of influence they have, and determine how they feel about the program. A stakeholder analysis tool such as the one shown in Figure 8.13 can be very useful.

Develop Stakeholder Management Strategy

Once the stakeholder analysis is complete, a program manager can develop a stakeholder-management strategy. The strategy should consist

Stakeholder name	Relationship to project	Level of influence	Allegiance to project	Resources					
				People	Money	Material	Facilities	Knowledge	Decisions
Sue Williams	Sponsor	H	⇑		X			X	X
Ajit Verjami	Functional manager	H	⇓	X		X			X
Steven Cross	Software vendor	L	▭	X			X	X	
Lynda Donovan	Core team member	M	⇑					X	

Influence key: L = Little or no influence
M = Some influence
H = Considerable influence

Allegiance key: ⇑ Positive alliance
▭ Neutral alliance
⇓ Negative alliance

Figure 8.13 Example stakeholder analysis tool.

of a communication and action plan for each of the important stakeholders. It should also keep the program advocates engaged, describe how they can be used to influence others, and plan how to win over or neutralize the stakeholders who are not current advocates.

The stakeholder strategy should consider the following aspects:

- What is wanted or needed from each stakeholder?
- What is the message that needs to be delivered to each stakeholder?
- What is the best method and frequency of communication with each stakeholder?
- Does the strategy reflect the interests and concerns of each stakeholder?

Manage Stakeholders

The final step in the process, managing the stakeholders, involves implementing the stakeholder-management strategy and plans developed in the previous step. It involves engaging the important stakeholders in communications, building an honest working relationship with them, and negotiating a series of solutions that are acceptable to the stakeholder and beneficial to the program. For the stakeholders who are difficult to influence, a significant amount of negotiation and bargaining may be required.

Program Implementation

Stakeholder management on a program is a continual and cyclical process. It is continual because stakeholders have to be managed throughout the PLC; it is cyclical because stakeholders can come and go during the course of a program. This requires repetition of the entire process at various points in the program. The number of repetitions is dependent upon the duration of the program and the amount of stakeholder turnover that occurs.

Good stakeholder management practices can be utilized as an effective defense mechanism to neutralize the effects of corporate politics surrounding a program. Program management is not a good career choice for individuals who are adverse to organizational politics. Because of the politics that surround a program, the program manager is forced to participate in the political maneuvering to prevent actions outside of his or her control from negatively impacting the program outcome.

The program manager, therefore, needs to be politically sensible. This means that he or she understands and is sensitive to the concerns of the powerful stakeholders.[16] Utilizing stakeholder-management techniques is an effective approach for identifying the potential sources of political tactics and for developing a strategy for countering effects on a program.

SUMMARY

The PLC is a framework that can be used to synchronize the work, deliverables, and decisions of a program. The schedule management process is used to manage the horizontal collaboration between the project teams on the program. The financial management process is used to manage the financial aspects of a program, such as cost estimation and control, financial feasibility assessments, and cash flow. The risk management process can be used to identify and proactively deal with the various risk factors surrounding the program before they become problems that have to be managed. The change management process is a program manager's greatest weapon to combat program scope creep. Finally, stakeholder management is a process with which a program manager can manage the communication, political, and conflict resolution aspects of a program.

The Principles of Program Management

▼ The PLC is a framework for decision making and synchronization of the program team.

▼ Time scheduling in the program drives resource estimation and allocation, develop budgetary information, monitor execution of work elements, and measure the timeline.

▼ The financial management process is a framework to set and achieve the financial targets of the program.

▼ Risk management is a preventive process that allows the program manager to identify potential problems *before* they occur and put corrective action in place to avoid or lessen the impact of the risk.

▼ The change management process is critical to control the scope of a program.

▼ Stakeholder management increases a program manager's acumen in managing the political, communication, and conflict resolution aspects of the program.

Program Management in Practice

The Budica Program
Diane M. Yates and Dragan Z. Milosevic

Prologue

This program management in practice example focuses on program processes and how they interact with metrics and tools, as well as other key business practices. This is the only open-ended industry example we present, meaning that we don't disclose the outcome but leave the reader to choose what he or she believes is the likely outcome.

Digital Solutions is a global leader in the electronic test and measurement industry and is known for its customer-centric business strategy. For example, most of the functionalities built into the instruments are actually recommended by their customers. The Budica program is intended to deliver a new derivative product as requested by a small but strategically important customer. The program is geographically dispersed with the core team divided between the United States and the United Kingdom, with the customer located in China. The program finds itself in deep trouble, failing to meet the program strategy intended.

The reader will recognize how Digital Solutions uses standard program management practices for defining, planning, and executing the program (see Chapters 6 and 7). The program team applies a traditional phase/gate development process, with key decision checkpoint approvals provided by senior management. Additionally, monthly program reviews are conducted to review program status with senior management.

This example also demonstrates how the Budica program team uses program management processes to manage the details of the program. Processes include the PLC, schedule management, financial management, change management, and risk management. This example also shows how trade-off decisions between project deliverables and program milestones are made and why.

The effective use of program processes are important for successful planning and execution of a program. However, as this example also shows, they are not always instrumental in guaranteeing business success when external market or environmental factors change significantly. The program manager must also look beyond the operational aspects of a program and focus on the "big picture" that surrounds every program.

A Business Opportunity Missed

George Wellington, program manager for the DPWO260 Millennium digital oscilloscope division at Digital Solutions Inc., remembers the day he realized that the program he had been managing became untenable and would not meet its goals. Not only was it late and over budget, but also nearly the entire market had shifted to a competing product.

"When I took the program over for Jeremy Williams (the previous program manager)," Wellington said, "I had my hands full just coordinating the program team. Every time I talked to one of the project managers in the United States they complained about how difficult it was to coordinate the work with the UK team. When I first talked to Ian McClellan, one of my project managers in the UK, he said that the marketing team hadn't provided his team with a firm set of specifications or features to work with. Worse, he said his team was not motivated and was functioning more as individuals who simply had a job to do instead of a team who owned the project and was invested in its outcome.

One of the first changes I made as program manager was to give the UK team the authority to make decisions locally. This increased their motivation and made the difference with regards to project ownership." However, according to Wellington, the problems didn't end there. "We couldn't get our customers to agree on the features and functions required, and there were problems understanding the standards that the Japanese customers were requiring."

Some programs are doomed for failure before they even reach the planning stage of the PLC. The key for program managers and senior managers is to recognize this as early as possible and make corrective action or terminate the program to minimize the losses. This program is an example of a poorly defined and executed program that was allowed to expend precious company resources beyond what was feasible from a business perspective.

Program Background and Business Goals

As mentioned Digital Solutions is a leader within the test and measurements industry and is known for its ability to obtain customer input and feedback and then use that information for its products. The program Wellington was referring to, named Budica—was one of several programs in the digital oscilloscope business unit.

The business unit's primary objective was to create competitive advantage in the digital oscilloscope industry. It's strategies to obtain this objective were to establish a presence in a new market segment, maintain market leadership in the market segments it currently serves, and introduce new technologies within its products. It also emphasized a commitment to customer satisfaction, both throughout the product development life cycle and lifetime support of the final product. The customer vision was incorporated into the unit's business strategy and enabled the company to obtain a competitive advantage in the industry, because most of the functionalities built into the instruments were actually recommended by customers.

Budica was initially seen as a program that would meet its business unit's objectives. The product was classified as derivative, meaning that incremental changes were added to an existing product. Customers were requesting additional features and interfaces that were consistent with market and technology trends. Although it used some technologies that were new to the company—for example, wireless technology—they were not new to the industry.

Wellington said, "Normally we don't compete in the wireless technology space, but the decision was made that we would do so in order to prevent our existing customers from looking to our competitors for solutions to their needs." Bill Walsh, the Vice President of strategic business planning, felt that the successful implementation of this program would create a new market in the United States and prevent competitors from making inroads in the Asian market. The program strategy was primarily defensive in nature; it would provide a compelling product that would prevent current customers from moving to a competitor's product.

"I preached strategy to my project managers," Wellington said. "I made sure they knew the importance of this program and how it aligned with our business goals. I was told time to market was the first priority, and if features had to be removed or postponed in order to meet schedule, then that was what should be done. I left it up to my project managers to coordinate schedules and make sure their respective teams bought into their projects and that key deliverables were completed on time."

Defining the Program: Functional and Business Alignment

Trina Tektondi, marketing manager for the DPWO260 Millennium oscilloscope, said that one of the features clients wanted was a wireless interface to their standard digital oscilloscope. A wireless interface would enable users to port waveforms and records from one oscilloscope to another without having to save the data on intermediate media, even if the other oscilloscope existed in another part of a building or at another location entirely. Additionally, the new product provided waveform monitoring capability that was more economical than the competitors' oscilloscopes.

Tektondi had clients in both the U.S. and Asian markets. One goal of the program was to meet expectations of both markets, but implement only one set of standards. However, from the beginning, there were discrepancies over how to implement the wireless interface and what standards were going to be used. Tektondi, located in the United States had sales people and technical marketing engineers working for her in Asia, who complained that meeting the needs of the Asian market was proving to be challenging.

"The problem, as I see it, started right from the beginning," Tektondi said in a conversation between meetings. "Product definition was a problem because wireless-interface standards were proving difficult to define. Lengthy communication with our customers put us behind schedule from the very beginning. But because of the short time to market target this program had, we were forced to align the product definition with some customers and not with others. Not having the product details nailed down in the beginning really cost us," she lamented. "I'll give you an analogy: We thought our customers wanted a chocolate chip cookie, but in the end, they really wanted a blueberry muffin."

Business Strategy and the Role of the Program Manager

"I feel my role as program manager was compromised from the start," Wellington mused. "I was brought in during the planning phase of this program, so much of the program definition work—summarizing market data, assessing the customer's needs, analyzing the program feasibility, defining the macro plan, performing a financial analysis—had already been completed. Not only that, but the teams had already been assembled by the previous program manager. He hand picked the PCT and had the functional project managers pick the most suitable people to form the project teams." Because of the short-cycle time of the program caused by the time to market goals, Wellington had little time to review all of the work and analysis that had been done previously.

"Still, I feel responsible for the outcome of this program," said Wellington. "It's my job as program manager to make sure that our company's business strategies and the objectives of our program are aligned. I should have caught on sooner that certain things in the product weren't well defined. This very important shortcoming most likely cost us the market and should have been identified in the risk analysis during the defining phase. Incredibly, there were no specific risk areas identified for this program because it was expected to be a relatively short program with established technology."

Digital Solutions has an excellent program management office within the company. "All of our program managers are trained in the latest program management techniques and have the latest processes and tools at their disposal," Sherri Woodward, director of program management said. "Furthermore, whenever one of our program managers is having trouble with a program, support is available to help him or her until the predicament is resolved and the problem taken care of. No one is left to struggle on their own."

According to Woodward, what Digital Solutions expects in a program manager is a skill set that includes good interpersonal skills, a working knowledge of the functional aspects of the program he or she will manage, good business acumen, market and customer knowledge, and strong leadership qualities.

Digital Solutions program managers are not expected to understand every aspect of the individual functional departments in detail, but they should have a broad working knowledge of the technical aspects of each. Having a technical background in a discipline such as engineering is not necessary but is helpful. Since programs are designed to be cross-project and interdependent in nature, managing programs successfully is more than managing multiple projects. Each project deliverable is dependent upon the successful delivery of all project deliverables and is an element of the total solution, or whole product.

"The role of the program manager is to provide a focal point for ownership and accountability for business results," said Woodward. "He or she is responsible for championing product development and is responsible for achieving specific business objectives. This individual works closely with the product line manager to make sure that business objectives for the program align with strategic objectives of the business unit and company."

Wellington had been with the company for a long time and successfully managed several other key programs before being assigned to the Budica program. "We should not have originally assigned an inexperienced program manager to a program like this, due to the unique challenges it presented," stated Woodward. "I think the communication problem between the U.S. and UK project teams, plus the short duration of this program, made it extremely challenging for anyone to manage. Unfortunately, there was not much George could have done once he was brought on board."

Program Implementation

Like many companies today, Digital Solutions uses established company processes and procedures for managing its programs. It is the program manager's responsibility to establish the program vision based upon the objectives established by senior management and create the appropriate links to the firm's strategic plan. The program vision is one of the key tools used by the program manager to provide focus and motivation for the cross-discipline program team.

Program Budica followed a traditional phase/gate development process, with key decision checkpoints at the completion of each phase. In addition to the formal decision checkpoints, monthly program reviews were conducted to review program status with senior management.

Work schedules were developed in the planning phase by the program manager and the core team and were integrated in a bottom-up process. Trade offs between project deliverables and delivery milestones were made during the integration process.

The program manager and his cross-functional project managers estimated the cost throughout the program's phases, including the defining, planning, designing, and ramping-up phases. A final financial performance was analyzed at program closure.

Risk analysis for Digital Solutions programs begins in the defining phase and continues through all phases of the development cycle. Enough cannot be said about the importance of performing an adequate risk analysis. Not following the established risk management process was a key factor in the Budica program's execution challenges. Having a cogent analysis would have caught the cross-project confusion and weak-product definition near the beginning of the program when it could have been dealt with more successfully.

"We were asked to do a risk assessment, but honestly, our team didn't have much to go on because the technology was new to us, and the duration of the program was so short that we basically had to try to do an assessment on the fly," said Wellington. This situation is a strong indication that aggressive risk management is needed on a program. When a program has a short cycle time and new technologies are being introduced, the program team should immediately establish a high-risk level for the program. Then, as more is learned about the program unknowns and mitigation plans are put in place, the risk level of the program can be lowered, if appropriate. "It was a corner that shouldn't have been cut. Like I said, everything was coming together really fast—product definition, risk assessment, cost—you name it," stated Wellington.

The Budica PCT also discovered that their WBS was incomplete and, as a result, product testing ended up being problematic near the end of the development phase. Testing wasn't started early enough in the engineering cycle to allow time to fix problems that were uncovered. Also, the lack of a robust change management process contributed to quality and functionality problems that emerged late in the program. Product definition continued well into the design phase and the lack of requirements change management resulted in the lack of test cases being written for some key functions of the product.

Tools are used in program management to facilitate processes and measure progress through all phases and are a mixture of procedures, software, and techniques. Some of the tools used on the Budica program were the WBS, Gantt and milestone completion charts, and spreadsheets to track costs and risks. A program strike zone was also established to help the program manager and his team monitor progress toward achievement of the program goals and business objectives.

Team culture is also important for effective management of programs. The program team members may be from several geographies and site locations. In the case of the Budica program, the teams were located in the United States and UK and at four different sites. The program manager was responsible for ensuring there was close alignment with the entire core team, no matter how many sites and geographies were involved. This was accomplished through a well-structured vision and set of objectives for the program. Additionally, well-managed and focused meetings, follow up of action items, and follow through by team members for completion of the action items were critical aspects of effective program execution at Digital Solutions.

Facing a Difficult Decision

For programs like Budica that have a short life, it is important to accelerate time to market. "Adopting the program management model helps our company manage complexities such as this," Wellington said. "The model encourages concepts like project ownership and concurrent development. It has worked well for this company in the past, and I'm sure it will continue to work well in the future."

However, one of the drawbacks with concurrent development is that problems are often not discovered until late in the program. This is where misalignment between the work of project teams tends to show up, as was the case with the Budica program. As discussed earlier this is also known as the big bang event. Because concurrent development involves functional project teams working simultaneously, the lack of integrated planning, clear project ownership, poor cross-project communication, and late product definition was problematic.

Digital Solutions was faced with the decision to finish the program or terminate it prematurely. Program metrics presented at the latest program review, using the program strike zone tool, revealed the following status:

- The program was four months behind the planned schedule.
- Current program expenditure was currently $500,000, compared to $470,000 total budgeted.
- Final cost was estimated at $1,000,000.
- A number of customers had moved to a competitor's product and had indicated that the Budica product features were not what they wanted.

In evaluating the program status, it was clear to senior management that the Budica program had gone dramatically off track. The finance manager for the Budica program strongly advocated for its cancellation due to the high-cost overrun and estimated cost to complete. The marketing manager, however, advocated continuation of the program because a strategic customer had the potential to become a multimillion dollar contract in the future and pulling out at that time would alienate them and cost the company future revenue.

The senior management team was faced with a tough decision, as both the finance and marketing managers made strong arguments in support of their respective positions. In the end, the senior executives took the correct approach. They evaluated the business objectives driving the program to make their decision.

To reiterate, the program strategy was to provide a compelling product that would prevent existing customers from moving to a competitor's line of oscilloscopes. This was in support of the business strategy to maintain existing share in Digital Solution's current market segments.

Clearly, the Budica program was failing to meet the program strategy intended—existing customers had already begun to cancel their orders for the oscilloscope and were instead ordering a competing product. Customers were indicating that the product features and functionalities were not in alignment with what they had requested. Additionally, even though the product still had the opportunity to capture a market segment that was new to Digital Solutions, the market was not large enough to warrant additional expenditure in company resources to complete the product development.

Program Completion or Cancellation?

Wellington understood the challenging state that his program was in, as well as the positions of both the Budica finance and marketing managers. Wellington also believed in his program and what the resulting product could offer his customers and did not want to leave the program unfinished. After all, he was the leader of the program team who had worked diligently and tirelessly on the program, despite significant challenges.

He was due to give his senior managers his recommendation for the future of the Budica Program. Should the program be cancelled or completed?

REFERENCES

1. Project Management Institute. *A Guide to Project Management Body of Knowledge.* Drexell Hill, Pa.: Project Management Institute. 2004.
2. Kemerer, Chris. *Software Project Management.* Boston, MA: McGraw-Hill, 1997.
3. Smith, Preston G. and Donald G. Rinertsen. *Developing Products in Half the Time: New Rules, New Tools*, 2nd edition. Hoboken, NJ: John Wiley & Sons, 1998.
4. Milosevic, Dragan Z. *Project Management Toolbox: Tools and Techniques for the Practicing Project Manager.* Hoboken, NJ: John Wiley & Sons, 2003.
5. Goldratt, Eliyahu M. *Critical Chain.* North River, MA: Great Barrington, 1997.
6. Keogh, Jim, Avraham Shtub, Jonathan F. Bard, Shlomo Globerson. *Project Planning and Implementation.* Needham Heights, Ma.: Pearson Custom Publishing, 2000: pp. 41.
7. Weiss, Joseph W. and Robert K. Wysocki,. *5-phase Project Management.* Cambridge, MA: Perseus Books, 1992.
8. Martinelli, R. and Jim Waddell. "Managing Program Risk". *Project Management World Today* (September-October 2004).
9. Smith, Preston G. and Guy M. Merritt. *Proactive Risk Management.* New York, NY: Productivity Press, 2002: pp. 32.

10. Frame, J. Davidson. *The New Project Management*. San Francisco, CA: Jossey-Bass Publishing, 1994: pp. 186.

11. Milosevic, Dragan Z. *Project Management Toolbox: Tools and Techniques for the Practicing Project Manager*. Hoboken, NJ: John Wiley & Sons, 2003: pp. 300.

12. Smith, Preston G. and Guy M. Merritt. *Proactive Risk Management*. New York, NY: Productivity Press, 2002: pp. 109.

13. Frame, J. Davidson. *The New Project Management*. San Francisco, CA: Jossey-Bass Publishing, 1994: pp. 49.

14. Keogh, Jim, Avraham Shtub, Jonathan F. Bard, Shlomo Globerson. *Project Planning and Implementation*. Needham Heights, Ma.: Pearson Custom Publishing, 2000: pp. 116.

15. Pinto, Jeffrey K. *Power and Politics in Project Management*. Newtown Square, PA: Project Management Institute Publishing, 1998: pp. 27.

16. Pinto, Jeffrey K. *Power and Politics in Project Management*. Newtown Square, PA: Project Management Institute Publishing, 1998: pp. 146.

Part III

Program Management Metrics and Tools

Part II explained how program managers navigate programs through the phases of the PLC with the use of proven practices and processes. Even though these practices and processes are revealing, they don't constitute the whole picture for one reason: They don't include program management tools and performance measures that help to implement the practices and processes. Part III explains how performance measures and tools enable the effective execution of program management practices and processes.

In Chapter 9, we will shed light on performance measures that we term metrics for program management. Their intent is to focus attention on the business results of the organization and serve as the critical success factors in programs and projects. Given both their strategic and tactical nature, we emphasize the need to *carefully* engineer and install metrics. This means that they are compatible (or mutually aligned), balanced, consistently implemented, and tiered.

To help visualize a system of metrics and how they are tiered, we offer the Program Management Value Pyramid, which includes the following four levels:

- Strategic management metrics
- Program portfolio management metrics
- Program management metrics
- Project management and team execution metrics

The pyramid helps integrate the four levels and guide business improvement efforts. We stress the point that each company should develop a menu of metrics that respond to its own specific situation. Metrics also need to support the company's business strategy.

Chapter 10 specifies five management tools that complement one another in implementing the strategic planning process. These tools enable program managers to fathom the nature and position of their programs in relation to all current and future programs and to the companies, business strategies. Each of the five tools has a clear-cut aim. The program business case defines the program, tests the feasibility of the program against the business objectives and operating environment, and sets the program success criteria. On that basis, a program is checked for alignment with the business strategy by means of the program alignment matrix and, if aligned, is added to the portfolio of programs. There, the program portfolio map is put to work in evaluating, prioritizing, and balancing the portfolio of programs from multiple strategic criteria identified by the executive team of the organization. The program road map indicates the timeline and dependencies of programs in the portfolio as they exist in the company's development pipeline. The program complexity assessment helps determine the skill level that is required of the program manager and PCT to run a program.

For the tactical process of program planning and execution, tools are also needed—we call them operational program-management tools. In Chapter 11, we present 5 of these complementary tools. The aim of these tools is to allow the program manager to effectively plan, monitor, report, and control progress of individual programs. He or she uses the program strike zone to identify the program's critical success factors and develops the program map to find the critical cross-project dependencies that need to be coordinated. The program manager also uses the P-I matrix to find and focus on the program areas of highest risk. When the program execution unfolds, the program review is applied to check phase gates and other major program events. Throughout execution, the program dashboard helps report program status. It is recommended that program managers use the strategic and tactical tools as a tool set, although each may be employed individually.

An example that covers both metrics and tools is provided at the end of this section. It demonstrates how a high-technology organization aligns metrics and tools use with its business strategy.

Chapter 9

Program Management Metrics

Humans know that measures such as the level of their blood pressure, cholesterol, blood sugar, and white blood cells reflect the state of their health. Similarly, program managers know that measures such as time to money, development cost, gross profit margin, and profitability index reflect their programs' health. This chapter deals with performance measures that senior managers, program managers, and project managers need to understand to assess the health of their programs.

It is commonly understood today that one of the key rationales for using metrics is that "what gets measured gets improved." In particular, using performance metrics will help program managers and their sponsors understand how well a program is performing, where and why a program has problems, and tailor actions to eliminate the problems. This will, in turn, improve the program and bring it closer to its goals. Therefore, devising and employing appropriate metrics should aim to improve business results of the organization.

Program metrics not only measure the health of individual programs but also show the effectiveness of program management-related processes, such as strategic management and portfolio management. In this manner, program management metrics are an effective means to integrate and synchronize strategic, portfolio, planning, and execution activities.

We first explain detailed reasons why a program manager needs to utilize metrics with a description of the design requirements for metrics, or what kinds of metrics are needed. We then present the program management value pyramid (PMVP) to develop a menu of metrics, which is conceptually grounded in the integrated management system from Chapter 3. After explaining how metrics can help improve business results, we delve into what may be the most important aspect of metrics—customizing metrics to support a company's business strategy. Hence, the purpose of this chapter is to help practicing and prospective senior managers, program managers, and other program stakeholders, accomplish the following objectives:

- Understand why program management metrics are needed
- Utilize the PMVP to determine the types of metrics to use
- Choose value-added program management metrics from a menu of options
- Understand how to customize metrics to support a company's business strategy

RATIONAL FOR USING METRICS

Business understanding and experience leads to the following question: *What is the business rationale for using metrics?*

There are at least two logical and often cited reasons for using metrics, with one less obvious reason explained later. First, as stated earlier, what gets measured gets improved. Second, metrics measure performance toward the achievement of the business results intended, as well as other program critical success factors.

Metrics measure achievement of the critical success factors of the program. Programs using comprehensive metrics to measure and monitor performance will have fewer problems, hence, higher success in accomplishment of program goals. For example, recent studies cite metrics as a key to programs and project success.[1] Additionally, a recent study by one of the authors involving 229 programs and projects found that companies using consistent, tiered, balanced, and mutually aligned metrics outperformed companies who used sporadic, nontiered, schedule-oriented, and nonaligned metrics. The use of program metrics is an institutionalized practice of leading companies. They focus on regular, periodic measurements of program and business performance. Other companies still favor schedule-only metrics that are sporadically utilized, normally when a program is in trouble. Additionally, leading companies do not apply the same metrics throughout the PLC.[2] Programs have different priorities and emphases in different phases of the life cycle of the program.

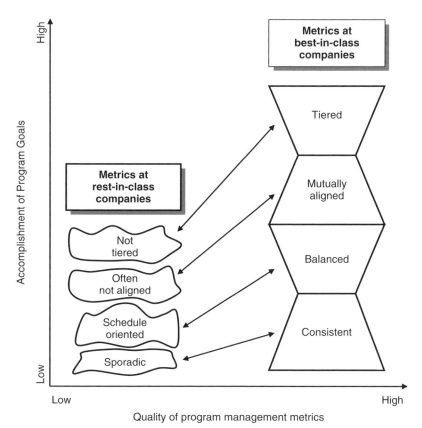

Figure 9.1   Best versus rest-in-class in the use of program management metrics.

Figure 9.1 demonstrates the key differences between companies that methodically design their set of program metrics (which we refer to as best-in-class companies). There are also those that take an ad hoc approach (which we refer to as rest-in-class companies). Best-in-class companies experience both a higher rate of program goal accomplishment and produce a higher-quality metric set—quality that is quantified by usefulness.

A detailed comparison between the best-in-class and rest-in-class companies revealed that the best companies engineered and installed a system of metrics that were balanced and mutually aligned. Balanced metrics are those that cover all dimensions of a program, including schedule-oriented metrics, as well as those for financial, customer, process, and human resource utilization. Balanced metrics also include both leading and lagging metrics. Leading metrics are forward looking, such as a projected finish date based upon the *rate of milestone completion*. Lagging metrics, such as percentage of deliverables completed, are most valuable for a retrospective view of program performance. Both leading and lagging metrics are important and useful.

Mutually aligned means that metrics are compatible, using the same baseline information. For example, *performance-to-planned schedule, probability of completing the program by a certain date*, and the *cumulative percentage of milestones that are accomplished* are all based on the same baseline—the program schedule.

Literature claims that this behavior, using balanced and mutually aligned metrics, is not by chance but by design and is aimed toward enhancing success in terms of accomplishing the program goals.[3] Consequently, best-in-class companies translate program success criteria into specific metrics the program team can act on and also create incentives for accomplishment of the metrics. For example, the executive team at Lucent Technologies studies the best time-to-market performance measures in the industry, which are turned into targets for their own teams to beat. This is a culture of continual improvement by making program targets highly aggressive.[4]

To reflect how metrics vary, successful companies create metric tiers. Tiering allows the companies to categorize a diverse continuum of programs and projects, while concentrating on crucial features. One company uses the following three tiers for its research and development programs: basic research programs (tier 1), technology development programs (tier 2), and specific programs (tier 3). *Quality of the research* and *percent of goal fulfillment* are examples of tier-1 metrics. *Number of deployable technologies* and *breakeven after release* exemplify tier-2 metrics. When it comes to tier 3, good examples are the *number of customers who found defects* and *time to market*.

In summary, best-in-class organizations carefully build metrics systems that align program execution with the organization's business strategy. In that effort, the emphasis is on using metrics to measure performance on a consistent basis from inception to completion of the program. They insist on measuring multiple facets, balancing metrics to obtain a holistic picture of program health, and selecting metrics that are aligned and compatible. Finally, by dividing metrics into tiers, leading companies define metrics for use by all levels of management.

MEASURING PROGRAM MANAGEMENT EFFECTIVENESS

This brings us to the third, and less obvious, reason for using program management metrics. One of the most fundamental duties of senior management is to account for, and use as effectively as possible, the corporate assets which he or she oversees. To date, senior managers have not designed a methodology that enables satisfactory evaluation of their program management assets, except in a passive fiduciary way. Hence, top corporate managers have limited capability to evaluate one of the strongest competitive weapons a corporation may build—its program management discipline.

The real value of the program management discipline is only obvious when we take a close look at the role program management plays in the development of product, service, or infrastructure capabilities. However, one cannot judge the intrinsic value of program management by just looking at the value of recently developed new products, services, or infrastructure capabilities. Measuring program management performance and effectiveness requires a shared understanding between all stakeholders and their participation in the evaluation process. Participative roles will vary between stakeholders, but, in general, only a high level of interaction between these members can lead to the establishment of credible metrics of value-producing performance.

We have constructed an approach that enables stakeholders to get involved in a relevant and direct fashion to ensure that program management performance and effectiveness are adequately measured—the PMVP. All stakeholders, including CEOs, senior management, program management directors, functional managers, program managers, and project managers will all benefit from use of the PMVP.

PROGRAM MANAGEMENT VALUE PYRAMID

In Chapter 3, we explained the integrated management system and its subsystems that describe the program management capability of a company—strategic management, portfolio management, program management, project management, and team execution. The subsystems are used to build the PMVP, as shown Figure 9.2. Metrics related to each subsystem allow the PMVP to serve as a tool to analyze the performance of

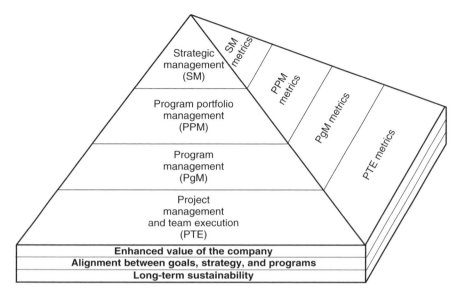

Figure 9.2   The program management value pyramid (PMVP).

the program management function and to be used as a guide for continual improvement efforts.

The PMVP's foundation is built upon three premises that should be in place to create and sustain program management as a competitive advantage for an enterprise:

- The program management discipline should enhance the value of the firm.
- There must be alignment between the firm's business goals, strategies, and programs.
- The program management discipline is sustainable over the long term to continuously generate useful products, services, or infrastructure capabilities.

These premises indicate the following: First, the ability of program management to create value for the corporation should be measurable; second, how well programs and program management are aligned with the business goals and strategies should be measurable; and third, how well program management executes its function should be measurable. If program management performance can be effectively measured in these areas, one has the ability to evaluate the overall value of program management within the firm.

The bottom level of the pyramid, project management and team execution, contains the tactical elements of the PMVP. As explained in Chapter 3, project management and team execution form the basis for planning, implementing, and delivering the interdependent elements of the product, service, or infrastructure capability. This is where the hands-on work gets completed. The functional project managers are responsible for the detailed planning and execution of the project deliverables pertaining to their respective operating functions (for example, hardware engineering, software engineering, and marketing). Metrics that assess the adequacy of the tactical elements would be those that permit measurement of parameters, which are important to practices that support the program management capability.

The second level in the pyramid relates to program management. As described in Chapter 3, it is not by accident that program management is in the center of the integrated management system and PMVP. Program Management is the "glue" that translates business strategic objectives into tangible products, services, or infrastructures that are developed by integrating the work efforts and deliverables of multiple project teams. Effective program management metrics focus on the program's ability to transcend across strategic planning and execution.

The third level is program portfolio management. Portfolio management is an effective process for choosing the most valuable programs, balancing and prioritizing them, and making sure they are aligned with the strategic goals of a company. Various metrics are available for evaluating the effectiveness of the portfolio management process within an organization.

The highest level of the PMVP is strategic management. This level deals with the question of whether program management is a viable part of a company's business goals and strategy. Some appropriate metrics are covered in the PMVP that can be used to answer this question.

THE PMVP MENU OF METRICS

Table 9.1 presents a menu of the most commonly used metrics for each level of the PMVP, as they pertain to program management. The metrics are categorized by what is being measured (financial return, strategic alignment, and so on) and by the PMVP level (strategic management, portfolio management, and program management, and project management). A single metric may relate to more than one PMVP level, which is indicated in the table.

Table 9.1   Program Management Menu of Metrics

Metric	Definition	Usage
1. Projected value of the program road map PMVP Level: Strategic Management and Program Management		
Dollar sales from program road map	Fraction of future sales by year projected from programs on the program road map over multiple years; the probability of accomplishing each program goal is also provided.	Many companies tend to project future sales and income from programs on the road map. The cumulative value of these programs is equal to the value of the program road map in total. When this is taken seriously, assumptions about probability of projections and related sales/income are taken seriously too. That means biases for individual programs that affect the assessment are not allowed.
Projected future income from program road map	Fraction of future net income by year projected from programs on the program roadmap over multiple years; the probability times the net income for accomplishing each program goal should also be provided.	The forecasts (leading metrics) of the measures are made quarterly and are compared to the original forecasts.
2. Strategic alignment of programs PMVP Level: Strategic Management and Program Portfolio Management		
Alignment of programs to business-unit strategic goals	Percentage of total program portfolio that is compatible with documented business-unit strategic goals.	It appears difficult to find a program that does not support specific business-unit goals. But if so, an explanation should be provided. This is a lagging metric.

Table 9.1   Program Management Menu of Metrics (*Continued*)

Metric	*Definition*	*Usage*

3. Program portfolio distribution
PMVP Level: Program Portfolio Management

Program portfolio distribution	A way to express fractions of the total program portfolio among various dimensions that are important to program stakeholders.	The intent of these metrics is to ensure that program investments are well balanced among various dimensions. These lagging metrics should show balance across dimensions. If the programs are not in balance, the metrics help determine how to modify the program portfolio.
	For senior management, the following list shows how programs within the portfolio are distributed • Reward v. risk • Product lines of the business unit • Creation of new business, expansion of current business, maintenance of current business • Derivative, platform, and breakthrough products • Current markets, markets new to the business unit, and markets new to the world	
	For directors of program management, the following list shows how programs are distributed:	

Table 9.1   Program Management Menu of Metrics (*Continued*)

Metric	Definition	Usage
	• External v. internal development • Core v. new technologies • Resource distribution per technical discipline • New v. existing core competencies	

4. Portfolio risk
PMVP Level: Program Portfolio Management

Metric	Definition	Usage
Portfolio risk index	The total risk associated with all programs in a company's portfolio.	This is an emerging leading metric—not standardized, not universally recognized as a metric, and probably needs some time to reach an accepted form and formula but valuable
	Two often used forms are as follows: • Additive form, in which risks of individual programs are summed up • Multiplicative form, in which risks of individual programs are multiplied	The additive form assumes that individual program risks are independent of one another, which is not true. The multiplicative form assumes that individual program risks depend on one another. That means, for example, that a risk occurring in one program may trigger a risk in another program.

5. Customer satisfaction surveys
PMVP Level: Program Portfolio Management and Program Management

Metric	Definition	Usage
External	Average value of ratings given by key external customers, on a Likert scale of 1 to 5, with 5 being the highest value. It measures various dimensions such as	External customer surveys about program performance, service, quality, and competence of personnel provides a lot of information about customer satisfaction.

Table 9.1   Program Management Menu of Metrics (*Continued*)

Metric	*Definition*	*Usage*
	timeliness of the program completion and customer value of the program output.	
Internal	The same average value of ratings for internal customers such as engineering, marketing, etc. An example is timeliness of deliverables.	Similarly, internal customer surveys are important to understand satisfaction of internal customers in the program. To paraphrase an executive, external customer satisfaction begins with internal customer satisfaction.

6. Profitability index
PMVP Level: Program Portfolio Management and Program Management

Portfolio profitability index	The portfolio profitability index is the net-present value of all expected cash flows for the future in all portfolio programs, it is divided by initial cash investment into the program portfolio.	The profitability index is also known as the benefit-cost ratio. Typically, this metric is used as a program selection tool. Once a program is selected, companies periodically reestimate the index to see if the program is still promising. If it is not, the program may be terminated.
Program profitability index	The program profitability index is the net-present value of all expected cash flows for a future program; it is divided by initial cash investment of the program.	In a program selection process, if an index is greater than one for a program, it means the program may make more profits than it costs—in which case the program becomes a candidate for acceptance. Some companies estimate the index to visualize the profitability of the entire portfolio. If the index value

Table 9.1   Program Management Menu of Metrics (*Continued*)

Metric	Definition	Usage
		is greater then one, the portfolio as a whole will make money. This metric may help to set the direction for modifying the portfolio and adding more profitable programs. The forecasts of the metrics are made monthly or quarterly (these are of the leading nature) and compared to the original forecasts. The actual values of the metrics are lagging.

7. Milestone completion
PMVP Level: Program Portfolio Management, Program Management, and Project Management

Metric	Definition	Usage
Percent of the program milestones accomplished	The percent of all program milestones in the portfolio of programs achieved within a week of projected achievement date (or within appropriate time for an industry).	A good leading metric to reflect the in-process timelines of the program portfolio, individual programs, and projects within a program. The metric acts as an early warning signal for a company's time management system. For example, having this metric at the level of 40 percent, may mean that a lot of final program completion dates will be missed and other cycle-time dates will be delayed. It takes some time for a company to learn what levels of the metric can be tolerated. For example, one company considers any value of the metric below 80 percent unacceptable.

Table 9.1   Program Management Menu of Metrics (*Continued*)

Metric	*Definition*	*Usage*
Percent of program milestones accomplished	The percent of all milestones in a program achieved within a specified time period of the projected achievement date.	
Percent of project milestones accomplished	The percent of all milestones in a project achieved within a specified time period of the projected achievement date.	

8. Development system adoption
PMVP Level: Program Management and Project Management

Program development system adoption	The percentage of programs and the projects making up a program in the total portfolio; this follows the established program management system that has standard decision checkpoints.	This lagging metric determines the percentage of the company utilizing a common development system. The metric value will be dependent upon the expected adoption strategy of the company. If, for example, the company includes technology development programs, the metric value will be small because many technology development programs are exploratory and commonly do not follow a standard development system.

9. Product, service, or infrastructure quality
PMVP Level: Strategic Management, Program Management and Project Management

Reliability / defect rate assessment	At the company level, it is the percentage of the company's product,	Each company will have its own style of measurement. These lagging metrics serve

Table 9.1   Program Management Menu of Metrics (*Continued*)

Metric	Definition	Usage
	service, or infrastructure output that meets or exceeds the accepted quality standards. At the program level, percentage of a program's output that meets or exceeds the accepted quality standards. At the project level, percentage of a project's output that meets or exceeds the accepted quality standards.	the purpose of knowing the level of quality provided to customers.
Customer fall out	At the company level, a fraction of the company's product, service, or infrastructure output that has defects captured by customers. At the program and project level, a fraction of a program's output that has defects captured by customers.	

10. Program cycle time
PMVP Level: Program Management

Program time to market	The time between the point when the customer's need was identified until the point when commercial sales or use of the program output begins.	These are all classic lagging metrics. The first one is used to measure how fast the program reaches the phase of commercialization or implementation.
Program development time	The time between the point when the program was established until the point when the launch or go-live decision is made.	This metric was established to measure achievement of the speed of program results promised in the program business case.
Program time to results	The elapsed time from the point when a program was	Typically used by the real fans of fast time-to-market

Table 9.1   Program Management Menu of Metrics (*Continued*)

Metric	Definition	Usage
	established until the point when results (such as market share or cost-reduction goals) from the program business plan are achieved.	philosophy. Attempting to achieve fast time to profit brings a risk of focusing on programs that can have fast time to profit, such as derivative programs, and ignoring programs that are more risky and typically not as fast, such as breakthrough programs.
Program time to profit	The elapsed time from the point when a program was established until the point when the profit goals of the program are achieved.	Another speed metric focusing on how fast the program investment was recouped and first profits were made.
Program time to breakeven	The elapsed time from the point when a program is established until the point when all program investments are equal to the program net income.	The beginning of program implementation denotes the point when the program begins to incur expenditures for labor and materials. As the program progresses through its life cycle, it continues to incur expenditures. When the net income from the sales (or cost savings) from the output becomes equal to the expenditures for development and production, the point of breakeven is reached. The logic is that shorter time to breakeven equates to faster recuperation of the program expenditures.

11. Profit margin
PMVP Level: Program Management

Metric	Definition	Usage
Gross profit margin	Gross profit is a percentage of sales for program products.	These are leading indicators, in which forecasts of the profit margins are made

Table 9.1   Program Management Menu of Metrics (*Continued*)

Metric	*Definition*	*Usage*
		periodically and compared to the original forecasts. The actual profit margin values are lagging metrics.
Net profit margin	Net profit margin is equal to net sales minus cost of goods sold.	

12. Market share
PMVP Level: Program Management

Market share	The percentage of the market that will buy or use the output of a program. The probability of accomplishing that market share goal should also be given consideration.	All projections must be done with a clear understanding of assumptions behind probabilities. Personal biases in favor or against the program should not be allowed to affect the probabilities. The forecasts of the metrics are made periodically (leading metrics) and compared to the original forecasts. The actual values of the metrics are lagging.

13. Resource Management
PMVP Level: Program Management and Project Management

Program staffing level	Number of actual person hours per program (over a specific period) divided by the planned number of person hours over the same period.	These metrics will measure if a program and its projects are staffed for success and address the practice of understaffing programs. Ideally, the value of these metrics is one (all planned resources are actually assigned to a program), assuming that the planned number of the resources was accurate.

Table 9.1   Program Management Menu of Metrics (*Continued*)

Metric	*Definition*	*Usage*
Project staffing level	Number of actual person hours per project (over a specific period) divided by the planned number of person hours over the same period.	These metrics do not indicate if the program is staffed with people who have the appropriate skills and experience.

14. Program cost
PMVP Level: Program Management and Project Management

Program development cost	Total cost of a program from formal initiation to launch or go live. This metric includes all costs associated with the development, including nonrecurring engineering, design, and development labor in all operating functions participating in the program—all parts and materials used during development and any associated indirect and allocated costs deemed appropriate by the accounting department, in accordance with generally accepted accounting practices.	This metric is useful in calculating the financial returns on the program. Additionally, tracking the development cost of each program and the total for all programs; this assists senior management with managing overall development spending to the available budget set by the firm.
Manufacturing cost	This reflects the total cost to manufacture a product or capability. It includes the labor, materials, overhead, and any designated company allocations of indirect costs deemed appropriate by the accounting	This metric is used together with the rest of the costs and benefits to determine the overall financial returns on the program. It also helps senior management determine future continual improvement efforts to drive further margin and

Table 9.1   Program Management Menu of Metrics (*Continued*)

Metric	*Definition*	*Usage*
	department, in accordance with generally accepted accounting principles.	profitability gains on the product.
Nonrecurring engineering cost	This represents that portion of the total development cost related to the direct engineering and design effort for a specific new product, service, or infrastructure capability. This measure does not include any follow-on engineering effort that is applied for sustaining or support.	This is a useful metric to assist senior management with managing the effective usage of resources and budget. Also, by monitoring this at the program level and in total for the organization, management can ensure that the available dollars and resources are applied to the highest-priority programs based upon the firm's business objectives and strategy.

15. Technology innovation
PMVP Level: Program Management

Metric	*Definition*	*Usage*
Percentage of technologies new to the world	The fraction of technologies in the program output that are new to the external ecosystem, including customers and competitors.	These metrics measure the technical complexity and technical risk of a program. The higher the percentage, the greater the technical complexity and risk. They also give an indication of the technical competitive advantage that a program may provide.
Percentage of technologies new to the company	The fraction of technologies in the program output that are new to the company.	

The menu was developed from our experience with specific companies; therefore, it is not a comprehensive list of all possible metrics. Rather, we present metrics that we have found to be the most generally representative and useful for the managerial functions in the PMVP.

The heading in column one of the table, metric, identifies the metric being presented; the heading in column two, definition, provides the definition of each metric; and the heading in column three, usage, explains the purpose and how to use the metric.

In practical terms, it is highly unlikely that any one firm would, or could, use all the metrics presented in Table 9.1. It is presented as a menu of metrics. The listing should be seen as a restaurant menu that presents many dining options. From a restaurant menu you choose and order only a subset of the available menu items—seldom all of them. Similarly, from the menu of metrics you choose only those metrics that you need. We will explain later how to customize a metrics set for a specific company's needs.

We'd like to remind the reader once again that Table 9.1 is a menu of common metrics used in practice and does *not* represent that all the metrics should be used on a program. Each senior management team and PCT must select a set of metrics for their particular measurement needs. A common question we are asked is how many metrics do we need? The answer to this question is that it depends. A study by *BusinessWeek* and The Boston Group shows that the sweet spot is somewhere between 8 and 12 metrics, which was Chu Woosik, a senior vice president of Samsung Electronics confirmed.[5] See the box titled, "How Many Metrics Do You Need?" for additional guidance.

How Many Metrics Do You Need?

The answer to this question of course depends on the nature of the program. The strategy, culture, and stakeholders' values of the company, as well as how big, complex, and technically and commercially risky a program is, all play a major role in deciding how many metrics you need. Additionally, accelerated time to money objectives for the program will require special attention to how the program is managed and what metrics will best assist management in keeping it on track. As a general rule, one needs a sufficient number of metrics to maintain good control of a program, and as few as possible to still stay nimble. How many is that for an average program?

Some examples may provide guidance. In one successful high-tech company, program teams have eight standard metrics agreed upon with senior management. In another company, programs use between 10 and 14 metrics. In a leading semiconductor company, very complex, fab building programs evaluate about 20 metrics each month. Our personal experience is that about ten metrics are sufficient, if an effective metrics system is constructed.

THE NATURE OF THE PMVP

In developing the PMVP, our focus was on those measurements that relate to program management's contribution to building and sustaining a business-unit's competitive advantage and the improvement of business results. The PMVP provides a holistic approach for evaluating the strategy for obtaining business objectives, how well a company's development funds are invested, the value of program management to the company, the effectiveness of program management practices, and how individual programs are progressing.

The PMVP is designed to function in a top-down fashion, level by level, to create a balanced and aligned set of metrics that cover multiple dimensions of product, service, or infrastructure development. The value pyramid is also a dynamic model, meaning the metrics set created will dynamically change as a program progresses through the development life cycle, as different metrics are needed at different points in the life cycle.

The PMVP is applicable to both centralized and decentralized forms of program management structure. In a centralized form of program management, there exists an organizational home for all or a majority of program managers. In this case, the use of the PMVP will be determined between the director of program management and senior management, and the commonality of metrics across programs is established. The decentralized form means that all or a majority of the program managers are assigned to one or more business units, and metrics will be based upon the specific business needs of the business unit.

The PMVP is an effective tool for measuring and guiding continual improvement in an organization. However, utilizing the PMVP does not mean automatic improvement. Metrics can only provide data. It is up to the senior management team of a business to interpret the data presented by the metrics and turn it into information that can be used to develop the necessary improvement actions. Let's look at some of the key questions and considerations for management at each level of the PMVP.

Strategic Management Metrics

At the top of the pyramid is the strategic management domain along with the firm's business objectives. This level is, therefore, the primary driver of the PMVP. Strategic management metrics are used as an input for business planning, management reviews, and for addressing the following questions:

- Will the current portfolio of programs achieve the strategic objectives?
- What is the cost/benefit ratio for achieving the strategic objectives?
- How long will it take to achieve the strategic objectives?
- Are we getting returns on program management as we expect?

Strategic management metrics related to program management are solid indicators of the return the business is gaining from using a program management model, whether enough is being spent on program management and if the business can expect to receive further positive returns from program management.

Portfolio Management Metrics

When a business designs a balanced and prioritized portfolio of programs, if develops the strategy for achieving the key business objectives it has defined. Additionally, by converting the portfolio into a program road map, it further develops the strategy by creating a time-phased plan for achieving the objectives.

Metrics that measure the effectiveness of program portfolio management generally indicate if a business's development dollars are invested properly for achievement of the business returns desired. They also indicate how the investment is balanced across multiple dimensions such as reward versus risk and new versus core technologies.

Program Management Metrics

Program management practices are a mirror of the culture of the program management organization and are risky to neglect. Like any cultural element, they can degrade or improve quickly. For this reason, program management practices need to be continually monitored and measured to become and remain a competitive weapon for an organization.

These metrics measure the program-level practices of the program management discipline within an organization. They are designed to give a business an indication of whether program management is helping to achieve its goals, if programs are cost effective, and if quality and customer satisfaction is sufficient. These metrics also indicate the effectiveness of the program management discipline by measuring elements such as whether the program management acumen and core competencies are improving or declining within a business.

Program management metrics also measure the performance of the program team on a particular program and give the program manager and his or her senior management team an indication of whether the program will attain its critical success factors.

Project Management and Team Execution Metrics

Project management and team execution helps deliver critical elements of the integrated management system and subsystems of the whole product. These metrics measure the hands-on performance of the project teams. The functional project managers are responsible for the detailed planning and execution of the project deliverables pertaining to their respective operating functions. Each project manager, along with his or her respective functional team specialists, will develop a specific project plan and project schedule, budget, and performance goals. Project metrics are designed and used to measure the functional team performance to the goals. Additionally, metrics used to measure the effectiveness of the cross-project coordination and collaboration are also needed at both the project and program levels.

CUSTOMIZING A METRICS SET

As we mentioned earlier, each company will need a distinct set of metrics, depending on its business strategy. Therefore, the set of metrics will vary by the type of industry the company is part of and by its business strategy (for example, differentiation, cost leadership, or best cost). The PMVP does not provide a mechanism to automatically indicate which metrics to choose; this is the job of senior management.

When a business unit determines its own distinct set of metrics, it should do so in collaboration with company stakeholders. As noted in the following list, each stakeholder will emphasize different metrics, according to their needs:

- The board of directors, financial community, and CEO will primarily be interested in the strategic metrics. See the box titled, "I Have Only Three Minutes a Month" for an example of the metrics one CEO wanted to see.
- Business management will show a strong interest in metrics that assess the strategic health of the business unit, the alignment of development programs with the business strategy, and the balance of the program portfolio relative to the business unit needs.
- Program management directors will be concerned with metrics from all levels of the PMVP pyramid, with most interest in metrics assessing the portfolio of programs and the management of individual programs.
- Members of the PCT and project teams will be most keenly interested in metrics about project management and team execution.

I Have Only Three Minutes a Month!

Al Petroff, CEO of DirectConnect, a world premiere producer of interface cable, was very expressive, almost rude: "Look guys, my time is very expensive, and I am sick of wasting my time reading poor reports. I need you to design a report showing the monthly status of my 40-plus programs going on at any time. It must be a one-pager showing the most important things about my programs; it need not contain words, only numbers and graphical symbols, and I must be able to read it in three minutes because I have only three minutes for that purpose a month. Is it clear?" A consultant that Petroff was talking to nodded and said, "Yes, it is clear."

Seven days later the consultant was back with a one-page report. He began by saying, "We included five metrics, covering program management-related measurement from strategic management, portfolio management, and program management: (1) program management return, (2) projected future income from program road map, (3) program portfolio distribution, (4) external customer satisfaction survey, and (5) percent of the program portfolio milestones accomplished.

Then, he went on to say, "Each metric shows one number for the month and one for the cumulative value, where applicable. Quality of the monthly status, cumulative value, and overall trend are shown by colors. A green status signifies progress as planned, a yellow status indicates a heads up to management of a potential problem, and a red requires management intervention. We have tested the time needed to read and interpret the report with our executives, and they need an average of three minutes." Petroff took a long look at the one-page report, paused, and said, "This looks good, let me test it."

In addition to selecting from a set of metrics, thought has to be given to how all metrics from the various levels of the PMVP interconnect and which are compatible with one another. It is the job of the senior stakeholders of the business to provide the interconnection and alignment strategy of the set of metrics. Let's be reminded that business strategy dictates the configuration and focus of a company's program management function. This means that each element of program management needs to be aligned with the business strategy—program management metrics included. To illustrate, we'll look at three companies with three different business strategies as defined by Porter's model, and demonstrate how the strategy drives the selection of the metrics.[6] Company A in Table 9.2 utilizes a differentiation strategy to offer its customers something different from their competitors. In particular, Company A focuses on technology innovation and fast time to market to achieve differentiation and gain higher profits and market share. Secondarily, Company A strives to achieve high quality and doesn't allow runaway cost. To measure performance to their strategy, Company A selects metrics for measuring time to market, percentage of new technologies developed, profitability index, and market share. Additionally, they choose customer fall out and development cost metrics to measure their secondary strategy of providing high quality and cost containment.

Company B focuses on a low-cost strategy aimed at establishing a sustainable cost advantage over its rivals. The intent is to use the low-cost advantage as a way of underpricing rivals and capturing market share. This strategy will leave company B with a small-profit gross margin per product, but profit can be achieved through a large volume of sales and lean staffing. Metrics to support this business strategy should measure development and manufacturing cost, market share, program-staffing level, gross profit margin, and sales volume. Secondary objectives are to bring the product to the market in an average industry standard time and with an average quality. This is complemented with the second priority metrics of development time and defect assessment rate.

The best-cost strategy pursued by Company C combines upscale features with low cost. Company C aims to become a low-cost provider of IT products that have high-quality features. A secondary strategy is to develop innovative information technologies within reasonable time frames. Primary metrics to measure this strategy should include defect

Table 9.2 Strategy drives metric selection

		Company A	*Company B*	*Company C*
Company Business		High-tech electronics	Conventional manufacturing	Automobile company IT
Business Strategy		Differentiation	Cost leadership	Best cost
Type of programs		New product development	New product development	Infrastructure development
Metrics			**First Priority**	
		Time to market	Development cost	Defect rate assessment
		Percentage of new technologies developed	Manufacturing cost	Customer fall out
		Profitability index	Market share	Development cost
		Market share	Staffing level	Manufacturing cost
			Gross profit margin	
			Sales volume	
			Second Priority	
		Customer fall out	Development time	Development time
		Manufacturing cost	Defect rate assessment	Percent of new technologies to company
		Development cost		

rate assessment, customer fall out, development cost, and manufacturing cost. Secondary metrics could include percent of new technologies developed and development time.

These examples illustrate how business strategy drives the choice of metrics. We showed only three examples of companies with different strategies and the corresponding sets of metrics chosen. Each company in the world will have its own unique business strategy and set of metrics to measure the effectiveness and achievement of its strategy (see box titled, "The Control Chart as a Metric for Program Team Charter").

The Control Chart as a Metric for Program Team Charter

Customizing your set of metrics needs to be done for all levels of the PMVP. For the program level, here is an example of customization in a high-tech firm. The GM of several $10 million-plus programs asked program managers to develop and use a metric to measure how each PCT member views the extent to which their program manager drives the team to deliver the program charter requirements. They designed and deployed an aggregate metric consisting of 14 items, each item being one requirement from the charter, including schedule, budget, communication, and enforcement of the company's values.

The degree of accomplishment for each of the 14 requirements was measured by means of a Likert-type survey on 1 to 5 scale, with 1 being "not at all per program charter" and 5 being "always per program charter." Each month, five of the core team members were randomly chosen for charting the metric. For each of the five evaluations, a mean of ratings for the 14 requirements was calculated.

The calculated mean is the metric charted on a control chart with the purpose of tracking the performance of a specific program manager. It determines whether his or her performance is consistent with the program charter and, if not, indicate what improvement action is needed.

This example shows us that when a customized form of a metric is needed, you can find a way to create the customization, even if the metric is new and atypical.

USING METRICS TO IMPROVE BUSINESS RESULTS

All metrics should help improve business results. To demonstrate this point, we choose and evaluate the use of a set of metrics from the PMVP, as follows: 1) *Projected future income* from program road map from the strategic management and portfolio management levels, 2) *portfolio risk index* from the portfolio management level, and 3) *percentage of program milestones accomplished* from the program management and project management levels.

Projected future income from program road map: A metric that shows the fraction of future net income anticipated from programs on the program road map over a multiyear period. For discussion sake, let's assume that the metric shows that the projected income is lower than the stated goal. The question from senior management will then become "why?". A close look at the supporting detail of the metric reveals the income from many programs is low. An even closer look shows that the majority of programs are for derivative products, not new architectures. Derivative products serve to extend the life of a particular product line but do not provide much income due to ever-decreasing profit margin caused by competitive pressures. A deeper analysis shows that the only way to achieve the income goal is to terminate some of the derivative programs with low-revenue generation and shift investment to new platform programs that will provide higher levels of net income. As a result, the use of this metric highlighted a serious flaw in the execution of this organization's business strategy and helped to resolve it, leading to improved business results in the form of higher-net income and profit.

Portfolio risk index: A program portfolio-management level metric indicating the total risk of all programs in a company's portfolio. For example, let's say that the portfolio risk index indicates a low-risk level for a business unit's portfolio. The conventional wisdom tells us that it is good to keep risk levels as low as possible. But it also says that low risk typically means low gains. Further analysis of the portfolio shows that most of the individual programs face low risk and low-projected profits. These types of programs are commonly called "bread-and-butter" programs. They are typically the mainstream of a business, use existing technology, and target markets and customers that are well known. The portfolio team realized that they were selling their future by doing mainly bread-and-butter programs. To invest in the future, the company needed to terminate some of these programs and instead invest in programs that create new technologies, bring higher-profit margins, and are also more risky. This metric demonstrated an unbalanced portfolio (in terms of risk) and helped the company choose a different program portfolio that promised improved business results in the form of higher profits and new products and markets.

Percentage of program milestones accomplished: A metric that indicates percent of all milestones in a program achieved within a specified time period (such as a week) of the projected achievement date. To understand how it can help improve business results, let us look at an example of a program schedule that is structured with 300-lower-level milestones over the 11-month program duration. The first month does not have any milestones; the remaining 10 months have 30 milestones each. At the end of the third month of the program, only 40 percent of the 60 milestones planned to that point were accomplished. What does this indicate? This metric shows that the rate of milestone completion was poor compared to the plan, and that the final program completion date is likely to be significantly delayed. This is an early warning signal for management to take comprehensive corrective actions that will lead to timely completion of the program. This metric's function is to warn the business insert problems in the management of the program schedule so that corrective actions can be implemented to remove the barriers and achieve the business results intended.

SUMMARY

The purpose of metrics is to help improve business results of the organization. To serve this purpose, metrics need to be carefully developed to ensure they are compatible (or mutually aligned), balanced, consistently utilized, and tiered.

We provide a menu of commonly used program metrics that can be configured as a customized set of metrics by the management team of a business. Each business should construct its set of metrics based upon its business strategy and the specific industry in which it operates.

One convenient way to visualize a system of metrics is the PMVP that has the following four levels: strategic management, program portfolio management, program management, and project management and team execution. This pyramid provides an integrated and holistic approach for selecting and utilizing a metrics set, which evaluates the performance of the four levels of the integrated management system.

The Principles of Program Management

▼ The fundamental purpose of program management metrics is to help improve business results.

▼ The PMVP is an integrated and holistic way to customize a metrics set.

▼ A set of metrics should be balanced, mutually-aligned (compatible), and tiered.

REFERENCES

1. Hauser, J. and F. Zettelmeyer, "Metrics to evaluate R&D," *Sloan Working Paper # 3934, MIT*, (October 1996).
2. Tipping, J. W., E. Zeffren, et al., "Assessing the Value of Your Technology," *Research Technology Management*, Vol. 38, No. 5 (1995): 22–39.
3. Hauser, J. and F. Zettelmeyer, "Metrics to evaluate R&D," *Sloan Working Paper # 3934, MIT*, (October 1996).

4. Meyer, Chris. "How The Right Measures Help Teams Excel," *Harvard Business Review* (May–June 1994): pp. 95–103.

5. McGregor, Jena. "The World's Most Innovative Companies," *Business Week Online* (April 24, 2006).

6. Porter, M. E., *Competitive Strategy: Techniques for Analyzing Industries and Competitors*, 1st edition. New York, NY: Free Press Publishers, 1998.

Chapter 10

Strategic Program Management Tools

The topic of this chapter is tools to support the strategic aspects of program management. To be clear, it is not the program manager who develops and owns the strategic tools. The strategic tools are owned by senior managers who use them to supervise and manage programs at the strategic level. Why, then, do we bother to discuss these tools in the section of the book dealing with management of a single program, a level that is not the purview of senior management but of program managers? The significance of these tools is that program managers need to comprehend the information obtained from the tools to communicate with senior managers about their programs from a strategic perspective.

The conventional wisdom holds that tools are enabling devices to reach an objective or, more specifically, a deliverable. Like any other tool, strategic program management tools include procedures and techniques by which a deliverable is produced. Their core—either of a qualitative or quantitative nature—is in their systematic procedure.

We explain the following strategic program management tools in this chapter:

- business case
- alignment matrix
- portfolio maps
- road map
- complexity assessment

The purpose of this chapter is to help practicing and prospective program managers achieve the following objectives:

- Learn how major strategic program management tools are developed and utilized to increase program efficiency and effectiveness.
- Be able to describe how strategic program management tools are selected to match the program situation.

Mastering the information contained in the tool description is essential for program managers. The information can help program managers understand how the company's senior management utilizes these tools to execute the organization's business strategy, program portfolio, and senior management-owned program decisions. This will enable program managers to comprehend the nature and position of their program in the context of all programs and of the company's business strategy.

For a user-friendly presentation of the tools, we designed the chapter in a way that enables maximum understanding of the development and utilization of each tool. Presentation of each tool will be segmented into sections that discuss the following:

- *Description of the tool*: A brief description of the characteristics of the tool and how it is used by the firm.
- *Developing the tool*: Steps that are involved in constructing or developing the tool are detailed in this section. They are made up of a series of substeps that describe specific activities.
- *Utilizing the tool*: Multiple elements are involved in this section. The section called *When to Use* explains situations in which the tool can be applied. How much time is needed when utilizing the tool is described in *Time to Prepare*. The *Benefits* element specifies what value the tool creates for the user. By contrast, the *Advantages and Disadvantages* element generally concentrates on the simplicity/complexity and ease-of-use issues of the tool.
- *Summary and highlights of the tool*: At the end of each tool section, a summary reminds the reader of the purpose, use, and benefits of the tool—offering highlights for appropriately using or structuring it.

At the end of the chapter, we offer a brief set of situations in which the strategic program management tools are used.

PROGRAM BUSINESS CASE

The program business case is a start-up document used by senior management to assess the feasibility of a program from multiple business perspectives. It establishes the program vision by describing a business opportunity in terms of alignment to strategy, market or customer needs, technology capability, and economic feasibility. It also provides a balanced view of business opportunity versus business risk. The program business case is used for the following purposes:

- Gaining agreement on program scope and business success criteria.
- Obtaining approval of funding and resource allocation for program planning and implementation.
- Evaluating a program against others in the portfolio of programs.
- Obtaining approval to proceed from the define phase to the planning phase of the PLC.

The program business case consists of the following five major elements: (1) Description of the business opportunity and product, service, or infrastructure concept; (2) program alignment to strategic, market, and technology goals; (3) business success criteria; (4) cost versus benefit analysis; and (5) risk analysis.

Developing the Program Business Case

Step 1: Prepare information inputs: The business case for a program must be correct based upon the knowledge available at the time it is created, unbiased and clear. This requires quality information about the following:

- The business environment
- Customer requirements
- The business strategy
- Technology road map
- Program targets

The business environment and customer requirements information is necessary to build the foundation of the business case. Understanding the needs of the customer, as well as the state of the environment within which the business operates, is needed to position and differentiate the product, service, or infrastructure capability being proposed. The business strategy specifies the strategic objectives that the organization is striving to achieve. The technology road map shows both internal and external technologies that will be available for the proposed program. The program targets consist of high-level directives from senior management to gauge initial feasibility of a program to meet the business needs. These typically consist of cost, schedule, and primary feature or functionality targets.

Step 2: Describe the business opportunity and product, service, or infrastructure concept: There are two steps to this action. First, a description of the opportunity that the program will fulfill is provided, with a focus on business opportunities. Second, a description of the product, service, or infrastructure capability concept is documented, with focus on how the concept addresses the business opportunity that is presented.

Step 3: Define program alignment: The purpose of programs is to serve as basic building blocks for the execution of an organization's strategy.[1] The premise is that if programs are aligned with the firm's strategy, they will better support the objectives of that strategy. Since one tangible way to express the strategy is to define its specific, measurable, attainable, relevant, and time-based objectives, one can use the stated objectives to assess how well an individual program supports them. This step describes how the program aligns to and supports achievement of one or more strategic objectives of the organization. Additionally, a description of how the program result fulfills documented customer and market needs is included. Finally, a description of any new technologies that will be included in the product, service, or infrastructure capability being developed is provided in this step.

Step 4: Define the program success criteria: Identification of the critical success factors for a program should begin during the define phase of the PLC. The critical success factors ensure that the product, service, or infrastructure capability under development supports key business objectives such as profitability, time-to-money, productivity gains, and technology advancement.[2] This step forms the genesis of the program strike zone, which is described in detail in Chapter 11.

Step 5: Perform the cost versus benefit analysis: The heart of the program business case is the feasibility assessment that results from the cost versus benefit analysis of the program. The cost/benefit analysis should identify both tangible and intangible benefits with the benefits expressed in quantifiable terms such as dollars gained or saved, hours saved, and gross margin increase. The cost versus benefit analysis should answer the following questions:

- How much will this program cost to implement?
- How much will this program contribute to the company bottom line?
- Is the program worth investing in terms of achievement of specific business objectives?

Step 6: Analyze program risk: In this final step, all potential risk events that may affect the business success of the program are identified. At this stage of the program, it is a high-level look at the known risks. The risk events are then analyzed, and a plan to minimize the impact and probability of occurrence for the high-level risks is developed. The risk analysis should answer the following questions:

- What is the probability of success for this program?
- What will be done to maximize the probability of success?
- How will the known risks be avoided or mitigated?
- Does the level of risk prevent investment in the program?

Utilizing the Program Business Case

The program business case is developed in the define phase of the PLC. Presentation of the business case to senior management stakeholders normally signifies the end of the define phase. The program business case, along with the market or customer requirements document, forms the basis on which the program plan is developed. The business case is updated as needed and reviewed as part of the implementation plan approval meeting. Prior to entering the implementation phase, in which maximum resources are expended, the business case needs to be reviewed for validity. During program implementation, the business case needs to once again be reviewed prior to releasing funds for large expenditures, such as factory tooling. Finally, the business case information is used during the program retrospective to evaluate if the program was successful in achieving the business objectives intended.

Time to Prepare

The time to prepare a program business case varies greatly, depending upon the size and complexity of the program; the type of product, service, or infrastructure capability being developed; and the company's industry. Depending upon these variables, duration to prepare the business case can range from two weeks to several months. Having stated this, however, one of the greatest opportunities to reduce development cycle time is in this arena. The more efficient a company can become in collecting and analyzing the inputs needed to develop the business case, the more efficient it will become in preparing the business case for a program. Any time saved on the front end of a program is an opportunity for decreased time to money.

Benefits

The benefits of creating a good business case for a program are many. The primary benefits are fourfold. First, the business case answers the following critical question: "How will this program help our company meet its business and strategic goals?". In that way it helps senior management of a firm make sound decisions when considering development investment options. Second, it establishes alignment between strategic goals and program execution based upon multiple business perspectives. Third, consistent use of the business case for all programs helps to make the portfolio management process more effective by enabling the evaluation of programs within the portfolio on a consistent set of criteria. Finally, it establishes the vision—or future state—to effectively plan, execute, and deliver the output of the program.

Advantages and Disadvantages

An advantage of the business case is that all crucial information for the program decision making is conveniently included the case. However, time necessary for preparation of the business case can be a real disadvantage. In particular, the preparation can be time consuming, if discipline is not applied to effectively collect the data inputs as well as effectively analyze the cost, benefit, and risk information. The opportunity for *analysis paralysis* is great, and when it happens, it consumes critical time-to-money advantage.

The program business case is a critical tool used to assess the feasibility of investing in a program based upon cost, benefit, and business risk factors. Once prepared, the program business case provides the vision to guide program planning and execution and becomes the means to align execution to business strategy.

Program Business Case Highlights

Check to make sure that the program business case is properly completed prior to presentation to senior management. It should do the following:

- Describe the business opportunity and product, service, or infrastructure capability being proposed
- Demonstrate alignment of the program to strategic, market, and technology goals
- Clearly identify the critical business success criteria for the program
- Present a concise cost versus benefit analysis and a thorough assessment of business risk

PROGRAM ALIGNMENT MATRIX

The alignment matrix helps establish the degree to which the program is aligned with the organization's business strategy (see Figure 10.1). This is, of course, also balanced against the specific customer needs that the firm is attempting to meet. The first column of the matrix contains the list of the organization's business objectives that serve as criteria to align programs with the organization's strategy. Then, in each of the remaining columns, the degree of alignment of individual programs with each objective is assessed using a qualitative scale. As an outcome, a qualitative goal-by-goal alignment evaluation for each program is generated that may be used for different strategic purposes.

Developing the Alignment Matrix

Step 1: Prepare information inputs: The assessment of a program's alignment with an organization's business strategy calls for quality information that typically comes from three inputs, as follows:

- Approved business strategy
- The portfolio of programs
- Program business case (or preliminary business case information)

The approved business strategy provides a list of the organization's business objectives that the strategy is striving to accomplish. To assess the degree of alignment of individual programs to the strategic objectives, a list of current and future programs is needed. This information is typically found in program portfolio documents. Finally, to understand how well each program is aligned with the strategic objectives, the program business case is needed.

Step 2: Identify the organization's strategic objectives: Strategic objectives are defined by the organization's senior management—sometimes formally in strategic plans but other times informally. In either case, the objectives should be used for the alignment matrix assessment. Because each organization is a unique entity, the strategic objectives found in the alignment matrix will be unique as well. Additionally, as the strategic objectives of an organization are updated and modified, the objectives in the alignment matrix need to be updated accordingly.

Step 3: Identify the programs: There are two steps to this action. First, the names of the new and existing programs are entered into the columns of the alignment matrix (see Figure 10.1). As stated earlier, the list of the programs is normally part of the portfolio of programs documentation. If a formal portfolio of programs does not exist, an active program roster will suffice. Second, the program strategy and goals should be developed

Example of organizational business objectives	Program 1	Program 2	Program 3	Program 4
Features are clearly differentiated from competing products	P	F	F	P
Enables performance increase in high-speed devices	F	N	F	P
Falls within the time-to-market window	F	F	P	F
Supports the common platform architecture	F	F	P	F
Reduces manufacturing cost	N	F	N	P
Legend: F = fully supports P = partially supports N = does not support				

Figure 10.1   An example of an alignment matrix.

and documented to secure the information needed to assess alignment of programs to strategy.

Step 4: Define the alignment scale: Scales of course vary, and the choice of scale is specific to the organization. We believe that a simple, qualitative scale is completely adequate and provides the value we want from this matrix. An example of a simple qualitative scale is a three-level scale shown in Figure 10.1. The scale includes the following rankings: *fully supports* for the highest degree of alignment, *partially supports* for the medium-level alignment, and *does not support* for no alignment of the program with a specific goal.

Step 5: Assess the degree of alignment: Now is the time to use the adopted scale of assessment and assess each program's alignment with each organizational business objective. Typically, decision makers who do the assessment should use a workshop format, in which information from multiple perspectives is exchanged and the assessment decisions are shared and consensual.

Utilizing the Alignment Matrix

Typically, the alignment matrix is prepared for the program portfolio reviews. At that time, program selection and risk balancing, the first two goals of the program portfolio process, are completed first. What remains to be done are "strategic buckets," the third portfolio goal, which aligns programs with strategic objectives of the organization.[3] Based on the alignment information obtained, by means of a tool like the alignment matrix, the preliminary selection and risk balancing of programs may be changed, which may provide better alignment of programs with the strategic objectives of the organization.

Time to Prepare

Filling in the basic information of the matrix takes very little time, usually only a few minutes. However, a team of decision makers, usually the senior managers owning the program portfolio process, may easily spend a few hours developing and agreeing upon the level of alignment data for the alignment matrix. The true value of this tool is in the discussion that it generates between senior managers; therefore, we recommend a workshop format to facilitate the discussion.

Benefits

The alignment matrix enables an organization to refine the selection of the preliminary portfolio of programs by pointing to programs that are best aligned with the organization's business objectives.[4] Based on that, one can eliminate some programs that are not strategically aligned; one can also add new programs that are better aligned. That's the matrix value, which is strategically precious given how difficult it is to select the most valuable programs that are risk balanced and aligned with the strategic objectives of the organization.

Additionally, it forces the senior managers of an organization to develop and document the organization's strategic objectives. The alignment matrix aids the program manager in understanding how well his or her program supports the strategic objectives of the firm. This knowledge will be useful, for example, when a program manager requests additional resources from a less strategically aligned program.

Advantages and Disadvantages

The alignment matrix is visual and easy to read. However, it is not easy to compare the alignment of multiple programs across multiple objectives, in the absence of one alignment score. This busy aspect of the matrix can be disadvantageous.

The alignment matrix helps to establish the degree to which a program is aligned with the organization's business strategy. This makes it possible to refine the portfolio of programs and for the program manager to fully understand the strategic significance of his or her program.

Program Alignment Matrix Highlights

Check to make sure that the alignment matrix is properly completed. It should do the following:

- Show the organization's business objectives
- List all existing and new programs
- Have a scale to measure the alignment of each program with the strategic objectives of the organization
- Indicate the degree of alignment of each program with each organizational business objective

PROGRAM PORTFOLIO MAPS

Portfolio maps—we refer only to the bubble diagram form of portfolio maps—are information displays that visually show key parameters necessary to successfully balance a portfolio of programs. Typically, the *x* and *y* axes of the diagram represent two of the key parameters of a program, as depicted in Figure 10.2. The bubbles in the diagram indicate the position of the program within the two dimensions, while its size and color may point to additional parameters such as program size, percent complete, or program type. The distribution of the programs on the diagram, visualized as bubbles, provides senior management with information needed to decide how to distribute their investment in the programs as they pertain to the parameters selected.

Developing the Portfolio Maps

Step 1: Prepare information inputs: Balancing a portfolio of programs calls for quality information about the following:

- Strategic and tactical plans
- Program selection criteria
- Program roster with programs' alignment to strategy

Step 2: Select the type of chart: Making a choice of the chart (see box titled, "This Is Not the Old BCG Matrix") begins with a clear understanding of the parameters on the axes of the portfolio map, such as the following:[5]

- Program importance versus ease of execution (in scoring models-based charts)
- Adjusted net present value versus the probability of technical success (in risk-return charts)
- Program costs versus life cycle phases (in other bubble diagrams)

No single chart can comprehensively characterize a portfolio. By utilizing several charts with varying program parameters, one can accurately capture major strategic requirements of the portfolio.

Step 3: Chart programs: Draw the *x* and *y* axes (see Figure 10.2), label them with the chosen parameters from step 2, and enter the scale. This

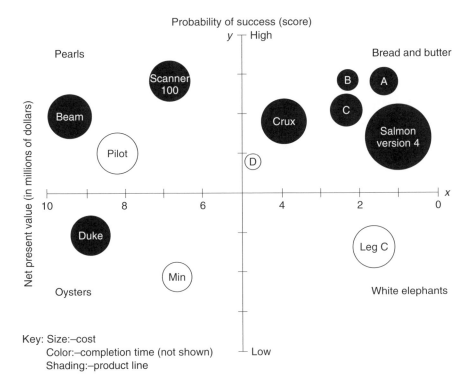

Source: Adapted from Strategic Decisions Group model

Figure 10.2   An example of a portfolio map that shows value versus risk.

type of diagram is based on scores from program selection, and in this example, each parameter's scale is a numerical average of three constituent scales. For program value, these include strategic importance to company goals; organizational impact (for example, profitability impact); and economic benefits (for example, dollar savings). However, ease of execution relates to cost of execution; program complexity (for example, difficulty of implementation); and resource availability. The intent here is to assess the program in a comprehensive, multicriteria fashion.

Step 4: Interpret the diagram: Identify programs that are located in favorable quadrants (the upper, left in Figure 10.2), dissect those in the low importance and difficult quadrant (lower right in Fig. 10.2), and strive to balance ease and importance of programs. For example, in Figure 10.2, we interpret the information in the diagram as follows:

- Easy-to-do programs are in short supply, just 3 out of 12. We need more of these "low-hanging fruit" programs, seeking a balance between easy and difficult programs.
- There are only three programs in the most sought-after upper, left quadrant, an intolerably low number.
- Another problem looms on the horizon with too many programs—seven precisely in the right quadrants. This is the most unpleasant place, and one should think about which to kill.
- Two programs are of high importance but difficult to do. Perhaps, there are ways to remove some barriers to ease implementation.

In this relatively simple example, the bubble diagram demonstrates its forte, the ability to display program information well.

This Is Not the Old BCG Matrix
When first exposed to bubble diagrams, many people are quick to assume that they are the same as the Boston Consulting Group matrix (or GE or McKinsey matrices) of the 1970s.[6] Actually, they are not. Old matrices were used to plot the existing strategic business units on market attractiveness versus competitive position matrix, capturing the present state of the units. Program bubble diagrams add newly selected programs to the existing programs, focusing on the future. In addition to this difference in unit of analysis and time horizon, dimensions of the diagram are also very different, ranging from probability of technical and commercial success to the *ease of program execution*. Many dimensions may be used.

Step 5: Balance the portfolio: Managers, not the portfolio map, need to make a decision to act and balance the portfolio. For the example shown in Figure 10.2, one can suggest that the following balancing actions be taken:

- Increase the number of easy-to-do programs
- Increase the number of the most sought-after programs
- Kill programs with the lowest importance
- Increase the ease of implementation of all programs in the high value, risk quadrant

Utilizing Portfolio Maps

Portfolio maps are versatile tools that can be employed in multiple ways. One way may be to start at the top of the company (a portfolio map for top strategic programs) and cascade through portfolios of different organizational levels. Good examples are new product, information technology, and manufacturing groups.[7] In this context, the portfolio maps are employed for periodic reviews of the program portfolios, quarterly or semiannually.

Time to Prepare

Provided that information about programs is available, constructing a portfolio map is a quick action. For a portfolio diagram consisting of 20 programs, a group of decision makers may need only a few minutes to create a portfolio diagram. The large portion of time consumed is in analyzing the position of programs in the diagram and making decisions to balance the portfolio of programs. These steps may take several hours to a full day.

Benefits

The basic value of the portfolio map is its ability to act as a meaningful information display, helping discern programs in less-desired quadrants. Additionally, the sense for the utmost importance of balancing programs across the desired quadrants comes across clearly. Such a visual demonstration of the program portfolio is unmatched by other tools.

Advantages and Disadvantages

Portfolio maps offer substantial advantages through simplicity, with user-friendliness and a data-based view. The data is viewed and factual. The problems with portfolio maps may result from information overload, when too many diagrams are used.

This section is about portfolio maps in the form of bubble diagrams, which is information that visually shows key parameters necessary to successfully balance a portfolio of programs. The bubble diagrams cannot suggest actions to balance an unbalanced portfolio of programs. However, they can indicate the importance of balancing programs across the desired quadrants. The essence of constructing the diagrams is highlighted below.

Program Portfolio Maps Highlights
Be sure to adequately construct the program portfolio map. It should show the following:

- Chosen dimensions on the x and y axis.
- The scale for dimension.
- Programs as bubbles in quadrants that are balanced by management.

PROGRAM ROAD MAP

The Program Road map is an information display that visually shows the time phasing of the programs within the portfolio. Typically, the road map is segmented into product, service, or infrastructure type, or into market and customer segments. This is represented by the vertical grouping of programs, as shown in Figure 10.3. The horizontal axis of the program road map shows the distribution of program completions over a predefined time horizon, which may be months, quarters, or years. It should be noted that only programs selected through the portfolio management process are included in the program road map.

Developing the Program Road Map

Step 1: Prepare information inputs: Developing the program road map requires accurate information about the following:

- Program portfolio prioritization and balancing
- Product, service, or infrastructure segmentation scheme
- Estimated program launch dates

The results of the portfolio management selection and balancing process should be complete to accurately depict the programs that have been approved for planning and execution. A segmentation scheme by product, service, or infrastructure type (or by market segment) adds clarity to the program road map structure. Estimated program launch dates for product, service, or infrastructure availability drives the timeline for program completion dates.

Step 2: Assess market and customer need: Complete a thorough market analysis to determine the appropriate market or channel segments for

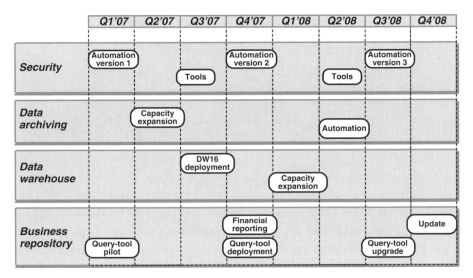

Figure 10.3   An example of the infrastructure program road map.

each of the programs on the road map and the customer expectations of availability dates for the products, services, or infrastructure capabilities to be developed.

Step 3: Utilize the portfolio analysis results: The portfolio management process yields a prioritization of programs selected over a predetermined time horizon. It is recommended that the programs are added to the program road map in similar prioritization. In other words, the highest-value programs are given priority with respect to time phasing of resources.

Step 4: Assess the organizational development capacity: This step includes an analysis and understanding of the firm's capacity in the following areas:

- Availability of adequately skilled resources
- Planned availability of existing and new technologies
- Availability of funding

Most firms' development appetites normally exceed their available resources. Part of the exercise of developing the program road map is to balance the timing of approved programs with the available resources so that the road map reflects what is possible and practical.

Step 5: Finalize time phasing of programs: Based upon the organization's development capacity, market or customer demand, and portfolio management prioritization, adjust the targeted timeline for each program on the road map accordingly. It should be noted that this is many times an iterative process.

Utilizing the Program Road Map

The program road map tool is dynamic in nature and is prepared or updated in conjunction with the portfolio management process. Once the portfolio of programs has been updated, any programs added or removed from the portfolio are also added or removed from the road map in similar fashion. Program timeline targets are then adjusted accordingly. The road map represents the best and current view of the firm's product, service, or infrastructure schedule targets.

Time to Prepare

The estimated time to prepare the program road map depends upon the number of programs undertaken by the firm and the complexity of the markets served. Preparation time can range from a few hours to a few days.

Benefits

This tool provides an organized and disciplined approach to characterize a firm's development strategies by balancing the timing of program demands with the availability of resources to maintain reality for what is practical and achievable. Additionally, it provides both senior and program managers with a view of the portfolio of programs over time. Also, this tool provides the completion and delivery targets from the market and business perspectives to help guide the program planning process. And, finally, the tool shows the grouping of programs by market segment or product, service, or infrastructure type to facilitate more effective resource allocation across the programs.

Advantages and Disadvantages

The program road map is visual and easy to read. However, developing and updating the program road map requires discipline that many organizations lack to keep the information current and relevant.

This section describes the program road map as an organized and disciplined approach to illustrate the time phasing of the programs within a portfolio. It is an effective tool for senior managers and program managers alike. For senior management, the tool provides a means to balance the market needs for products, services, or infrastructure capabilities with availability of development capacity over a predetermined time period. The program road map also provides target delivery dates that program managers can use to guide their program planning activities and end results.

Program Road Map Highlights

If the program road map is properly constructed and used, it should show the following:

- The grouping of programs into product, service, or infrastructure type, or into market and customer segments, on the vertical axis
- The distribution of program completions over a predefined time horizon on the horizontal axis
- Only programs selected through the portfolio management process

---

PROGRAM COMPLEXITY ASSESSMENT

The program complexity assessment tool helps determine how a new program stands on multiple dimensions of complexity. To accomplish this purpose, the structure of the program complexity assessment tool features several parts. First, the tool includes the dimensions of complexity that reflect the program management strategy of the company from multiple perspectives, including program business environment, program strategy, and program characteristics. Each dimension of complexity is assessed on the anchor scale (second part), and when the complexity scores of each dimension are connected, a line called the complexity profile (third part) is obtained. The complexity profile is a graphical representation (Figure 10.4) of a program's multifaceted complexity used by management to make several major decisions, as described in the following benefits section.

Complexity dimension	Low complexity	1	2	3	4	High complexity
Business climate	Stable			X		Uncertain
Market novelty	Derivative			X		Breakthrough
Financial risk	Low		X			High
Objectives	Clear		X			Vague
Technical requirements	Clear		X			Vague
Organization	Assembly			X		Array
Technology	Low tech		X			Super hi tech
Speed to market	Regular		X			Blitz
Geography	Local				X	Global
Team members	Experienced	X				Inexperienced

Figure 10.4   An example of program complexity assessment.

Developing the Program Complexity Assessment

Step 1: Prepare information inputs: Determining realistic program complexity takes quality information about the following:

- Business environment
- Program strategy
- Program characteristics

The business environment information is necessary to establish and evaluate complexity dimensions related to the environment such as the business climate, market novelty, and financial risk of the program. Objectives and features set (part of technical requirements) are defined in the program strategy. Other dimensions such as organizational, technology and speed-to-market complexity are determined from program characteristics identified in the program business case.

Step 2: Identify complexity dimensions: What constitutes a complex program? According to Edmonds, a complex program can be defined as one whose variety and/or connection of parts or aspects have increased in multiple dimensions.[8] In particular, a program that features newer technology, faster speed to market, and total-market novelty is more complex than the one including mature technology, regular speed to market, and a known target market. In selecting the complexity dimensions with which to assess a program, choose those that systematically characterize the programs from varying angles. As illustrated in Figure 10.4, the dimensions should also reflect the company's business environment and strategy.

Step 3: Define complexity scales: First, choose a scale for each dimension. Alternatives for the scales abound; in Figure 10.4 we chose a scale in which the complexity for each dimension is measured on a simple four-level scale (1 is the lowest complexity and 4 is the highest complexity). Next, define anchor statements for each level of the scale. Anchor statements help build consistency when assessing the complexity for each dimension (see boxes titled, "Some Anchor Statements May Be More Detailed" and "Some Anchor Statements May Be Less Detailed"). Without anchor statements, each assessor may evaluate the levels differently, leading to inconsistent complexity evaluations. Well-defined anchor statements help to ensure all assessors approach the scale for each complexity dimension from a consistent frame of reference.

Step 4: Assess complexity: Relying on quality information inputs and using the established scales, decision makers are able to assess the complexity of each dimension. For example, in Figure 10.4, speed to market is assessed as a level-2 complexity, (fast and competitive). Once all complexity dimensions are assessed, connect the obtained scores for each dimension to produce the complexity profile, which helps to visually depict the overall program complexity. The profile in Figure 10.4, for instance, shows that the program is of medium complexity, with all dimensions at levels 2 and 3, except team members who are experienced (the least complex). The profile is important in making some decisions described in the following Benefits section.

Some Anchor Statements May Be More Detailed
Here's an example of detailed anchor statements for the technology complexity scale.[9]

Level 1: Low tech: The program uses technologies that at program initiation were well established both in the industry and the company.

Level 2: Medium tech: The program uses technologies that at program initiation were new to the company but existed in the industry.

Level 3: High tech: The program used technologies that at program initiation were all or mostly new to the company but existed in the industry.

Level 4: Super high tech: The program used technologies that at program initiation were nonexistent in the industry and the company.

Some Anchor Statements May Be Less Detailed
Here's an example of less-detailed anchor statements for the speed-to-market complexity scale.[10]

Level 1: Regular: Delays are not critical.

Level 2: Fast competitive: Time to market is important for business success.

Level 3: Time critical: Completion time is crucial for meeting the window of opportunity.

Level 4: Blitz: Crisis program in which immediate solution is necessary.

Utilizing the Program Complexity Assessment

Typically, the program complexity assessment tool is prepared during the program define and plan phases of the PLC to serve purposes discussed in the benefits section that follows. However, this tool should be utilized dynamically and updated periodically in high-velocity environments where the program scope and business climate may frequently change. It is crucial that the senior management team of an organization who manages the portfolio of programs uses this tool to approve the overall level of complexity for each program.

Time to Prepare

Once the information inputs are defined, it typically takes less that one hour to create and utilize the program complexity assessment tool. If you complete the complexity evaluation without the basic knowledge of the program and business environment, it will require significant time on the part of the assessors. It is highly recommended that the informational inputs be collected prior to assessing a program's level of complexity.

Benefits

The program complexity assessment tool's value is multifold. First, knowing the program complexity helps balance the portfolio of programs with a mix of low- and high-complexity programs. Further, the complexity assessment aids in the planning process, indicating how to adapt one's management style to the level of complexity of the program. For example, a high-tech program that experiences scope freeze in the third quarter of the development cycle will require a rigorous change management process due to the high number of interdependencies that a change may affect. The program complexity assessment tool also helps the senior management team determine the level of skill and experience needed for the program manager and the key leadership positions on the team to successfully define and execute the program. Additionally, the tool may influence how much contingency buffer to build into the program budget and schedule—the more complex the program, the bigger the buffer. Finally, the tool may help identify the categories of risk and the level of robustness you will need in your risk management plan.

Advantages and Disadvantages

As an advantage, the program complexity assessment tool is easy to read. The senior management team and program manager can quickly get a feel for the level of complexity of each program within the organization. However, when there are too many complexity dimensions identified, or the scales are too vague, the tool can become unreliable and disadvantageous.

The program complexity assessment tool helps evaluate complexity of the program from multiple dimensions, spanning from program business environment to program strategy and to program characteristics.

Understanding the level of complexity of a program helps to balance the portfolio of programs from a complexity perspective; aids in the determination of the skill set and experience level required of the PCT; guides the implementation of key program processes such as change management, risk management, and contingency reserve determination; and helps the program manager adapt his or her management style to the level of complexity of the program.

Program Complexity Assessment Highlights

If the program complexity assessment tool is properly structured and used, it should include the following:

- Program complexity dimensions
- Anchor scales and statements for each dimension
- The program complexity profile—a line connecting scores for each dimension

---

SITUATIONAL USE

Details about situations in which to use each of the tools are described in Table 10.1. Specifically, we've identified several program situations and marked how each situation favors the use of the tools. (If additional situations are needed, brainstorm to identify them). Indicate which tools will support each program situation and then choose between the options.

IMPROVING BUSINESS RESULTS

Companies use strategic tools to support achievement of program deliverables, but the ultimate essence of using them is to help improve business results related to programs. Let's see how every tool from this chapter can be employed to improve results. This will help us better comprehend why program management elements' such as strategic tools' are so important for good business.

Consider the immense value of the information that the business case provides. First, information about how a program helps the company meet its business and strategic goals; second, information about alignment between strategic goals and program execution based upon multiple business perspectives; third, information about evaluation of programs within the portfolio on a consistent set of criteria; and fourth, information about the program vision to effectively plan, execute, and deliver the output of the program. Simply, the better the information contained within the program business case, the easier it is to determine if a program will help achieve the business results intended.

Table 10.1   A Summary Comparison of Strategic Program Management Tools

Situation	Favoring Program business case	Favoring Program alignment matrix	Favoring Program portfolio maps	Favoring Program road map	Favoring complexity
Show future company business strategy			✓	✓	
Show program launch timeline				✓	
Show program business and marketing viability	✓				
Show program technical feasibility	✓				
Show program aligned with business strategy	✓	✓			
Display program portfolio visually		✓	✓		
Show skill level required to manage the program				✓	✓
Need to adapt program management style					✓
Indicate the level of contingency reserves					✓
Indicate robustness of the risk management plan					✓

The alignment matrix enables senior management to select the most valuable programs that are risk balanced and are also aligned with the strategic objectives of the organization. Business results will be improved if the most valuable programs are chosen.

The program road map provides an organized and disciplined approach to characterize a firm's development strategies by balancing the timing of program demands with the availability of resources to maintain reality for what is practical and achievable. Additionally, it provides both senior and program managers a view of the portfolio of programs over time. Also, this tool provides the completion and delivery targets from the market and business perspectives to help guide the program planning process. Finally, the tool shows the grouping of programs by market segment or product, service, or infrastructure type to facilitate more effective resource allocation across the programs.

The basic value of the portfolio map is its ability to act as a meaningful information display of how well a portfolio of programs is balanced, according to multiple business dimensions. Such a visual demonstration of the program portfolio is unmatched by other tools.

A program map equips an organization with a characterization of its development strategies by balancing the timing of program demands with the availability of resources to maintain reality for what is practical and achievable. Logically, then, better program maps will provide more effective resource allocation across the programs, which leads to better business results.

The business results hinge on the following factors: (1) How well balanced the portfolio of programs is with a mix of low- and high-complexity programs, (2) knowing the level of skill and experience needed for the program manager, and (3) in what way to adapt one's management style to the level of complexity of the program. We can fairly state that higher-quality factors mean better business results. The quality of all three factors is determined by the quality of the program complexity assessment; therefore, we can conclude that the assessment can improve the business results.

SUMMARY

The strategic program management tools enable program managers to comprehend the nature and position of their program in the context of all current and future programs in relation to a company's business strategy.

Each of the five tools is designed with a distinct purpose. The program business case is used to clearly define the program and program output, test the feasibility of the program against the business objectives and operating environment, and define the program success criteria. On the basis of the program business case feasibility, a program gets funded, is checked for alignment with the business strategy by means of the program alignment matrix, and then added to the portfolio of programs. The program portfolio map is utilized to evaluate, prioritize, and balance the portfolio of programs from multiple strategic vectors identified by the executive team of the organization. The program road map shows the time and sequence relationship of the portfolio of programs as they exit the company's development pipeline. The program complexity assessment is then utilized to gain an understanding of the skill level required of the program manager and PCT, of how to adapt the program management style to respond to the program's complexity, and to identify risks associated with the complexity detail.

The Principles of Program Management

▼ The program business case assesses the feasibility of investing in a program and obtains the means to align execution to business strategy.

▼ A program is aligned with the business strategy by means of the alignment matrix.

▼ Portfolio maps display key parameters necessary to successfully balance a portfolio of programs.

▼ The program road map provides a means to balance the market needs for programs with availability of development capacity over a predetermined time period.

▼ Program complexity assessments to help determine the skill set and experience needed by the program manager and program team.

REFERENCES

1. Cleland, David I. *Project management: Strategic design and implementation.* Blue Ridge Summit, PA: TAB BOOKS, 1990.
2. Martinelli, Russ and Jim Waddell, *"The Program Strike Zone: Beyond the Bounding Box".* Project Management World Today (March-April 2004).

3. Cooper, Robert G, et al. *Portfolio Management for New Products*, 2nd edition. Cambridge, MA: Perseus Publishing, 2001.

4. Archer, N.P. and Ghasemzadeh F. "Project Portfolio Selection Techniques: A Review and a Suggested Integration Approach". *Innovation Research Working Group*, McMaster University, (1996).

5. Benko, Cathleen and F. Warren McFarlan. *Connecting the Dots: Aligning Projects with Objectives in Unpredictable Times*. Boston, MA: Harvard Business School Press, 2003.

6. Schwalbe, Kathy, *Information Technology Project Management*, 4th edition. Cambridge, MA: Course Technology, 2005.

7. Cooper, Robert G., et al. *Portfolio Management for New Products*, 2nd edition. Cambridge, MA: Perseus Publishing, 2001.

8. Edmonds B., "What is Complexity?" in F. Heylighen and D. Aerts *The Evolution of Complexity*, Dordrecht: Kluwer, 1996).

9. Shenhar, Aaron J. "One size does not fit all projects: Exploring classical contingency domains," *Management Science*, Vol. 47, No. 3 (2001), 394–414.

10. Shenhar, Aaron J., D. Dvir, et al., "Toward a NASA-Specific Project Management Framework," *Engineering Management Journal*, (2005).

Chapter 11

Operational Program Management Tools

The topic of Chapter 11 is program management tools of the operational nature, those that program managers need on a daily or frequent basis for managing a single program in collaboration with project managers on their core team. The operational tools are enabling devices for both program and project managers to reach an objective or, more specifically, a program deliverable. As stated in Chapter 10, the tools do not make decisions; program and project managers make the decisions using information provided by the tools. The operational program management tools, just like strategic program management tools, include systematic procedures and techniques by which they provide program information.

We explain the following operational program management tools in this chapter:

- Program strike zone
- Program map
- P-I Matrix
- Program review
- Program dashboard

Most of these tools are used throughout the PLC. Two of the tools, the program dashboard and the program review, are also useful for senior management's use to evaluate multiple programs simultaneously under way in the organization.

The purpose of this chapter is to help practicing and prospective program and project managers, as well as other program stakeholders, achieve the following objectives:

- Learn how to use major program management tools of the operational nature
- Select program management tools that match their program situation

Mastering these skills is crucial to the management of a single program in creating a repeatable process of successful program definition, planning, coordination, monitoring, and control.

Note that there is a difference in the development and use of the operational and strategic program management tools. As mentioned in Chapter 10, program managers may not get involved in the development of the strategic program management tools. They use them primarily as a source of information that helps them comprehend the nature and position of their program in the context of all programs and the company's business strategy. In contrast, program managers spearhead the effort of developing and using the operational tools. However, it is worth emphasizing that the project managers on the program supply much of the information for most of the tools. This is very typical, considering that most of the program focus is on cross-project collaboration and dependencies. The case is the same in the use of the operational program management tools—the program manager and project managers jointly use the tools. Therefore, as some experts argue, knowing how to use these operational program tools is a factor of success for the program manager and project managers.

For a user-friendly presentation of the tools, we designed the chapter in similar fashion to Chapter 10, with the goal of maximum understanding of the development and utilization of each tool. Presentation of each tool will be segmented into the following sections: (1) Description of the tool, (2) developing the tool, (3) utilizing the tool, and (4) summary and highlights of the tool.

At the end of the chapter, we present a brief set of situations in which the operational program management tools are used.

PROGRAM STRIKE ZONE

The program strike zone is utilized to identify the critical success factors of a program, to help the organization track progress toward achievement of the key business results, and to set the boundaries within which a program team can successfully operate without direct management

Program strike zone			
Critical success factors	**Strike zone**		**Status**
Value proposition • Increase market share in product segment • Growth within six months of launch • Growth one-year post launch	**Target** 10% 5%	**Threshold** 5% 0%	Green
Schedule • Definition approval • Program plan approval • Initial "power on" • Validation release • Product launch	March 15, 2008 June 15, 2008 October 1, 2008 March 1, 2009 June 30, 2009	March 30, 2008 June 30, 2008 November 1, 2008 April 7, 2009 August 15, 2009	Red
Resources • Team staffing commitments complete • Staffing gaps	June 30, 2008 All project teams staffed at minimum level	July 15, 2008 No critical path resource gaps	Yellow
Technology • Technology identification complete • Core technology development complete	April 30, 2008 Priority 1 and 2 technologies delivered at Alpha	May 15, 2008 Priority 1 technologies delivered at Alpha	Green
Financials • Program budget • Product cost • Profitability index	100% of Plan $8500 2.0	105% of Plan $8900 1.8	Green

Figure 11.1   An example program strike zone.

involvement. It is an effective tool for ensuring the program is planned with the correct set of success criteria and that the program stays within those boundaries throughout the PLC. As shown in Figure 11.1, elements of the program strike zone include the critical success factors for the program, target and control (threshold) limits, and a high-level status indicator. The thresholds also serve as the dividing lines between program team empowerment and executive management intervention.[1] A green status indicator signifies progress is as planned, yellow status indicates a heads up to management of a potential problem, and red requires management intervention to proceed.

Bill Shaley, a senior program manager for a leading telecom company, described the program culture within his company this way: "Managing a program is like having a rocket strapped to your back with roller skates on your feet; there's no mechanism for stopping when you're in trouble." Sound familiar? The program strike zone is such a mechanism that is designed to stop a program, either temporarily or permanently, if the negotiated threshold limits are breached. At which point, the program is evaluated for termination or reset for continuation.

Developing the Program Strike Zone

Developing an effective program strike zone is a critical activity for ensuring that the program manager, executive management, and program team all understand and agree upon the critical success factors of the program. It is also critical for establishing the boundary conditions that will drive effective decision making on the program. Developing the program strike zone consists of the following steps:

1. Prepare information inputs
2. Identify critical success factors
3. Set the target and threshold values
4. Negotiate and ratify the final target and threshold values
5. Adjust the program strike zone as needed

Step 1: Prepare information inputs: Defining a meaningful program strike zone requires quality information from the following sources:

- Approved business case and objectives
- Program road maps
- The team's capabilities, experience, and past track record
- The program's complexities and risks

The initial set of success factors are derived directly from the approved business case and program road maps. To be able to establish and later negotiate the control limits for each success factor with the program manager, the GM needs to know the program team's capabilities, experience, and past track record. Then he or she must use that information to balance thresholds against the new program's complexities and risks.

Step 2: Identify critical success factors: Identification of the critical success factors begins during the define phase of the PLC. The factors represent a subset of the metrics normally tracked by a program team. The program strike zone should include only the few critical success factors that represent business objectives. Each organization will have a unique set of factors, based upon its products or services, and way of measuring success. The factors deemed as "must haves" often include market objectives; financial and schedule targets; value proposition of the program output (for example, product, service, or infrastructure capability); and technology objectives.

Step 3: Set the target and threshold values: The target and threshold control limits shown in Figure 11.1 form the strike zone of success for each critical success factor. The target value for a critical success factor is based on the program business case and baseline plan. The threshold value represents the upper or lower limit of success for a particular critical success factor.

Step 4: Negotiate and ratify the final target and threshold values: Once the program manager and PCT establish the recommended target and threshold values for each critical success factor, the program manager presents the information to the senior executive sponsoring the program. Based upon the complexity and risk level of the program, and upon the capability and track record of the program manager and team, the executive sponsor will adjust the values accordingly. For example, on a program that is low complexity, low risk, and is being managed by an experienced program team, the range between target and threshold values will be opened up to allow for a higher degree of decision-making empowerment for the program manager. Conversely, on a program that is of higher complexity, risk, or is being managed by an inexperienced program manager and team, the range between target and threshold values will be tightened-up to limit the decision-making empowerment of the program manager and team. Once the targets and boundaries are negotiated, the team is empowered to move rapidly, as long as they do not violate one of the strike zone threshold values.

Step 5: Adjust the program strike zone: The program strike zone is dynamic in nature and may be adjusted or changed as the business or environment evolves during the life of the program. As stated earlier, the program strike zone is initiated in the define phase to align the program business case to the critical success factors that represent key business objectives. During the planning stage, it is expanded to contain a full set of critical success factors that align the program plan to business objectives. During the implementation phase, the program strike zone is utilized to guide the decisions of the program team and to gauge overall program progress. Finally, during program closure, it is utilized to determine program completion and success. Throughout the PLC, changes to the operational or business environment may require alterations to the data elements in the program strike zone.

Utilizing the Program Strike Zone

The program strike zone is used by executive managers and program managers. It establishes executive management's role in the program decision-making process and provides a catalyst to keep them actively engaged in the success of the effort. Executive managers utilize the program strike zone as a "no-surprises" tool to ensure that a new program's definition supports business objectives; it is also used to establish control limits and check that the program team's capabilities and performance are in balance with the complexity of the process. As long as the program progresses within the strike zone of each critical success factor, the program is considered on target and the program manager remains empowered to manage the program through its life cycle. However, if the program does not progress within the strike zone of each critical success factor, the program is not considered on target and the executive managers must directly intervene. The executives can either reset the critical success factor target or thresholds; modify the scope of the program to bring it within the current targets; or, in the worst case, cancel the program to prevent further investment of resources.

Program managers utilize the program strike zone to establish the program vision based upon the critical success factors for the organization; to negotiate and establish their empowerment boundaries with executive management; to communicate overall program progress and success; and to facilitate various trade-off decisions throughout the PLC (see the box titled, "Using the Program Strike Zone"). As indicated in Figure 11.1, the structure of the tool allows the program manager to communicate program progress with a dashboard or status-light reporting method that quickly summarizes program status for executive management.

Using the Program Strike Zone

For an example of how the program strike zone is used, refer back to Figure 11.1. A product manager, who is part of the PCT, has submitted a product feature change request that she recently received from a key customer. The customer has stated that the feature change is a "must have" for his company to commit to the number of products that they have agreed to purchase upon product launch.

After evaluation of the impact of the change by the program change control board, it was determined that the "initial power-on" threshold date of November 1, 2008 would be surpassed by a week. The program has now moved outside the strike zone for this critical success factor and status is changed to *red*. The program manager is no longer empowered to make the decision to include the feature change and must escalate the issue to executive management.

Executive management must now intervene in the program to decide if the power-on date should be extended to accommodate the request, thus, changing the control limits of the strike zone; refuse the request and hold to the current schedule; or take other actions to attempt to preserve the November 1, 2008 date, such as remove other less desirable features from the product.

In this example, senior management decided to accept the requested change, adjust all threshold dates for the schedule criteria, and alter the program plan accordingly.

Time to Prepare

Development of a program strike zone can be accomplished within an hour, provided the information inputs are available. Establishing the critical success factors and negotiating the threshold values can be more time consuming and easily take a full workday to complete.

Benefits

The program strike zone is an important communication and management tool that helps program managers keep a program aligned with the business objectives as it progresses throughout the PLC. Its major value is setting and communicating the vision for the program team by identifying the elements of program success as gauged by a business unit. The program strike zone also fosters a "no surprises to management" behavior by increasing the flow and relevance of information between the program team and executive management. This results in an efficient means of elevating critical issues and barriers to success for rapid decision making and resolution. Finally, the program strike zone provides a mechanism for stopping a program, if it is no longer aligned to the objectives of the company.

Advantages and Disadvantages

The program strike zone is user-friendly in terms of its simplicity and graphical appeal for its high-visual power. It is flexible from a usage standpoint, in which both senior managers and program managers consider it a useful tool. A disadvantage may arise when too many critical success factors are fed into it, causing the tool to lose its simplicity.

A challenge historically plaguing business management has been the ability to convert comprehensive strategic objectives into effective execution that delivers tangible results. The program strike zone bridges the chasm that can exist between strategic planning efforts, program planning, and execution output within an organization. If implemented properly, the program strike zone will help to keep senior management and the program team focused on issues critical to the success of the program. This tool is used to identify the critical business success factors of a program, help the program and executive managers track progress toward achievement of the key business results desired, and set the boundaries within which a program team can operate without direct senior management involvement.

Program Strike Zone Highlights
Check to make sure that the program strike zone is properly structured and used; it should include the following:

- Target values for the critical success factors.
- Threshold values for the critical success factors.
- Status indicators.
- Information that was negotiated between the general manager and program manager.

PROGRAM MAP

The program map is a tool that indicates critical cross-project dependencies with respect to time (Figure 11.2). In particular, the map shows the critical deliverables of each project team throughout the PLC. It then illustrates through the use of arrows the project teams that are dependent upon each deliverable. This mapping of deliverables from one team to another helps the program team to determine and fully understand the cross-project interdependencies that exist on the program. As an example, in Figure 11.2, power control SW is a deliverable generated by the software development project team, which is then delivered to the enclosure project team for further development. In turn,

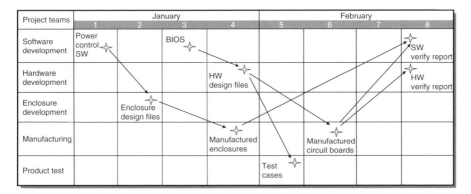

Figure 11.2   An example of a partial program map.

the enclosure design files deliverable is generated by the enclosure development project team and delivered to the manufacturing project team. This mapping of deliverables shows the critical interdependency between the software, enclosure, and manufacturing project teams of the program.

Developing the Program Map

The crux of the program map is to show the big-picture view of the cross-project dependencies throughout the PLC, one that will serve as the critical target for the program deliverables. However, this big picture hinges on the knowledge of many details and the application of the following steps:

1. Prepare information inputs
2. Identify project deliverables
3. Map the deliverables
4. Establish cross-project dependencies

Step 1: Prepare information inputs: Successful mapping of cross-project deliverables is dependent upon the availability and full understanding of the following program information:

- Key elements of the program strategy and requirements
- Project managers' inputs
- Knowledge of the program technology

Clearly one needs to understand the program strategy and the details of the product, service, or infrastructure requirements in terms of the program feature set; program scope (preliminary- or mature-program WBS); and program goals (for example, schedule goals) to be able to identify the projects that make up the program and their critical deliverables. The project managers, in turn, use this information to identify their teams' deliverables and estimate the time and resources required to develop each deliverable. In addition, to fully understand the definition and work to generate a deliverable, a firm understanding of the technology employed on the program is also a critical input to the mapping process.

Step 2: Identify project deliverables: Using the information contained in the program WBS and the detailed requirements, each project team identifies its primary deliverables to the program. Obviously, this is a case in which the quality of the requirements and the WBS will affect this step. Poor requirements will lead to poor deliverable identification, which will ultimately lead to a poor plan and program execution. The message here is to take time to gather and analyze the program requirements *before* the program map is created, or be prepared to repeat the exercise at a later date. For each deliverable, document the critical information needed for the mapping process—deliverable name, time to develop the deliverable, and dependencies required to complete the deliverable.

Step 3: Map the deliverables: We have found that the most effective method for creating a program map is to use a team workshop format that is led by an independent facilitator. Necessary participants include the program manager and the PCT. It is important that the program manager does not take on the facilitation duties. It is best if he or she is a full participant in the activities and discussion during the workshop. We do recommend, however, that the program manager be prepared to supply plenty of coffee, donuts, pizza, and maybe dinner to keep the team engaged and willing to participate in a potentially lengthy exercise.

During the mapping session, utilize a large work area, such as a white board or wall, to create swim lanes along the horizontal axis that represent the various project teams on the program. Then create vertical partitions that represent the lowest level of schedule tracking that will take place on the program (day, weeks, months, or quarters), as shown in Figure 11.2. On the appropriate dates, enter the deliverables for each project team into the swim lanes of the program map. Using the example shown in Figure 11.2, the BIOS is a software project deliverable due in week three, and the manufactured circuit boards is a manufacturing project deliverable due in week six. Continue the mapping exercise until all program deliverables are entered on the program map.

Step 4: Establish the cross-project dependencies: This is the crucial time when the team sequences the deliverables and sets up the interdependencies among them. Essentially, cross-project dependencies will determine the sequence of development, interfaces, and responsibilities, as well as the initial timeline. For example, in Figure 11.2, the arrow from enclosure design files to manufacturing means the enclosure design files are delivered to the manufacturing team in week two. Multiple rounds of iteration are normally needed to refine the deliverables, interdependencies, and delivery dates. As a note of caution, once a timeline is established through the mapping process, it should not be considered the program schedule. At best, this exercise yields a schedule with a 50-percent confidence level. This should only be considered a first-pass timeline for the program (see Chapter 8).

Utilizing the Program Map

For any program, small or very large, the program map is highly recommended to fully understand the deliverables and interdependencies between the project teams on the program. The best time to create the program map is during the planning phase of the program, after the program scope and requirements are well defined. To repeat, the team workshop approach for creating the program map yields information-rich exchange, cross-project discussion, face-to-face interactions, and reciprocal iterations; it also has no equal when it comes to developing a quality program map.

Time to Prepare

If the facilitated workshop approach is used to develop the map, one workday is a typical time duration. However, large programs can take multiple days. If the facilitated workshop method is not utilized, allow a full work week to complete a program map for small- to medium-sized programs and multiple work weeks for large programs.

Benefits

The value of the program map lies in showing cross-project dependencies and enables benefits that stem from knowing the dependencies. These enabling benefits include a clear sequence of the program implementation, established commitments for each project team, and critical interactions between PCT members.

Advantages and Disadvantages

As an advantage, the program map gives a clear visual representation of the cross-project deliverables and dependencies. As stated earlier in the book, a critical role of the program manager involves managing the cross-project dependencies on the program; the program map, therefore, establishes his or her plan for management of those dependencies. Creating and revising the program map can be a large investment of program resources; thus, it could be disadvantageous if the prerequisite information is not available and reasonably stable prior to creating the mapping process.

The focus of this section is the program map, which indicates critical cross-project dependencies over the program timeline. This tool helps establish a clear sequence of the program implementation, commitments, interactions, and responsibilities of critical team members.

Program Map Highlights

To ensure that the program map is appropriately constructed, include the following:

- Individual projects and their deliverables
- Program timeline
- Cross-project dependencies and connecting project deliverables

---

P-I MATRIX

The program P-I matrix is a tool used to identify program risks, assess their probability and potential impact, and provide a representation of risk severity to facilitate effective program decision making (Figure 11.3). The matrix is usually divided into red, yellow, and green zones, which

Probability (P)	Risk score = P + 2 x I				
Nearly certain = 5	7	9	11	13	15
Highly likely = 4	6	8	10	12	14
Likely = 3	5	7	9	11	13
Low likelihood = 2	4	6	8	10	12
Very unlikely = 1	3	5	7	9	11
	Very low = 1	Low = 2	Medium = 3	High = 2	Very high = 1
	Impact(I)				

High severity
Medium severity
Low severity

Figure 11.3   Sample P-I Matrix dividing risks into low, medium, and high severity.

represent high-, medium-, and low-severity risks respectively, based on the organization's thresholds for risk severity. The position of a risk in the matrix determines its ranking or severity. The higher the value of a square, the higher is its rank and the more severe is its impact. To be effective, the matrix must be realistic (as to the severity of the risk), timely, and brought to life by the joint effort of the program and project managers who own the program matrix.

Developing the P-I Matrix

Making decisions is perhaps the toughest job for program managers and the PCT. Many times decisions are made with incomplete information and uncertain outcomes, which is one reason why decision making can be so difficult. The P-I matrix is a tool that can be developed and utilized to provide a more complete representation of information in support of program decision makers. Development of the P-I matrix involves the following steps:

1. Prepare information inputs
2. Identify program risks
3. Assess risks
4. Establish risk response strategies

Step 1: Prepare information inputs: Ideally, preparation of the P-I matrix should start with the following rigorous inputs: [5]

- Risk management policies
- Program planning outputs
- Historical information
- Project managers' risk assessments

Risk management policies provide recommended methods for dealing with risk throughout the program's life. The second input, program planning outputs, include performance baselines such as program goals and specification baselines, which is what is at risk on a program. Having full knowledge of the baselines is crucial in developing the P-I matrix. Because risk management tends to be data intensive, reliable historical information such as past program records, postmortems, or published sources (for example, benchmarking studies) is vital. It is also important for all project managers to express their assessment of cross-project risks because these risks make up a large portion of the P-I matrix (see Chapter 8).

Step 2: Identify risks: The purpose of this step is to identify all the potential risks that may significantly influence the success of the program. As discussed in Chapter 8, program risks are those that either jeopardize the critical success factors of the program or affect multiple projects. The latter type of risk is referred to as cross-project risks. Risk events that are contained within a single project are referred to as project-only risks, and should not be included in the program P-I matrix.

It is crucial that the program manager and the project managers collaborate to identify the program-level risks with which to populate the program P-I matrix. To begin this exercise, the team reviews each project-risk list to determine which of the identified project risks need to be elevated to the program level. A brainstorming exercise by the PCT is recommended to identify any other program-level risks with which to populate the program P-I matrix.

Step 3: Assess risks: At this stage of the risk management process, it's likely that a large number of risks have been identified, depending upon the type of program. The challenge for the program manager is to identify those that have both the highest impact on the program and those that are most likely to occur. Therefore, the impact, probability, and severity of each risk needs to be analyzed. In this assessment, we tend to use a nonnumeric probability scale. For example, on a five-level scale, the ratings are $1 =$ very unlikely, $2 =$ low likelihood, $3 =$ likely, $4 =$ highly likely, and $5 =$ nearly certain. [2] Consequently, one will qualitatively assess each risk's probability of occurrence. When this is completed, the next step is to assess the potential impact of each risk, again on a discrete scale. One example is a scale rating impact, such as $1 =$ very low, $2 =$ low, $3 =$ medium, $4 =$ high, and $5 =$ very high. To illustrate its use, let's assume a risk that will be assessed has a simple program schedule slip impact. The scale can define the levels of impact, as shown in Table 11.1.

After all risks are assessed in this manner, it is time to use a formula to combine the probability and impact of each risk to establish a measure of severity. Although nonlinear formulas can be employed, linear formulas such as $severity = probability + N$ x $impact$ are easier to apply. For example, n can be equal to two, meaning that impact is twice as important as probability in establishing risk severity. In this case, the assessed probability and impact for each risk would be entered in the formula $severity = probability + 2$ x $impact$ and the obtained value input into the P-I matrix, which consists of 5×5 squares. This is the formula utilized for the risk severity calculation in Figure 11.3.

Some larger programs commonly focus on the top ten highest-ranked risks. In contrast, some smaller programs decide to manage the top three risks, arguing the lack of resources prohibit taking on a larger number of risks. Both approaches may be dangerous. So, what is a reasonable way out? The answer is in the P-I matrix—respond to the highest-ranked risks in the matrix, down to an agreed level. [3] For example, focus on handling risks down to risk score of 11 (see Figure 11.3) and treat other risks as noncritical. With this approach, one neither squanders resources

Table 11.1   An Example of a Five-Level Scale that Rates the Risk Impact on a Schedule.

Scale	1	2	3	4	5
	Very low	*Low*	*Medium*	*High*	*Very high*
Risk Impact on Schedule	Slight schedule delay	Overall program delay 5 %	Overall program delay 5–14%	Overall program delay 15–25 %	Overall program % delay 25

nor disregards significant risks. It should be noted that noncritical does not mean not important. Rather, it means that scarce program resources are not immediately needed to address the risk event but may be needed in the future.

Step 4: Establish risk response strategies: Risk response is not normally part of the P-I matrix development; however, developing the P-I matrix only makes sense if risk response planning is performed, and culminated into its most crucial action—determining the preventive action, trigger point, and contingent action for the critical risk events. The preventive action is the primary for strategy risk response, or plan A. When executed, however, it may or may not work. The point at which we establish that the primary strategy doesn't work is the trigger point. At that time, the backup strategy or contingency action, plan B, is taken to counter the risk.

Utilizing the P-I Matrix

Not surprisingly, the dominant mode of using the matrix is normally informal in practice. However, best practices have shown that it is advantageous to first develop the P-I matrix in the define phase of the PLC and present it as part of, or in conjunction with, the program business case. The P-I matrix should then be updated through the planning phase and utilized as part of the program plan. Periodic reevaluation of the matrix throughout the implementation phase of the program is also a best practice. The information contained in the P-I matrix is crucial to review during the program launch approval meeting (see Chapter 7). Although at times it may seem to be overly simplistic for large and complex programs, the P-I matrix is abundantly applied in these programs focused on the larger number of highest-ranked risk events, with more formality.

Time to Prepare

Many teams running small and simple programs can expect to expend one to a few hours of their time to develop the P-I matrix. This time proportionately rises as programs get bigger and more complex. Tens of hours may be necessary for a team in charge of a large and complex program to devise the matrix. However, periodic updates can be accomplished as part of the regular PCT meetings and require a minimal amount of time.

Benefits

The P-I matrix helps sift through the myriad of uncertainties and pinpoint and highlight the program areas of highest risk—both before work has begun and throughout the program. [4] This offers an opportunity to focus program resources and to identify effective ways of reducing risk events in a proactive manner, rather than being confronted by them if they turn into problems later in the program. In addition, the matrix generates information for more reasonable contingency planning.

Advantages and Disadvantages

The matrix's major advantages are visual presentation of risk information and simplicity. If you eliminate the advanced statistics, its quantitative part is still simple and adequate. But the matrix's major focus is on individual risk events, and it does not provide a reliable mechanism to deal with interacting risk events.

The P-I matrix helps pinpoint and highlight the program areas of highest risk and identify effective ways of reducing those risks in a proactive manner. It also enables more reasonable contingency planning and provides an early warning of risk.

P-I Matrix Highlights

Check to make sure that the program P-I Matrix is appropriately structured, including the following:

- Identified risks
- Risk probability and impact
- A method of prioritizing risks by severity

  Use the matrix to determine the following:

- Preventive action, trigger points, and contingent action
- Name of the risk owner

PROGRAM REVIEW

A program review is a formal event in which senior management evaluates the status and progress of a program. The program review can be used in many ways. From the individual program perspective, the program review can be used to assess a major developmental milestone as part of the traditional phase/gate process such as a business-case approval milestone or engineering release milestone. It could also be used to request management's attention to address a significant barrier outside the control of the program team. Finally, a program review can be used as a normal operational status and progress update to management. Senior management can use the program review format to evaluate the status and progress of all programs currently under way within the organization.

Developing the Program Review

Communication between the program manager and the senior management team is a critical element of program success. Early and often communication with senior management ensures they are actively involved in the program. The program review is an effective tool for facilitating good communication between the program manager and senior managers in the enterprise. Development of the program review involves the following steps:

1. Identify the type of program review
2. Determine the purpose of the review
3. Determine the attendees
4. Create the agenda
5. Conduct the meeting
6. Document the program review meeting results

Step 1: Identify the type of program review: Program reviews are most often defined as part of an organization's development process and can be several types, as follows: a formal PLC decision checkpoint, a progress and status update, or an unplanned review due to issues needing immediate attention. Formal PLC reviews are critical checkpoints in a program in which senior management approval is needed to progress from one phase of development to the next. Progress and status reviews are used to keep senior and functional mangers informed of program status. Depending upon the number and frequency of formal phase/gate reviews, a company may or may not employ progress and status reviews. As an example, Daimler-Chrysler's phase/gate process has ten formal gate reviews. Due to this high number of decision checkpoints, the program team is free to operate without progress and status reviews between gates. [5] By contrast, Intel has only four formal gate reviews in their PLC process and employs monthly progress and status reviews between gates. [6] Unplanned reviews are required when the program team encounters problems or barriers to progress and requires senior management intervention to proceed.

Step 2: Determine the purpose of the review: One of the biggest complaints about most organizations is that there are too many meetings that waste time and reduce worker productivity. One of the easiest ways to combat unnecessary program reviews is to clearly determine the purpose of the meeting. As noted in the first step, the type of program review should be identified to drive the overall purpose of the review. For example, the purpose of a formal decision-checkpoint review will be to make decisions and obtain authorization from senior management. The purpose of an unplanned review may be to solve a problem or choose between multiple courses of action. Once the overall purpose of the meeting is determined, a clear statement of the review objectives and desired outcomes should be established.

Step 3: Determine the attendees: An important part of the program review, as well as the objectives and desired outcomes, is to determine the attendees. Careful selection of the presenters, reviewers, and decision makers will go a long way toward achieving a successful program review. Generally speaking, the smaller the audience and the fewer the number of presenters, the better.

Step 4: Create the agenda: An effective agenda is the means to achieve the objectives and desired outcome of the program review. The agenda should consist of a list of activities that needs to take place in the review to achieve the outcome. It is a good idea to include the type of action (for example, decision, authorization, action assigned to someone) needed for each agenda item. Most organizations use a predetermined agenda for formal PLC program reviews (see box titled, "Predetermined Product Program Review Agenda) and less formal agendas for progress and status reviews. Agendas for unplanned program reviews need to be created based upon the problem, barrier, or decision driving the need for the review.

Predetermined Product Program Review Agenda

Tektronix a leading manufacturer of electronic test and measurement equipment, followed has a preset and consistent agenda that involves an overview of the entire product road map within a business unit, overview of the market, and detailed status and progress report for each program. An example agenda is as follows:

Each product line reviews the following:

- Product road map (product line manager)
- market update (marketing manager)

Each program manager within a product line reviews the following:

- Summary of program status (program manager)
- Schedule update
- Program financial update
- Risks and risk-reduction activities
- Resource status
- Key customer events and interactions
- Review the program strike zone

Attendees for the program reviews may include the vice president of the business unit, his or her senior staff, the director of program management (meeting chairperson), and the program managers for each program under review.

Step 5: Conduct the meeting: Begin the program review by communicating the purpose (objectives and desired outcome) of the review and stating the agenda items. Next, clarify the participant roles, identify the decision makers, and set the ground rules for engagement. The role of the individual chairing the review is to keep the discussion on target, ensure all points of view are heard, and make sure the desired outcomes are achieved. It is highly recommended that a common format is used and consistent information is presented for each program under review. Prior to closing the program review meeting, all decisions should be noted and documented and all action items assigned—relative to what is to be accomplished by whom and by what due date.

Step 6: Document the program review meeting results: At the close of the program review meeting, it is generally the responsibility of the meeting chairperson to document the results of the meeting. This person is responsible for keeping track of all the key decisions and actions directed by senior management relative to each program. A copy of the notes of this meeting should be disseminated to all parties who are being directed to take action, as well as a copy to all interested senior managers. The meeting chairperson should also follow up on all key actions with the specific owners to check that they have appropriately accomplished their action item. Unresolved issues from the program review meeting normally are addressed at the next review. See the example below for an example of monthly program review notes.

Program Review Meeting Notes

Description/Action	Owner	Due Date
Systems Programs		
- Road map is missing the new hybrid system in year 3. Correct by next review.	Bob Lane Product Line manager	May program review
- Quantum program: current phase/planning The schedule critical path and the SW development resources are not in sync. Resolve ASAP.	Harry Vargas program manager	4/25/0x
- Apollo program: current phase/launch Product demos are late by two weeks. Attempt to resolve schedule conflict by next week.	Craig Steinbrenner program manager	4/12/0x
Communications Product Line		
- Marketing update Trade show plan for latter part of year needs to include shows in Singapore and Taiwan.	Carol Simpson Product Line marketing manager	Next program review

Description/Action	Owner	Due Date
- Brightstar program: current phase/planning Recent key competitor move in market invalidates this product proposal. Program is to be cancelled at separate meeting.	Brad Raiden program manager	TBD
- Nebula program: current phase/concept ROI for this product proposal does not reflect the contract SW labor required from India. Product proposal milestone to be delayed until financial returns are corrected.	Jean Taylor program manager	4/6/0x
General Comments		
- Gary Rost, vice president of R&D has requested that the dashboard report be distributed at least two days prior to the scheduled program review. Director of PM to resolve before next meeting.	Todd Hampton director of PM	Next program review

Utilizing the Program Review

As pointed out earlier, there are several events that may require formal program reviews. Generally, they should be used as required by the formal development process utilized by the firm, or when unplanned events occur and either senior management or program management deems that the program needs intervention by senior management.

Time to Prepare

A consistent program review format can greatly reduce the amount of preparation time prior to the review. With this in place, preparation for the review can be completed within two hours. Time to conduct the review varies from organization to organization, but a good rule of thumb is 30–40 minutes is adequate to review a single program.

Benefits

The program review facilitates the flow of program information between the program team, senior management, and key stakeholders of an organization. It keeps senior management fully involved in the development process early and often during the PLC. It helps clarify the role of senior management as reviewers of program information, decision makers at key phase transitions, and problem solvers when risks and barriers are encountered by the program manager and his or her team. Also, it distills discipline by requiring that the program manager continuously knows and is able to communicate the current state of all aspects of a program; it is also an effective tool for communicating and resolving program risks and barriers to success.

Advantages and disadvantages

The advantages of the program review include the commonality and consistency of both format and content of program status. However, without commonality and consistency of both format and content, the program review can be time consuming and an overhead burden for the program manager.

The formal program review is a useful tool for program managers to communicate program status and progress of a program, or to deal with specific unplanned events in which management intervention is necessary. The program review helps to ensure senior management of an organization is actually involved with the program and fulfilling their role as decision maker and champion.

Program Review Highlights
To ensure that the program review is effective, it should do the following:

- Clearly identify the type of review being conducted
- Have a clear purpose, intended outcome, and agenda
- Be attended by the correct stakeholders
- Have consistent format of information
- Have a documented history of decisions made and actions needed

PROGRAM DASHBOARD

The Program dashboard highlights and briefly describes the status of the program by reporting on progress toward achievement of the major business goals of the program. Colors—green, yellow, and red—are used to indicate the progress toward reaching the program goals. Green indicates the program is on track to meeting the goals, yellow indicates a goal is in jeopardy, and red indicates a goal has been compromised and immediate action is needed to recover (see Figure 11.4). The use of colors to indicate program status reminds many users of the automobile dashboard, in which color lamps indicate the status of all major car systems, hence the name of this tool. [7] A program summary that includes status of work completed, significant accomplishments, and risks and issues currently being addressed by the program team may also be shown. At times the program dashboard predicts trends and states actions required to overcome issues and reverse negative trends. In this case, the program dashboard is used to predict the future state of the program based upon past and current performance. The program dashboard can be used for individual programs and also for multiple programs under way to give senior management a quick understanding of the status of all programs within their organization. This broader use of the tool is covered under variations in a later section.

Developing the Program Dashboard

The program dashboard can be developed to show a concise overview of a program's status and, in doing so, serve as a primary communication

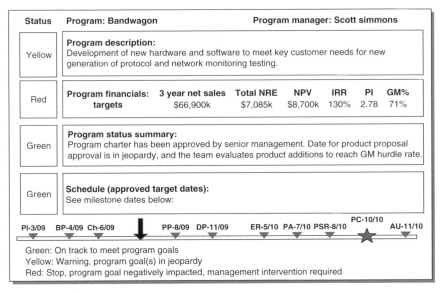

Figure 11.4   An example of a program dashboard.

device between the program manager and all program stakeholders. Development of the program dashboard involves the following steps:

1. Prepare information inputs
2. Design the dashboard
3. Determine the variance
4. Identify issues and risks
5. Predict trends
6. Specify actions

Step 1: Prepare information inputs: Producing a meaningful dashboard is dependent upon the following program information:

- Integrated program plan
- Work results and program records
- Proactive cycle of program control (PCPC). This is needed only in case of the proactive program dashboard (see Chapter 7)

The program plan with various baselines—business goals, cost, schedule, and so on—is necessary to compare actual progress of the program against planned expectations, assess performance, and indicate current status of the program. The actual state is derived from work results and other program records such as minutes of meetings and progress statements. [8]

Literally, the program dashboard's design is no more than an application of the proactive cycle of program control, a control method that includes the following five steps:[9]

1. What is the variance between the baseline and the actual program status (program status summary, program financials, schedule, and program cost)?
2. What are current issues causing the variance (program status summary)?
3. What risks may pop up in the future and how could they further derail the program (program status summary)?
4. What is the trend—the predicted completion date, budget, scope, and quality—if current issues and risks persist (program status summary)?
5. What actions should be taken to prevent the red light events from happening and deliver on the baseline plan (program summary status)?

Obviously, the majority of the questions are about the program future, attesting to our intent of giving the program dashboard a proactive nature. With all this information in hand, the PCPC can be utilized to provide the algorithm for processing information in the dashboard.

Step 2: Design the dashboard: The starting point for designing the dashboard is determining which stakeholders will receive the information communicated by the program dashboard (for example, business-unit general manager), and what information they wish to see. For example, they may use the program dashboard to measure the progress of a program toward its contribution to the company strategic goals. The stakeholders' informational needs drive the design in terms of the purpose, functionalities, frequency, responsibilities, and distribution of the program dashboard. Consistent design in the form of format and content enables comparison with current, previous, and future program dashboards for all programs.

Step 3: Determine the variance: This step determines the variance between the program baseline and the actual program status. A comparison of the baseline plan and the actual work results should easily yield the variance, which is the difference between the two. Once the variance has been determined, use a scale that translates the amount of variance into a status color. For example, a schedule variance of more than 5 percent behind the program plan yields red-light status indication. Referring to the example program dashboard in Figure 11.4, status indicators are shown for the program status summary, program financials, schedule, and cost.

Step 4: Identify issues and risks: If there is an unfavorable variance, the issues causing the variance need to be identified in the program dashboard. Additionally, the most critical risks for the program (determined from the P-I matrix) need to be identified. Current issues are meant to communicate the present problems that the program team is dealing with and what the impact is on the program. Program risks look into possible future troubles and assess the potential impact. In Figure 11.4, this would be shown in the program status summary box.

Step 5: Predict trends: What is the trend—the predicted financials, schedule, and cost—if current issues and risks persist (see box titled, "Forgetting Trend Analysis: Déjà Vu?"). Although forecasts of this type are notoriously vulnerable, their essence is less in their accuracy and more in creating early warning indications of where program results may end up if the issues and risks are not resolved and mitigated.

Forgetting Trend Analysis: Déjà Vu?
Western corporations invented quality management, forgot it, and relearned it in hard battles with their Japanese counterparts. [10] Are we seeing the same phenomenon of forgetfulness when it comes to trend analysis? The majority of program dashboards place an emphasis on historical data, leaving out trend forecasts such as predicted completion date. As experts wrote long ago, the purpose of program control is "no sudden shocks, please!". In other words, we predict trend because it is much better if we could know in advance what is going to happen, rather than just watch it happen. This makes trend analysis the single most important piece of information in program control.

Step 6: Specify actions: If trends are unfavorable, what actions should be taken to prevent them and deliver on the program baseline? With a look into the future, we need to specify corrective actions, assess the impact, and assign the owner. Along with the trend, the specification of corrective actions is perhaps the most valuable part of the dashboard.

Utilizing the Dashboard

All programs, small or large, can benefit from the use of the program dashboard to evaluate and communicate program performance. It is effective if the program dashboard is completed and presented as an element of the communication section of the program plan to establish a baseline for program status. Post-baseline plan approval, the program dashboard is the primary tool for determining and communicating program status throughout the remainder of the PLC. Pressed for resources, small programs can gain additional benefit from the program dashboard by using it as the only program status reporting tool.

Time to Prepare

One hour is normally sufficient for a program team—on a simple and small program—to prepare a program dashboard. Even as preparation time requirements increase with the size and complexity of programs, it is clear that a few hours of a large program team's time should suffice for the program dashboard production, with additional time required to develop backing data.

Benefits

If the time spent on the development of the program dashboard is viewed as an investment, then return on this investment can be lucrative in multiple ways. First, the program dashboard process ensures proactive cycle of program control. Second, the tool is a vehicle to preserve stakeholders' involvement in the program. [11] Third, the program dashboard instills discipline, takes out daily firefighting of program problems, and introduces the clear essence of program management—think, predict, act. Fourth, by covering program status on a single page, the program dashboard provides the crucial points of the program status, while enabling senior management to review performance progress and trend at a glance.

Advantages and Disadvantages

Major advantages include simplicity and visual power, enabling readers to easily glance and grasp where the program is and where it is going.

Preparation of the dashboard may be a disadvantage because it is time consuming for larger programs. This is a trade off worth taking when one is aware of the program dashboard's benefits.

Variations

A significant variation of the program dashboard used in many organizations is a dashboard targeted at multiple programs that are simultaneously under way in the organization. (see Figure 11.5). It contains many of the same elements as the individual program dashboard but is primarily focused on a quick understanding of the status of the program. Senior management can quickly determine if the program is tracking as planned or is in need of their intervention due to problems outside the span of control of the program manager and team. Generally, the summary dashboard is focused on the key success criteria, as described in the program strike zone.

The program dashboard is a tool that highlights the status of the program, predicts trend, and states actions to overcome negative trends. Several benefits may ensue. First, the process that results in the program dashboard development ensures PCPC. Also, the dashboard helps preserve stakeholders' involvement in the program and creates a historical trail and record. In addition, the program dashboard instills discipline in

Summary program dashboard				
Business unit	Product or platform	Program name	Green = Progressing to plan Yellow = Caution Red = Action needed	Action being taken
MBU	Product	Budica	Green	None
MBU	Product	Miss mercedes	Yellow	Schedule replan in process
OBU	Platform	Silverbow	Green	None
OBU	Product	TKS	Green	None
MBU	Platform	Shogun	Red Change in critical feature and launch date impacted	New plan has been submitted for approval

Figure 11.5   An example of a summary dashboard.

reporting program status. Below we highlight the key points about the dashboard.

Program Dashboard Highlights

Check to make sure that the dashboard is appropriately structured. It should include the following:

- Variance between the baseline and actual program/activity status
- Issues (or problems) currently causing the variance
- Risks that may emerge in the future and further impact the variance
- Trend resulting from the variance and risks.
- Actions required to reverse the negative trend or maintain the positive trend
- Back up data

Situational Use

Details on situations in which to use each of the tools are described in Table 11.2. Specifically, we've identified several program situations and indicated how each one favors the use of the particular tools. If they do not provide enough details to decide which tool to deploy, brainstorm to identify more situations. Indicate which tools will support each program situation and then choose between the options.

Improving Business Results

As is the case with strategic tools, operational program management tools support a specific program management deliverable. Fundamentally, everything accomplished on a program has a business purpose. Hence, the crux of employing operational tools is to improve business results related to programs.

By keeping a program aligned with the business objectives, setting and communicating the vision for the program team through the elements of program success, and fostering the concept of "no surprises to management," the program strike zone results in an efficient means of elevating critical issues and barriers for rapid decision making and resolution. This creates the climate for improved business results.

The program map with its cross-dependencies enables a clear sequence of the program implementation, established commitments for each project

Table 11.2 A Summary Comparison of Program Management Tools

Situation	Program strike zone	Program map	P-I matrix	Program review	Program dashboard
Indicate critical success factors	✓			✓	✓
Show how to achieve business results	✓			✓	✓
Indicate critical cross-project dependencies		✓			
Show implementation sequence		✓			
Need focus on most critical risks			✓		✓
Treat risks independently from one another			✓		
Show go, no go points, and major events				✓	
Provide built-in proactive approach			✓		✓
Display trend	✓				✓
Provide early warning signal	✓				✓
Short time to train how to use the tool	✓	✓	✓	✓	✓
Take little time to apply	✓		✓	✓	✓

team, and critical interactions between PCT members. For example, it is likely that target profitability for a program is not going to be achieved, if the sequence of implementation or which functional team provides what deliverables is unknown. In contrast, if the sequence and owners of deliverables are known and there is adequate management of the program, it is likely that there will be better business results.

When we use the P-I matrix it helps us sift through the myriad of uncertainties and pinpoint and highlight the program areas of highest risk. The purpose is to prevent the risks from turning into problems later in the program and diminishing the business results.

If we successfully deploy the risk plans and avoid the problems, we will accomplish better business results, such as faster time-to-money performance.

If there is a constant flow of program information between the program team, senior management, and key stakeholders of an organization, it is quite clear that the business results are likely to be better. For example, the information provided by the program review at key phase transitions or when risks and barriers are encountered by the program manager will definitely ensure better decision making and problem solving and, thus, improved business results.

The program dashboard is useful because it provides improved information of the essence of PCPC—plan, detect, predict, and act—and stakeholders' involvement in the program and the team cohesion. Therefore, the point is to have a high-quality program dashboard, to help improve business results that a program generates. A simple example is a dashboard that tracks the staffing levels for the program team. If the level is too low, and the dashboard indicates it, a triggered corrective action will improve staffing and, consequently, enable achievement of better-related results such as time to money.

Summary

The five operational program management tools—program strike zone, program map, P-I matrix, program review, and program dashboard enable the program manager to effectively define, plan, monitor, report, and control progress of individual programs.

Each of the five tools comes with a purpose. The program strike zone identifies the program's critical success factors. The program manager can use the program map to find the critical cross-project dependencies within the program that need to be coordinated, and the P-I Matrix is used to pinpoint and highlight the program areas of highest risk. When the program gets under way, the program review is employed to verify phase gates and other major program events. In between those major program events, the program dashboard helps report program status. Using all the tools as a toolset is preferred, although they may be employed individually.

The Principles of Program Management

▼ The program strike zone identifies the critical success factors and tracks progress toward achievement of the key business results.

▼ The program map indicates the critical cross-project dependencies between the project teams.

▼ The P-I matrix helps to identify the probability and potential impact of program risks.

▼ Program reviews enable senior management to evaluate a major developmental milestone, address a significant barrier outside of the control of the program team, and receive a normal status and progress update.

▼ The program dashboard highlights the status of the program and reports progress toward achievement of the major business goals of the program.

Program Management in Practice

Using Tools on a Mercedes
Sabin Srivannaboon and Dragan Z. Milosevic

Prologue

This industry example focuses on a company called RollingSys. This U.S.-based company is relatively small, with approximately 1000 employees, but is a market leader in the design and manufacturing of embedded computers. The company is highly advanced in its program management practices and provides an excellent example of the effective use of program management metrics and tools.

This example first demonstrates how business strategy should drive the selection of metrics and tools. RollingSys pursues an first-mover business strategy based upon technology innovation and fast time-to-market delivery. Priority selection of schedule performance, features, gross margin, profitability index, development expenses, manufacturing costs, market share, and resource allocation metrics support this strategy.

The alignment process of the Miss Mercedes program is also discernable. The business strategy begins with the strategic plan and continues with the selection of strategically focused tools such as the portfolio map and business case.

This example also demonstrates the use of many operational tools presented in Chapter 11. For example, the program strike zone and program map are used during program planning, and the program dashboard is used for program monitoring and control. We also present a tool used by RollingSys called the wiggle chart, which is an excellent tool that supports proactive program management practices by providing program trends.

Finally, this industry example presents the use of some common tools such as the work breakdown structure and the program schedule. However, details about these tools are not covered in this book, as they are readily available in existing project management literature.

Background

RollingSys is a privately held corporation that provides computer solutions to help customers design, build, deploy, and manage next-generation numerically controlled tool machines. These products require a lot of customization, resulting in each order being organized as a program. For this reason, the products are well known and have earned many outstanding awards.

RollingSys has seen an increase of orders from Asia and Europe, which puts a lot of pressure on their six program managers. After a long discussion, a decision was made to hire Keith Richardson, an experienced program manager. Mick Beggar, the director of the program management office (PMO) for RollingSys, prepared a plan for Richardson to transition to the new job, including familiarizing him with RollingSys's program management system. Following is part of the familiarization relating to the strategy and program management tools and metrics. Taking part in the discussion are Richardson; Beggar; and Charles Waters, a longtime program manager.

All Roads Lead to the Business Strategy

BEGGAR: I want to make something clear from the very start. RollingSys's choice and application of program management tools, like all managerial processes and actions, is driven by the business strategy. Therefore, we should first talk a bit about RollingSys's real strategy—not the company's public relations word on strategy but such things as the company's strategic uniqueness and what makes it successful.

As director of the PMO, I often interpret the strategy of RollingSys to my program managers. So, at this time, I feel obliged to put on the hat of the director of PMO and explain the business strategy. First of all, RollingSys is unique in terms of breaking the components of the business down and understanding what makes it successful and what doesn't make it successful. Our products are often recognized as the best products on the market. If you think of an airplane seat analogy, they are in the first class-section. Also, the ability of RollingSys to get customer input makes a huge difference. What our customers want to use the product for often becomes clear through the ability of the company to get engineers in front of customers. Our business is moving from the old traditional 'here is the computer we build' to "what are the features you need in the computer we are going to build?." Moreover, we have repeatability across different product lines. The repeatability allows the company to shift the program manager from one product line to another with an adequate understanding of performance criteria and his or her responsibilities.

What, then, is RollingSys's business strategy, and what is the role of program managers in the context of such strategy? RollingSys is often a first mover, so it is technology innovation and to be first in time to market. In that context, the program managers are responsible for more than just getting from program start to program finish. They are required to deliver the program on time, meet all the objectives of the program, and are responsible for business results. Therefore, you are expected, as any other program manager, to be a visible and seasoned business manager, aligning your programs with the strategy.

RICHARDSON: That having been said, can we now talk about tools? I would like you to take an example of a program and tell me how using specific tools in that program helped make it successful. Give me the background of the program.

Miss Mercedes

WATERS: I suggest using a program called Miss Mercedes as an example. The customer had all sorts of requirements, one being the program name. They said, 'In our country, a Mercedes car is the ideal of high quality. We want this program product to be of high quality, like a Mercedes car. In order for you to keep our high-quality expectations in mind at all times, let's call this program Miss Mercedes.' Basically, they wanted to have a capability added to RollingSys's existing product. RollingSys got an opportunity to win a large competitive sale in Spain, if the program was delivered on the particular date. So, RollingSys formed a team to execute this program. I was the program manager for it.

The program went from conception to completion in eight months, which is not normal. It is probably the best example of how our process works at optimum. More importantly, it brought the company several million dollars in revenue to date. In RollingSys, managers believe that the program was successful partly because it used all the tools of the program management knowledge base that's available at the company. So there is a belief that the program management system as a whole works very well here.

Selecting and Approving a Program

RICHARDSON: Can you explain what strategic tools were used in Miss Mercedes to make sure that it was well aligned with RollingSys's business goals and strategy?

BEGGAR: RollingSys uses both strategic and operational tools, divided per major program activities. But there is a word of caution here. Typically, program managers would not be involved in the development of strategic tools. Generally, executives do them. However, each program manager needs to know them well because he or she will use them in communicating with executives about program status.

The whole alignment process begins with the strategic plan and continues with portfolio maps and the business case. RollingSys's practice differs from some companies in, for example, the business case. The strategic plan drives the alignment. In other words, it is a tool or mechanism to ensure the quality of the alignment in RollingSys.

RollingSys's strategic plan usually includes the product road map, technology road map, customer-technology road map, and the business model for the next three years. It addresses things like mission, objectives, long-term strategy, market size, segmentation, competition assumptions, and market share. In particular, the product road map proposes products within the three-year time frame and addresses those currently in development. For each product, it includes start and completion dates, milestone dates, total nonrecurring expenditure, and the three-year sales forecast.

The strategic plan of RollingSys drives its formal portfolio management process, where programs are prioritized in terms of the program portfolio needed for customers and for the business. Miss Mercedes was no exception. It was a program selected from the product road map and prioritized in the portfolio process, and its implementation was sped up by the customer requirement.

Generally, executives focus on the strategic plan to analyze the growth plan and determine what the right markets are for the company, where the company is the most successful, and where the customer gets the greatest value for the products. Then, the programs are planned in alignment with the strategic plan over the next three years with their expected sales and profit dollars identified. By looking at them, we are able to see the growth from different directions, such as its existing business (extend or upgrade), new products in new markets, and pure technology transition. As a result, the importance of certain products becomes more obvious than others. Then, depending on product complexity, market pressure, and other significant factors, programs are initiated and selected into the program portfolio. We don't really use any specific tools for the selection of a program into the portfolio, but there is a lot of discussion. Once a program is selected into our portfolio of programs, we use a tool called a portfolio map to display all of our selections. Each one has different parameters on x- and y-axis, for example, net present value (x-axis). The portfolio map helps us to balance the selected programs by visualizing all selected programs, comparing them, and seeing where we have to intervene to achieve product balance. This is very important for our organizational success.

Now, we begin to develop the business case for programs close to their implementation time. Some companies, as I said, use the business case differently than we do. They use it to select programs into the program portfolio. We use it to approve a program for actual implementation in the concept phase. If we do a good job using this tool, i.e., the business case, we will give the go-ahead to good programs and kill poor programs. So, the business case tool is of make or break importance to our program success. Choose a poor program, and you are kind of doomed. Choose the right one, and you are given an opportunity to succeed. Our program managers tend to be assigned to a program shortly after a concept is approved and are responsible for successful completion of the program. They will make trade-off decisions on features to make sure that the plan is actually aligned with its objective.

Program Planning

RICHARDSON: Once you selected a program into the portfolio and a program manager was assigned to execute program, what tools did you use to ensure the quality of the alignment during program planning, and how did they contribute to the program's success?

WATERS: A lot of tools were used in Miss Mercedes. First, a tool called the program strike zone was used. The strike zone is simply a set of agreed upon program critical success factors and business results established to help executives and program managers monitor the programs by specifying quantitatively the boundary conditions under which the program can operate. Metrics such as time to market, target market, net present value and key milestone dates were included in the strike zone. In Miss Mercedes, the priority success factors were schedule, features, profitability. You see here how the business strategy (i.e., time to market) shapes the program strike zone, making schedule its first priority. Simply speaking, it was our primary alignment tool, and it helped me to develop a program plan and, at the same time, make sure that the program met the business needs. By doing all this, the program strike zone contributed to a successful program.

During a one-day workshop called Map Day, the core team developed a program map, which is a tool showing critical cross-project dependencies related to the program schedule and the critical deliverables of each project team throughout the PLC. The program map was used for two things:—to do a program preliminary work breakdown structure (WBS) and a preliminary master program schedule. Actually, the critical deliverables went into the WBS, which then served as a guideline to project teams to develop their project WBSs, which I took back and, after a thorough review with project managers, merged into a detailed program WBS.

Based on the critical cross-project dependencies from the program map, the initial program schedule was developed. Once the master program schedule was completed, it was decomposed into the seven project schedules. In this way, a hierarchical schedule with two levels was obtained—the first being the master program schedule, the second being the project schedules.

In terms of scheduling, standard project management tools were used for the master program schedule and the project schedules. The business strategy of being the first in time to market had a key role in determining which tool was chosen.

Program Monitoring and Control

RICHARDSON: What about tools used for monitoring programs?

WATERS: One of the tools we use is called the program dashboard. The dashboard is a management tool visualizing the status of programs, by using red, yellow, and green indicators. Red means management intervention is needed, yellow means warning, and green means that the program is progressing well. Tools like the dashboard and the program strike zone are commonly used everywhere in the organization to help executives communicate with program managers. The executives want to see if the programs are still aligned with the business strategy and determine if any corrective actions are needed to recover them from misalignment. In many instances, when there's an issue that pushes the program out of the success criteria limits, one of the program indicators will turn red. Then, executives and program managers will have to develop corrective actions and/or adjust the success criteria limits.

In Miss Mercedes, we also used the program review, which is a tool used to communicate program progress or to involve senior management when they to need to step in and make some tough decisions. We used periodic program reviews, in addition to the phase gate milestone reviews.

Lastly, a mandatory tool we use is the wiggle chart. The chart anticipates the expected rate of future program progress, focusing on predictions of major program events, like milestones and program completion. Let me show you an example (Figure 11.6). The vertical axis shows the team's predicted completion date for a specific program milestone (for example, the engineering release or product ship release), while on the horizontal axis we can see the actual date the prediction was made. The beginning point on the horizontal axis is the time when the schedule baseline is prepared and the high-level program milestone dates are marked on the vertical axis. After the beginning point, the program work is kicked off, and the horizontal axis represents the actual program timeline. The team reviews progress regularly and makes milestone predictions. By connecting all predictions for a particular program milestone into a line, we can obtain the milestone trend line. Should the line go upward, the program manager would know that there is a milestone schedule slip. Delivering the program milestone on time would produce a horizontal line. If we estimate an early program milestone completion, the trend line would go downward. Although it is effective in predicting milestone progress, the chart is even more effective if used to develop actions required to eliminate any potential deviation from the baseline schedule. In general, the wiggle chart acted like a compass, helping Miss Mercedes by warning us early of potential schedule slips so we could take corrective action and navigate toward success.

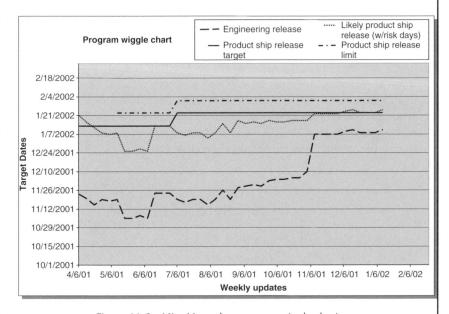

Figure 11.6   Miss Mercedes program wiggle chart.

Metrics

RICHARDSON: I read somewhere that man is a tool-using animal. In Miss Mercedes, you seem to confirm that. Now, what about metrics? Could you please elaborate more on the topic with regard to balancing the program and company needs, and how they helped the program success?

BATTS: Program performance is mostly measured by the ability of the program to meet major milestones, such as the launch target date. Since RollingSys wants to be a first mover and technology innovator, other metrics are important but not as important as schedule metrics. But there are some differences among programs and the metrics they focus on.

For example, in Miss Mercedes, the priorities of metrics were schedule, features, gross margin, PI (profitability index), development expenses, manufacturing costs, market share, and staffing levels. The schedule was so important that progress was measured by the ability to meet the milestones and delivery date, together with the feature sets requested from customers. In parallel with satisfying customer needs, the company's bottom line is of primary concern. So executives use the program strike zone to specify the boundary conditions that match the company's business needs, like ROI, in the form of the PI, or profitability index. The other metrics such as performance to development cost, manufacturing cost, and market share were of second level importance compared to time to market, feature set, gross margin, and PI.

Overall, what I am trying to say for Miss Mercedes is that customer needs, time to market, and feature set are balanced with our business needs, which is to make money—to be brutally honest. We combined those things together and created the business results. That also has another loud message: Our metrics, like tools, were driven by the business strategy. We did this properly, and metrics contributed to the program success by letting us know if we were heading to success and should stay the course, or if we needed to take corrective action.

REFERENCES

1. Martinelli, Russ and Jim Waddell. "The Program Strike Zone: Beyond the Bounding Box". *Project Management World Today* (March-April 2004).
2. Vose, David. *Risk Analysis: A Quantitative Guide*, 2nd edition. Hoboken, NJ: John Wiley & Sons, 2000.

3. Milosevic, Dragan Z. and P. Patanakul. "Standardization May Help Development Projects," *International Journal of Project Management*, Vol. 23, No. 2 (2005).

4. Meredith, Jack R. and Samuel J. Mantel Jr. *Project Management: A Managerial Approach*, 6th edition. Hoboken, NJ: John Wiley & Sons, 2006.

5. Concurrent Engineering User Group Benchmarking Study. Detroit: MI. (September 2005).

6. Martinelli, Russ and Chris Galluzzo. "Program Management and the Intel Product Life Cycle," *Conference on Project and Program Management*. Portland, OR. (July 2002).

7. Meyer, Chris. "How The Right Measures Help Teams Excel," *Harvard Business Review* (May–June 1994): pp. 95–103.

8. Wideman, Max. *Project and Program Risk Management*. Newton Square, PA: Project Management Institute, 1992.

9. Milosevic, Dragan Z. *Project Management Toolbox: Tools and Techniques for the Practicing Project Manager*. Hoboken, NJ: John Wiley & Sons, 2003.

10. Summers, Donna C. S. *Quality*, 2nd edition. Upper Saddle River, NJ: Prentice Hall, 2000.

11. Meyer, Chris. *Fast Cycle Time: How to Align Purpose, Strategy, and Structure for Speed*. New York, NY: Free Press Publishers, 1993.

Part IV

The Program Manager

To this point in the book we have explained how program management is an extension of business strategy; how a program is defined, planned, and executed; and what common processes, metrics, and tools are available for use by the program manager.

This part of the book focuses on the program manager. Specifically, the program manager's primary roles and responsibilities and the competencies that a program manager needs to effectively manage a program.

In Chapter 12, we begin by explaining that programs are the investment vehicles used to achieve ROI and other strategic objectives and why the business manager needs to delegate business responsibilities to a program manager for managing the ROI for a particular program. To achieve this responsibility, the program manager has two primary roles to fulfill: managing the business aspects of the program and leading the program team to success.

The first role, managing the business, begins with the focus on alignment of the program with business strategy and development of the program business case. Then, it continues with the management of program finances, resources, intellectual property, and business risks. The first role ends with monitoring the market and customers.

The second role of the program manager—leading the team—concentrates on establishing the program vision, empowering the PCT, and navigating change and risk through the collaborative work of the program team.

To successfully respond to these two roles, the program manager needs to be proficient in a broad set of competencies. In Chapter 13, we present the program management competency model. The model contains the necessary knowledge, skills, and organizational enablers to systematically support the recruiting, staffing, professional development, and career planning of the program manager. The competencies and enablers cover customer and market, business, leadership, and process and project management proficiencies.

Each company should adapt the mix of competency categories and tailor them according to its specific needs and business strategy. This may be needed if a company undertakes programs of diverse nature and complexity. Finally, we offer some advice on how the competency model can be of value in staffing program management positions.

An example from practice at the end of Part IV establishes two important points. First, the example shows how business strategy drives the mix of program manager competencies and how the competencies can be tailored to adapt to a company's strategy. Second, the example introduces a set of competencies needed for program managers faced with having to manage multiple programs concurrently.

Chapter 12

Program Manager Roles
and Responsibilities

This chapter turns the focus of attention to the program manager, where we comprehensively describe his or her primary roles and responsibilities. Numerous times in this book we have described programs as business vehicles to make money and achieve other strategic objectives and program managers as business leaders who are in charge of the vehicles.

The fact that we see program managers as business leaders portrays them as a special type of mover and shaker. And, they really are! Program managers make or break the business side of their programs, managing finances, monitoring the market, and battling business risks. In that way, their programs add to or take away from the value proposition to the company's portfolio of programs and improve or worsen the company's bottom line.

We present the subject matter through several steps. First, we show that the job of a business unit GM is to invest money to realize a set of business objectives and use program managers to accomplish that. We then describe the business responsibility of program managers in managing the investment. For simplicity, we split the responsibility into two groups: managing the business and leading the program team. This chapter seeks to help all members of an organization who are involved with a program understand the following:

- The role of the program manager is that of a business manager who performs the business function for his or her specific program
- How the program manager manages all aspects of the business on a program

- How the program manager manages the program business by leading a team of qualified functional specialists

Developing program managers who function as business leaders within an organization may be a significant paradigm shift for some organizations. However, it may be that this paradigm shift is needed to take a company further down the road of success.

INVESTING THROUGH THE PROGRAM MANAGER

Companies *survive* by generating enough income to pay for their operating expenses. Companies *grow* by investing a portion of their income in vehicles that generate more revenue and decrease operating expenses. Companies *sustain their growth* by developing and executing strategies to obtain long-term increases in profitability. Therefore, the income a company uses to fund its long-term growth is the investment that is used to generate an intended return. The programs within the portfolio of programs are the investment vehicles used to generate products, services, or infrastructure capabilities that become the means for achieving the strategic objectives. This investment model is illustrated in Figure 12.1.

This investment model for product, service, and infrastructure development is analogous to a personal investment financial model. As an example, let's look at the investment situation for a private investor named Shannon. Shannon's salary is the income she uses to pay her

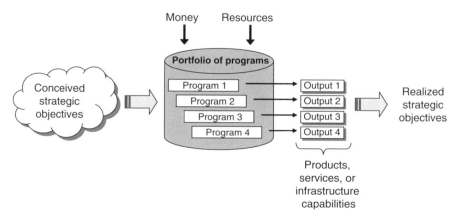

Figure 12.1   Investing in strategic objectives.

living expenses and fund her growth in net worth. Without investing a portion of that income in higher-growth investment vehicles, her net worth grows very slowly. If Shannon wants to accelerate the growth of her net worth, she looks for investment vehicles that will generate additional income over and above her salary. If she develops a portfolio of mutual funds, for example, each fund becomes a vehicle for generating additional income. Collectively the portfolio of mutual funds Shannon owns provides continual positive ROI, if managed properly.

A single mutual fund within the private investor's portfolio is analogous to a single program that a business manager invests in to generate a positive ROI for his or her business. Both a mutual fund and a program are vehicles for generating a higher rate of return. To extend this analogy, the private investor and the business manager both define short- and long-term goals they want to achieve and conceive investment strategies for attainment of the goals. They both turn their investment over to someone else to manage—the private investor employs an experienced mutual fund manager, and the business manager employs an experienced program manager. The mutual fund manager manages a collection of stocks that make up the mutual fund, and the program manager manages a collection of projects that make up the program.

The business manager, therefore, delegates the responsibility of managing the ROI for a program to the program manager. *Managing the business* on the program is the first primary role of the program manager. However, managing the business on a program is not a solo venture. To generate a ROI for a program, the product, service, or infrastructure capability has to be developed by a team of people, all of whom are responsible for developing some aspect of the whole solution. *Leading the program team*, therefore, is the second primary role of the program manager. Those two roles—managing the business and leading the program team—constitute the responsibilities of the program manager.

BUSINESS RESPONSIBILITIES OF THE PROGRAM MANAGER

A GM is responsible for managing the investment for his or her business portfolio of programs within a business unit. Three critical elements of a GM's responsibility for managing the investment for his or her business unit are: (1) defining the strategic objectives for the organization;

Figure 12.2   The GM's investment model.

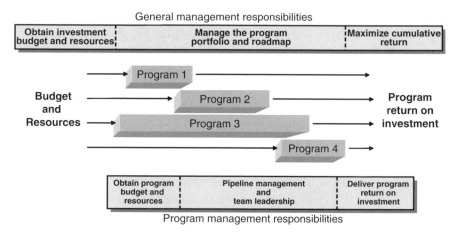

Figure 12.3   Shared business responsibilities.

(2) managing the return on his or her development investment (budget, resources, and capital expenditures) by selecting and prioritizing programs within the portfolio that generate the greatest return; and (3) developing a time-phased program road map (see Figure 12.2).

However, if we look at the portfolio of programs and the road map, it is clear that the GM cannot manage the business aspects associated with each and every program (Figure 12.3). As presented in earlier chapters, the responsibility for managing the business for the programs falls upon the program managers within the business unit. Each program manager operates as the GM's proxy, responsible for managing the business aspects of a single program.

The core responsibilities of the program manager in managing the business aspects of a program investment are described in the following section.

MANAGING THE BUSINESS

The first primary role of the program manager is to effectively manage all aspects of the business responsibilities bestowed upon him or her by the business unit general manager to achieve the business objectives and results intended. Accomplishing this is, of course, no easy task. As stated earlier, the program manager is acting as the proxy for the GM on his or her specific program. The successful program manager will perform this role in a mature and self-controlled manner. This role requires the program manager to be proactive and to ambitiously pursue the objectives laid out by senior management, while maintaining a great deal of flexibility to adapt as necessary during the life of the program.

There are several elements to managing the business of the program, as depicted in Figure 12.4. We now turn our attention to reviewing each of these elements in greater depth.

Aligning the Program to Strategy

The program manager is normally the primary business strategist on a program. He or she needs to continually focus on the strategic business objectives driving the need for the program and for its output. In the define and plan phases of the PLC, the strategic objectives guide the product, service, or infrastructure concept and the end state of the program plan.

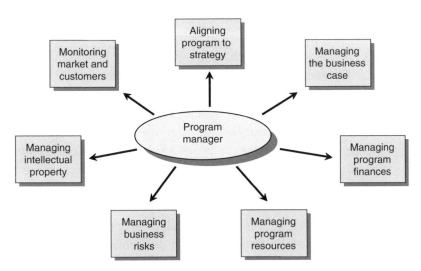

Figure 12.4   Elements of managing the business on a program.

A common pitfall within companies is that strong alignment to the strategies of a business is not established during the definition of a program and business case development. Instead, the alignment is attempted somewhere between business case approval and market launch (or system go live), and a misalignment develops between the program and strategic objectives. It is never a pleasant experience for the members of a business unit when they realize that the strategic return on their investment will not be realized due to a misalignment between strategy and execution.

During the tactical phases of the PLC (implementation, launch, and sustain), the program manager continually monitors the business environment within the company to ensure the program remains strategically viable. This is most important for programs with long development cycles which have a greater risk of strategic shifts due to environmental changes. At any point in which misalignment between the program objectives and the strategic objectives develops, the program manager has the responsibility for bringing it to the attention of the senior management team for appropriate resolution.

Developing and Managing the Business Case

The program business case establishes the business feasibility and viability of a program and is normally developed as the output of the define phase of the PLC. The program business case is the primary tool for the program manager to manage the business aspects of a program (see Chapter 10). It is, in fact, the business contract between the program manager and the senior management team of the company, as well as the basis of the program vision for the program team.

The role of the program manager is to drive the creation of the business case based upon the knowledge, information, and data available at the time. There are many unknowns to comprehend early in a program, so establishing and documenting the set of assumptions that the business case is built upon is crucial. For the business case to remain accurate and viable, the program manager needs to continually manage and update the business case as information and data becomes available throughout the PLC.

Part of the business case involves developing a robust set of critical success factors and metrics that are used to determine the viability of the program. Collectively, the critical success factors define business success for the program. They serve as the business objectives that the program team works to achieve, as well as the primary measure of checks and balances to guide decisions and to evaluate environmental changes.

At any point in the development cycle in which the business case no longer looks viable, the program manager must communicate that to the senior sponsors of the program. This, at times, can be difficult due to emotional and organizational culture factors, but as the GM proxy on a program, the program manager must manage the ROI. This usually means managing the program to success, but it can also mean cutting the losses as early as possible and recommending to cancel the program if the investment is no longer viable.

Managing the Program Finances

Once a program is approved or awarded, the investment funds are transferred to the program manager for management. In some cases, especially within firms utilizing traditional phase/gate development processes, funds may be released in conjunction with achieving specific key gates. The investment funds may come from various sources such as lending institutions, venture capital companies, shareholders, customers, or internal funds. This involves managing the program cash flow and managing the cost of goods sold. It is recommended that a program manager has a financial analyst on the PCT to help manage the financial aspects of a program, as managing program finances can be a time consuming activity. Historically, the relationship between the program manager and the financial analyst has been contentious due to competing agendas of the two job functions. The job of the financial analyst is to keep the program financially viable by controlling costs. Cost control, however, can be viewed as constraining to the program manager who is being challenged to manage and balance *all* of the competing objectives, including cost control. However, great benefits can be gained if the program manager forges a strong working relationship with his or her financial analyst, as shown in the box titled, "Knowing the Numbers."

Knowing the Numbers

Like most program managers, Michael, a product development program manager with Lockheed Martin, viewed financial analysts as a necessary burden or constraint. But a conversation during a routine meeting with Sarah, the financial analyst on his PCT, was the beginning of a million-dollar lesson that taught him the value of working with a financial analyst.

At one point during a discussion about cutting the cost of prototype or preproduction systems that were sent to lead customers, Michael inadvertently asked, "By the way Sarah, where does the revenue generated from the sale of the prototypes go?"

Because she had access to and understood the company's accounting system, Sarah was able to track the revenue. During a follow-up meeting, Sarah explained, "The money goes into the general sales and marketing department, which isn't tied to your department." Michael then asked Sarah if it were possible to route the revenue back to his department, and, better yet, back to his program. "I wanted to see if the money could be brought back to the business unit to use for developing more products, or brought back to my program to alleviate some of the financial constraints we had," Michael said.

Sarah was able to work within the company's financial system to have the money routed back to the program, once it was received from the customers. After the money had been collected and brought back into the program, Michael and Sarah presented their quarterly financial report to their management team. "For the current quarter, the program showed zero cost expenditure, and we were in the middle of the development phase where maximum expenditure was occurring," stated Michael. Sarah explained, "We collected enough revenue from the sale of the prototype products to cover our development expenses for the quarter."

Obviously, the senior management team was pleasantly surprised, and they required every program thereafter to follow the same financial process. As a result of this and further cooperation between the program manager and financial analyst, the program finished a million dollars under the planned budget.

"It was a learning experience for me," Michael added. "There is positive value in working closely with a financial analyst if you think beyond the constraint piece of it. They're just doing their job—not giving you an open checkbook—and that causes contention between you and the analyst at times. But they are definitely an asset."

Managing the Program Cash Flow

Managing the program cash flow involves more than just monitoring the total available program budget. It involves understanding and monitoring program expenditures and income on a periodic basis. All companies have to make money to stay in business, and the ability to maintain a positive cash flow is a critical component of an organization's success. This is especially true for smaller companies, start-ups, and any company that is not cash rich.[1]

Once a program is funded, management of cash flow for the program becomes the responsibility of the program manager. It is risky to monitor the program budget in total. If, for example, the budget was depleted three quarters of the way through a program, the team would suddenly be in trouble with respect to program finances, and most likely work would come to a halt (at least temporarily). For this reason, the program manager along with the program financial analyst should create a cash flow plan for the budget and then manage program finances to the plan. When a periodic cash flow plan is established (see Figure 12.5), deviations from the plan are much easier to react to and manage. This supports the 'no surprises to management' philosophy that we subscribe to.

Understanding the billing and payment terms of the company, its suppliers, and its customers should be understood and reflected in the cash flow plan. There is a time delay for both expenditures and income generated by a program due to billing cycles. For example, the organization sponsoring a program may have payment terms of 90 days after receipt of goods from its suppliers. This means that expenditures for the program are delayed by three months. Likewise, revenue generated by the program will also be subject to the payment terms of a company's customers.

Managing the program scope is one of the most critical aspects of managing the cash flow on a program. Increases in program scope without

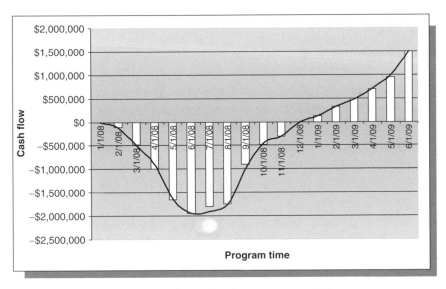

Figure 12.5   Example of a program cash flow.

increases in funding (commonly referred to as "scope creep") will negatively affect a program's cash flow. It is imperative that the program manager be aware of any work outside of the planned scope and of the financial impact of that work. For this reason, an effective change management process should be in place (see Chapter 8) to manage changes in the program scope.

The final aspect of managing program cash flow involves knowing when to ask for assistance from the senior management team of the business. Short-term and long-term program cash flow assistance can only be provided by senior management, and the program manager must be willing to work with the executive team to secure the finances needed to continue the program.

Managing the Cost of Goods Sold

The second aspect of managing the return on invested funds is the management of the cost of goods sold. The cost of goods sold includes the bill of materials and the cost associated with manufacturing, assembling, or otherwise physically creating the product, service, or infrastructure capability under development. The cost of goods sold is deducted from the revenue generated for each unit sold or used in the case of a service. Minimizing this cost through actions during the development cycle enables a company to retain a larger percentage of the sales revenue or cost savings generated once the product, service, or infrastructure capability is launched.

The program manager and his or her program team are responsible for driving the future cost of goods sold to a minimum as part of the development process. Cost of each component of the product, service, or infrastructure capability, as well as the total number of components, makes up the cost of the final unit. Many cost versus performance trade-off decisions are made during the development process to generate an optimum bill of material cost that achieves the performance requirements of the program. The program manager's role is to set aggressive, yet realistic, bill of material cost goals for the program team. Then he or she must measure progress toward achievement of the cost goals through the use of program metrics (see Chapter 9).

Like management of bill of material cost, managing the cost to manufacture, assemble, or create the final product, service, or infrastructure capability is the responsibility of the program team. The role of the program manager is to set aggressive, realistic goals to drive the most efficient and cost-effective approach for creating the product, service, of infrastructure capability under development.

Managing the Program Resources

Much of the funds invested in a program are utilized to pay for the resources needed to complete the work. This includes both human and nonhuman resources. Resource cost is by far the largest consumer of the investment budget, in many cases constituting more than 75 percent of the program budget. For this reason, effective management of the program resources is a critical aspect of the business management role of the program manager.

Developing a viable business case, aligning a program to the strategic objectives of a company, and managing the program finances are critically important aspects of the program manager's role; however, if he or she cannot secure and maintain the resources to complete the work, the strategic, business, and financial goals of a program will not be achieved. Therefore, resource management is a key factor in program success—equally important as the financial, time, and quality aspects of a program.

To be completely successful, a program must be adequately resourced. There is a high level of uncertainty with respect to resource utilization when program plans are initially generated. For that reason, resource management needs to be viewed as a continuous process that takes place throughout the PLC.[2] A common pitfall that we have observed is that a lot of effort goes into generating a program resource plan and gaining commitment from resource managers to fulfill the resource demand of the program. However, once the plan is approved, focus shifts to other aspects of the program, and resource constraints begin to mount. The role of the program manager is to manage resource utilization throughout all phases of the PLC.

Resource management involves the program manager understanding the skills, experience levels, and number of resources needed for a program. Then he or she must work with the organization's functional managers to adequately staff the project managers and individual contributors needed for the program. The program manager is first responsible for ensuring the core team is fully staffed. This means each project has a qualified project manager in place and represented on the core team. Further, each critical function supporting the program should have a qualified representative as a member of the PCT. Secondly, the program manager is responsible for working with each project manager to be sure that his or her team is adequately staffed. When resource constraints are identified on a project, the program manager assists the project manager in resolving the constraint as effectively as possible and, if necessary, adjusting the project plan and program plan to take the constraint into account. Resolution usually entails adjustments to program costs, resources, time, performance, or quality goals. At any point in a program in which resource gaps adversely affect the program critical success factors, the risk must be communicated to the senior management team of the business unit for assistance in resolution (see the box titled, "Resources Management that Hurts").

Resource Management that Hurts

Here's a view of human resources management from one experienced PCT member, Don Hallum of Rockwell Collins.

When taken on an individual basis, predicted program duration is determined by the critical path duration. However, when multiple programs are executed simultaneously, a given program's duration may be many times its critical path prediction, meaning too long.

One reason is contention for scarce resources. The scarcity of resources creates interdependence; the performance of a given program is not independent of the outcomes of other programs. As a given [when] development program approaches completion, managers assess that program via the work remaining to be done. Upon perceiving a gap, managers respond by allocating more resources to a specific program.[3]

Variable workloads usually arise because an organization takes on new programs whenever good marketing or technical opportunities present themselves. As a result, in some months many programs start and, in others, none do—a pattern that can create bottlenecks at crucial points in the development process.[4] This can cause loss of business revenues and opportunities due to late program execution and/or lack of development program capacity.

If human resource planning does not take into account these parameters inherent to managing multiple programs, it may hurt the company. A number of program schedules are created assuming that experienced (best case) resources will be assigned to key tasks, thereby compounding the problem. Many companies, while setting up timelines, only consider the best case scenario since the design was supposed to be shipped to the customer 'yesterday'.[5] Time after time large engineering programs are planned for perfect execution despite early warning from engineers and developers that live in the trenches.[6] Those programs deemed by upper management as 'critical' further compound the problem. Once their timelines are formalized, these programs have very high visibility and a very small hiccup along its progress is dealt with at utmost urgency. The customer is notified of every schedule-threatening issue, if the design team needs more time to solve them. These issues are usually customer-relationship breakers and are dealt with accordingly.[5] Overall, resource management of a program is a tough job.

Managing Business Risk

The financial world teaches us that all investment involves some level of risk. Developing products, services, and infrastructure capabilities are no different. Various levels of risk taking are necessary if a company is intent on being a leader, or even a strong contender in its industry. Market leaders understand that they must run directly toward, instead of away from risk to put distance between themselves and their competitors. This means that development programs can have a high level of business risk associated with them. Failure to identify and understand the risks, or uncertainties, can lead to substantial loss for the enterprise. The role of the program manager is to protect the company's investment in the program through effective management of the business risks involved.

Risk is inherent in every program—it comes with the territory. Good risk management will not make the risk associated with a program go away; it merely helps to ensure that the program team will not be blind-sided with a problem to which they cannot respond.[7] The program manager must be aware that the failure of a single project within a program can cause the entire program to fail. This requires that the program manager empower the project managers to manage the project-level risks and focus his or her efforts on identifying and managing the program-level risks (see Chapter 8). The role of the program manager is to be an advocate for effective risk management practices on the program and ensure that it becomes an institutionalized practice on the program. The program manager needs to keep the entire program team engaged in risk management practices, as risk management becomes useless if only a small number of program team members effectively practice it.

Managing Intellectual Property (IP)

IP may have significant strategic and commercial value to a company. Much of the IP generated by companies engaged in product, service, or infrastructure development is done so within the context of development programs. For these reasons, a crucial element of managing the business on a program involves the proper management of the IP created. This may include making sure that the proper patents, trademarks, and copyrights are obtained, depending upon the type of IP created (see box titled, "Intellectual Property of Another Kind").

The motivation for management of intellectual property is maximization of profit. For a high-technology company, IP is the engine of survival and growth. It is of utmost importance for managers to build a corporate culture that appreciates such a view.

IP is the know-how that is produced by our creative minds. In particular, it is the knowledge of doing something in a better way. It is also the knowledge to figure out that this better way may be very profitable if it can be commercialized. If a company patents its technology, for example, it would prevent others from using it without paying for its use.

It is not a job of program managers to know the ins and outs of managing intellectual property. Multiple players such as technologists, managers, functional managers, directors, legal experts, and officers all take part in managing IP. However, program managers must first understand that it is the IP portfolio that keeps profitable technology companies in their leading market positions. Second, they must have a developed sense for the following questions:

- *Is there a competitor who could use the IP developed on my program to compete against us?*
  It means that the program manager should know the market and customers and have a good understanding of the competitive analysis.
- *If so, what does that mean to my program's business?*
  Again, good knowledge of market, customers, and competitors, along with the program business case will help the program manager understand the answer to that question.
- *What can be done to protect the IP?*
  There are multiple ways to protect IP. For example, either patenting a new technology or idea or requiring nondisclosure agreements with all partners.

The program manager should use this knowledge to ensure that all players involved play the game in a way that benefits the program because, ultimately, the program manager owns the program success.

Intellectual Property of Another Kind

There are four main types of IP: patents for inventions, trademarks, design, and copyright. The first one—patents—relates to new products and processes that are capable of industrial application. Can a company's program management process then be considered part of its IP? Here's how one program management professional feels about that question.

A. Lois Lee, director of the PMO for Prdex, Inc., a high-tech manufacturer says, "One of the most important pieces of Prdex's IP is our program management process, although it was never legally protected as IP. Nobody has bothered to count, but I would suspect that we have made several hundred million dollars of profits through the use of our processes. Of course, I should have told you our history to understand my statement. We introduced our program management process 28 years ago. We have had five major revisions developed by a team of our best program managers on the basis of our collective learning. Further, all involved in programs-executives, engineering managers, program managers, program core team members, and extended team members received training on the program management process.

I guess that we have conducted 500-plus programs over the 28 years with the processes. So we probably made more profits than I previously stated. Program managers come and go, but our program management process stays. In a way, it connects all our program managers across the span of 28 years. It is a product of collective learning and know-how, and that's why we consider it IP."

Monitoring the Market

Besides a strong internal focus on the business of the program, the program management role also requires an external focus on the market and continual monitoring of market conditions. The market consists of the customers who buy the product, service, or infrastructure capability; the end users who will use it; and the competitors who are trying to sell their solutions to the same customers. There are many examples of program managers who have focused their attention solely on the effective use of their resources to achieve the goals of the program—all the while being unaware of shifts in the external market that have left their solution ineffective and unattractive to the customers and end users.

The best program managers continually monitor the state of the external environment to check if the business objectives and strategies that the program is based upon are still achievable. In particular, to keep a program feasible, from the business and strategic perspective, the program manager and other key members of the program team should be fully aware of the current state of the market (internal or external) in which the business operates and also be aware of emerging trends. To determine the current state and future trends of the market, the program manager needs to be engaged with the customer and bring the customer information back to the program team. It is crucial that program managers understand the wants and needs of their customers to obtain business success for the program.

However, it is even more important for them to understand the needs of the end-user. This is a critical distinction. In many cases, the customers and the end users are two different sets of people—the customer is the one who purchases the program output from the manufacturer, while the end user is the one who actually uses it. For example, a retail outlet is a customer of a digital camera manufacturer such as Kodak, but the person who buys it from the retailer and uses it for his or her personal use is the end user. It is the needs of the end user that must be transformed into the functional requirements of the product, service, or infrastructure capability. Therefore, the program manager and key members of the program team need to find ways to interface directly with end users. This may be accomplished by developing focus groups, directly observing end users, or developing prototypes for the end users to test. For an example of a creative approach to garnering end-user interface, see the box titled, "From the Design Room to the Showroom."

From the Design Room to the Showroom

Shane Plaxton, a Hewlett-Packard (HP) research and development manager in Vancouver, WA, knows first hand the value of interfacing with end users. A January 2006 article in the Oregonian explained HP's approach to putting its printer engineers and designers in direct contact with end users.[8] During the 2005 holiday season, HP sent more than 1,600 employees into electronics and office-supply stores to speak directly to end users, demonstrate how their products work, and offer their expertise.

Plaxton said the shopper feedback was a "little intimidating" because he is "not used to being out in front of 'customers.'" But he added that it was refreshing to step back from the details of printer design and see the big picture. "I spend all day looking at designs and product specs and details," he said. "This is a way of going out and seeing what end users really want to get."

The experience inspired him to take a fresh look at how HP can simplify its printer designs to make it easier for end users to choose the best model for them and to understand how it works.

Marketing experts say companies such as HP realize their product designers are too far removed from the end user and are working quickly to close the gap. Focus groups and customer surveys, the easiest and most popular marketing tools, are taking a backseat to more innovative approaches. Companies have discovered that observing end users and their decision processes in real-world situations gives a better picture of what buyers want.

HP workers each write a brief report on their store visits, highlighting end users' reactions to specific products. Jim Mury, the Vancouver manager in charge of HP's employee demonstration program, said, "The information gathered will be compiled into a larger report and delivered to each product team so that designers can incorporate customer feedback into the planning of future products. I truly believe that the information from our end users improves our products."

Program managers also need to understand who the competitors are and what they have to offer. To create a competitive advantage, it requires that competitors are known, that the competitor's solutions are understood, that the solution being developed by the program team can be compared to competitor's solutions, and that the differences between the solutions and the value proposition of the program team's solution can be explained to potential customers.

This exercise is often conducted by program teams but is treated as a one-time event either as part of the concept approval or the business case. However, to ensure competitive viability of the product, service, or infrastructure capability under development, this exercise needs to be repeated periodically—at least at every phase/gate review.

An appropriate level of marketing expertise on the part of the program manager is important for the following reasons:

- To achieve the business objectives intended, the product, service, or infrastructure capability must solve the problems and meet the needs of the customers and end users.

- Market knowledge improves the team's understanding of how the end users will utilize the program output.
- Market knowledge assists the program team in defining the product, service, or infrastructure quality expectations of the customers and end users. Quality goals that are too low result in customer dissatisfaction, and quality goals that are too high result in unnecessary cost to the business.
- Stronger knowledge of the market facilitates better decisions on the part of the program manager, the members of the program team, and the executive stakeholders.
- A much stronger business case can be developed when knowledge of the market is utilized—such as market size, market segmentation and sales potential of each segment, and customer and end user problems and needs. Market information will increase the probability of program and business success.

MANAGEMENT, LEADERSHIP, AND THE PROGRAM MANAGER

In program management, like any other discipline, there is a distinct difference between good program managers and great program managers. Good program managers are those that can manage the business and operational aspects of a program. Great program managers are those that have the ability to combine strong management skills with strong leadership skills. The primary differentiator between good and great is one's ability to lead the program team toward the achievement of the business objectives. The strong leader establishes a vision, or end state, and then guides the work of the entire team toward that vision, while helping the team to establish pride in their work as well as knowledge of how their work contributes to the program goals.[9]

A combination of strong management *and* leadership abilities is what separates the best from the rest. The program manager uses his or her management skills to utilize the physical resources (human skills, raw materials, and technology) of the enterprise to ensure that the work of the team is well planned out, is performed on schedule and within cost, is delivered with a high level of quality, and is executed productively and efficiently. However, because program managers rarely have resources directly reporting to them, they must use strong leadership skills to build relationships to influence, focus, and motivate the program team. In doing so, program managers utilize the emotional and spiritual resources of the organization to create and deliver great things.

So why is it important that program managers be both effective managers and effective leaders? The answer to that question is contained in Chapter 2—programs are complex by nature. The quickening pace of innovation has led to more complex solutions to meet the needs of both customers and end users. The high level of solution complexity has led to more specialization to breakdown the complexity into smaller elements that can be more easily managed. However, the challenge remains that the development team becomes fragmented into specialized work teams that are organized around the various elements of the product, service, or infrastructure solution. Pulling this group of functional specialists into a cohesive, effective team that is focused on a common set of objectives requires both leadership and management skills. Strong management skills are needed to organize the team and make sure the work is accomplished productively. Strong leadership skills are needed to ensure that the team members are performing as a single entity that identifies with the program more than their functional specialty and that they remain focused on a common vision of what they are trying to achieve.

It is fairly easy to recognize a program that is lead by a program manager who lacks strong leadership skills. It is normally characterized as a loose grouping of functional specialists who operate only within their area of specialty, functional agendas and goals take precedence over a common vision and goal, and there is a high level of dysfunction between the functional elements on the program. This results in a low level of interplay between the specialists and a high level of finger pointing and cross-functional accusation when the elements do not integrate into a common solution. If this situation sounds familiar, it is probably an indication that the second primary role of the program manager, that of leading the team, is not well understood within the business.

LEADING THE TEAM

Leaders take people where they need to go and help them understand how their efforts contribute to the goals of the organization. For the program manager, this means that he or she is responsible for keeping the team of functional specialists performing as a cohesive group to achieve the objectives of the program, while helping each project team understand how their element contributes to the creation of the whole solution. To accomplish this, the leadership role of the program manager consists of three primary elements, as follows:

- Establishing the program vision
- Empowering the PCT
- Navigating change and risk

Establishing the Program Vision

The foundational element of creating a cohesive, high-performing team is to establish a common vision for the program and then work through others to achieve that vision. The program vision is the end state that the program is trying to achieve. In most cases, it is the set of business objectives that the program is commissioned to achieve and is commonly referred to as "the big picture." Everything that happens on a program should do so in the context of the big picture, or program vision.[10]

One of the essentials of strong leadership is that the leader *pulls* upon the energy and talents of the team, rather than pushing, ordering, or manipulating people. Leadership is not about imposing the will of the leader, but rather about creating a compelling vision that people are willing to support and exert whatever energy is needed to realize it.[11] The program vision is the device that the program manager uses to pull the team together and to keep the project specialists focused on the cross-functional goals, rather than just the goals of their specific function. It is the basis of empowerment that allows the program manager to establish power of influence, rather than power of expertise. For an example of personal leadership principles one practicing program manager has developed, see the box titled "The Story of Leadership of a World-Class Program Manager."

The Story of Leadership of a World-Class Program Manager
This case happened in a world-class, market leading organization. One of the case participants is one of the authors, who is telling the story.

> Rich Paysinger was a program manager, who I have known for a long time as a repeat client. At the time, I was involved in a consulting project in his company. I was searching for a 'program manager from paradise' in this organization. More precisely, I needed to find an outstanding team leader, who would serve as a model program manager for the

consulting project. So, I asked Rich, a respected old-timer of the company's program managers, to introduce me to such a team leader. Rich told me that there was such a guy, Pali Pafgan.

So, I met Pali and interviewed him for 90 minutes. Simply, I asked him what principles of leadership philosophy he used.

Some of the major principles he used are summarized as follows:

Principle 1: *Obtain commitment:* Pali defined the role of each stakeholder, emphasizing that they are equally important. He tried to get all relevant stakeholders involved in his program as equal partners who build consensus whenever needed.

Principle 2: *Insist on transparency:* Pali showed an honest interest in others' opinions, insisting that everybody explicitly state his or her interest in the program. He worked hard to obtain information inside and outside the company and share it in a facilitated and informed decision-making model.

Principle 3: *Be a facilitator:* Pali created opportunities for his team members to hone their talents. He also insisted on empowering and emphasizing that the team develops its goals and work plan.

Principle 4: *Focus on human relations:* Pali concentrated on relationships within the team, trying whenever possible to select members based on their functional and people skills. His intent was to provide team building training to members who needed it and defined the team as the only place for decision making.

Principle 5: *Promote learning:* Pali encouraged members to behave like entrepreneurs and take risks. He also wanted members to feel safe in the program environment and to share their agendas and opinions for all members to scrutinize.

With his approach, he managed to become a leader who was respected by both people who worked for him and those who did not.

One of the biggest leadership challenges for the program manager is maintaining the program vision despite the large number of potential distractions that occur in any given day during the life of a program. These distractions create noise in the system that can cause the program stakeholders to lose focus on the goals of the program and forget what the program is all about.[12] It is important that the program manager remains diligent in maintaining and constantly communicating the program vision to all stakeholders, the program team, and sponsors in particular throughout the PLC.

Empowering the PCT

To lead, one must have a relationship with people who are willing to follow. For the program manager, those people are the PCT specifically and, generally, the entire program team. Every leader's potential is determined by the abilities and actions of the people that are closest to him or her. If the leader's inner circle of people are strong and capable, the leader can make a big impact within an organization.[13] This is true for the program manager whose inner circle consists of the members of the PCT. As discussed in Chapter 5, careful selection of the PCT members is critically important for this reason. First and foremost, program managers should select core team members who are experts in the function that they represent within the organization. The second critical criteria for PCT selection is a person's ability to step outside of their area of expertise to effectively collaborate with the other members of the leadership team toward the achievement of the program vision. One of the most significant ingredients for program success is the cooperation and collaboration between the project teams within the program, in which everyone understands that they cannot succeed unless everyone else succeeds.[14]

The program manager empowers the core team by giving away a portion of his or her own power to the project managers. Thus, the project manager will succeed by exercising control over his or her own decisions and resources. This is an important concept for a program manager to grasp. Nothing is more disempowering than giving the core team a lot of responsibility for program success, then not providing them with the resources and the ability to work autonomously. Effective leaders get the most out of the members of their inner circle by treating them in a way that bolsters their self confidence and provides them with the necessary resources to succeed.[15] A strong sense of self confidence makes it possible for the PCT members to take control of their portion of the program, which allows the program manager to focus on the big picture and the interdependencies between the teams.

Establishing an environment of "no fear" is also a critical element in developing a team's confidence, according to a senior program manager for a major aerospace company. As he told us, "A no fear environment is needed so people don't have to fear coming to the program manager with the *real answer*, especially if it's not a pleasant answer. You have to let people stumble a bit in order for them to learn and gain confidence."

Navigating Change and Risk

Leadership is directly tied to the process of creating innovative solutions.[16] Innovation by nature causes change, and change requires leadership to move people in a new direction.[16] It also requires a higher level of risk taking to try new ideas and solutions. Change, risk, and leadership are tightly intertwined. Good leaders will foster a climate of risk taking to empower their teams to step outside of their comfort zones and try untested ways to innovate.[17]

The role of the program manager is to be an agent for change on the program and to create a climate that supports risk taking. This empowers the program team to find innovative solutions for the solutions they are developing.

ATTRIBUTES OF A STRONG PROGRAM LEADER

Some program managers that we talked with believe that the sign of a strong leader is to have the right answer for every question and problem that comes their way, without having to escalate to senior management. In reality, as the leader of the program team, a program manager will not get rewarded for just being right. Rather, success is rewarded by leading a team of people toward the attainment of the business objectives that they are commissioned to achieve.

This is accomplished by effectively leading people to work at a high level of performance and to effectively collaborate as a cohesive team. To lead, the program manager must have a team of people who are willing to follow. People will only voluntarily follow someone who they believe is credible and has the ability to lead them to success. Strong leaders, therefore, must possess and demonstrate the following attributes:

- Vision
- Trustworthiness
- Credibility
- Competency

For program managers to create pull and influence the team to follow in the direction they have set, their vision must be clear, compelling, and achievable. The effective program manager continuously works to communicate the vision and to align the team in support of it. The program vision is the primary mechanism for pulling a team of functional specialists together to work in a collaborative manner.

At the heart of collaboration is trust. Trust is the fundamental cornerstone of human relationships, and without trust one cannot lead.[18] Program managers must continuously demonstrate that they are trustworthy and that they expect the same from all members of the program team. As John Maxwell states in the highly acclaimed book, *The 21 Irrefutable Laws of Leadership*:[10]

> People will tolerate honest mistakes, but if you violate their trust, you will find it very difficult to ever regain their confidence

Program managers need to possess a high degree of integrity and be consistently trustworthy. They need to take total responsibility for their own actions as well as the actions and results of their teams.

People will refuse to follow those they do not deem credible. To maintain credibility, the program manager must continuously do what they say they will do and always let people know where they stand on issues and stick to it. A program manager that fails to follow through on his or her commitments, or who vacillates on issues concerning the program, will have a difficult time convincing the team that they have the credibility to lead them to success.

Just as people will refuse to follow those they do not perceive as credible, they will also refuse to follow leaders they feel are not completely competent. Competency implies that the leader is organized and disciplined in all activities and will expect that of others. As the leader of the program, team members expect the program manager to lead by example and actively participate in accomplishment of the program goals.

SUMMARY

The programs within the portfolio of programs are the investment vehicles used to generate products, services, or infrastructure capabilities, which are the means for achieving ROI and other strategic objectives. A business unit manager is not able to manage the business aspects of all the programs within the portfolio. He or she, therefore, must delegate business responsibilities for managing the ROI for the programs to the program managers within his or her organization. This responsibility includes the following two roles: *managing the business* and *leading the team*. The first role, managing the business, includes the following:

- Aligning program with business strategy
- Developing and managing the business case
- Managing the program finances
- Managing resources
- Managing intellectual property
- Managing business risks
- Monitoring the market and customers

The second role, leading the team, involves the following:

- Establishing the program vision
- Empowering the PCT
- Navigating change and risk

To successfully fill these two roles, the program manager has to possess strong leadership and management competencies.

The Principles of Program Management

▼ The responsibility for managing the ROI and other business success criteria for a program is that of the program manager.
▼ The first primary role of a program manager is managing the business of the program.
▼ The second primary role of a program manager is leading the program team
▼ A highly effective program manager by demonstrates a combination of strong management and leadership abilities.

REFERENCES

1. Stagliano, A. R. "Cash is the Lifeblood of Every Contractor". *Building Profits Magazine* (2005).
2. Keogh, Jim, Avraham Shtub, Jonathan F. Bard, Shlomo Globerson. *Project Planning and Implementation*. Needham Heights, MA: Pearson Custom Publishing, 2000.
3. Repenning, N. "Resource Dependence in Product Development Improvement Efforts". *Cambridge, MA: Sloan School of Management, Massachusetts Institute of Technology* (1999).
4. Adler, P. A., Mandelbaum, et al., "Getting the Most out of Your Product Development Process," *Harvard Business Review* (March-April 1996).
5. Jagtap, M. "Analysis of Persistent Firefighting". *Portland, OR: Department of Engineering and Technology Management, Portland State University* (2005).
6. Herweg, G. and K. Pilon. "System Dynamics Modeling for the Exploration of Manpower Project Staffing Decisions in the Context of a Multi-Project Enterprise". *Cambridge, MA: System Design and Management Program, Massachusetts Institute of Technology* (2001).
7. Demarco, Thomas and Timothy Lister. *Waltzing With Bears: Managing Risk on Software Projects*. New York, NY: Dorset House Publishing Company, 2003.
8. "HP gets closer to users by design," *The Oregonian*, (Friday January 6, 2006).
9. Bennis, Warren G. and Burt Nanus. *Leaders: Strategies for Taking Charge*, 2nd edition. New York, NY: HarperCollins Publishers, 1997.
10. Maxwell, John C. *The 21 Irrefutable Laws of Leadership*. Nashville, TN: Thomas Nelson, Inc Publisher, 1991: pp. 58.
11. Kouzes, James M. and Barry Z. Posner, *The Leadership Challenge*, 3rd edition. San Francisco, CA: Jossey-Bass, 2003: pp. 143.
12. Kouzes, James M. and Barry Z. Posner, *The Leadership Challenge*, 3rd edition. San Francisco, CA: Jossey-Bass, 2003: pp. 114.
13. Maxwell, John C. *The 21 Irrefutable Laws of Leadership*. Nashville, TN: Thomas Nelson, Inc Publisher, 1991: pp. 110.
14. Kouzes, James M. and Barry Z. Posner, *The Leadership Challenge*, 3rd edition. San Francisco, CA: Jossey-Bass, 2003: pp. 250.
15. Kouzes, James M. and Barry Z. Posner, *The Leadership Challenge*, 3rd edition. San Francisco, CA: Jossey-Bass, 2003: pp. 307–322.
16. Kouzes, James M. and Barry Z. Posner, *The Leadership Challenge*, 3rd edition. San Francisco, CA: Jossey-Bass, 2003: pp. 187.

17. Kouzes, James M. and Barry Z. Posner, *The Leadership Challenge*, 3rd edition. San Francisco, CA: Jossey-Bass, 2003: pp. 207.
18. Kouzes, James M. and Barry Z. Posner, *The Leadership Challenge*, 3rd edition. San Francisco, CA: Jossey-Bass, 2003: pp. 224.

Chapter 13

Program Manager Core Competencies

What was your motivation for choosing the program management model for developing products? We posed that question to Gary Rosen, vice president of engineering for Varian Semiconductor Equipment. Rosen responded by stating the following:

> When observing the differences between poorly-run product development efforts and well-run efforts, I noticed the difference was that the well-run programs had a true program manager in charge. These people had a broad skill base that is needed in the capital equipment industry—good people skills, good business acumen, and good system (integration) skills. Unfortunately, not a lot of people have these broad skills.

It is a rare program manager that comes to the role totally qualified to fulfill all aspects of such a broad and encompassing set of skills and competencies. The successful program manager is constantly seeking to learn and broaden his or her knowledge and experience to take on more complex and critical programs. Senior management, in turn, needs to create a positive learning environment to encourage program managers to continually seek improvement and growth.

We developed the program management competency model, which addresses the breadth, depth, and complexity of the program management role. This model provides senior management with an excellent aid for developing program management as a true discipline that can provide competitive advantage through business efficiency. As the capability and experience of program managers grows, the greater the probability that they will successfully deliver the business objectives of the firm. The competency model is an excellent career development guide for program managers, project managers, and others aspiring to become program managers.

In this chapter we use the program management competency model to detail the knowledge, skills, and abilities needed for an experienced program manager to successfully perform his or her function. Additionally, we discuss the key organizational enablers needed to make the competency model fully effective and to adequately support the program management discipline. Finally, we show how the model can be customized to meet company needs. The purpose of this chapter is also to assist all members of an organization to understand the following:

- The core competencies that are needed for the experienced and successful program manager.
- Key organizational enablers that, if in place, assist the program manager in execution of his or her core competencies.
- The value of the competency set to the individual and the organization in pursuit of continual improvement for enhancing individual performance.
- The use of the program management competency model as a tool for career growth and performance evaluation of the program manager.

THE PROGRAM MANAGEMENT COMPETENCY MODEL

The program management competency model has been designed to provide the necessary knowledge, skills, and organizational enablers to systematically support the recruiting, staffing, professional development and career planning of the program manager (see box titled, "What is Competence?"). The information presented in this section is derived from companies that utilize program management as a true discipline and critical business function. These companies have devised comprehensive training and development approaches to support this key role. Although the technical aspects of product, service, and infrastructure development are critically important, much of the success of a program is behavioral and human-oriented in nature. Program management truly meets the traditional management attribute of getting things done through other people.

The program management competency model details the areas in which the program manager should gain proficiency to help the organization achieve expected business results. As mentioned, the model identifies the knowledge and competencies critical to success in performing the role of a program manager. The model also includes what we characterize as **key enablers**. Enablers are things that create or proactively encourage a positive environment to provide the maximum opportunity for success, learning, and growth to occur in program management. Enablers range from environmental factors to organizational and managerial culture, philosophy, and actions. Examples of some general enablers would include an organizational structure with clear roles and responsibilities defined for those involved in product, service, or infrastructure development and stable systems that can accurately and consistently measure program status and results.

What Is Competence?

The word competence means the ability to do something well. With respect to management, we found that the term competence is highly overdefined. Take for example the following definitions from Morris and Pinto:[1]

- "an underlying characteristic of a person in that it may be a motive, trait, skill, aspect of one's self-image (or social role), or a body of knowledge that he or she uses." While comprehensive, this definition may not be very user-friendly or practical.
- "the characteristics of a manager that leads to the demonstration of skills and abilities and results in effective performance within an occupational area." We find this definition more practical in nature, but it doesn't address the fact that competent managers need a strong knowledge base to effectively apply their skills and abilities.
- "the knowledge, skills, and qualities of effective managers used to effectively perform the functions associated with management in the work situation." We feel this definition best describes the competencies needed by practicing program managers to effectively manage complex and demanding programs in the contemporary business environment.

The following simple algorithm sums it up best:

Competence = Knowledge + Skills + Personal Qualities + Experience

---

The objective of the model is to provide a comprehensive set of competency criteria that can be used for the continued development and career growth of program managers. It also should be noted that each of the four competency areas includes its own specific technical competencies and tools related to its function. The term **technical** applies to the critical knowledge and tools necessary to become proficient in a particular discipline. Technical problems that occur within a program are not the direct responsibility of the program manager. These types of problems will normally originate within one of the functional disciplines, such as engineering. They should be addressed and solved by the project managers and the technical specialists of each functional discipline represented on the team. When, and if, a problem is not appropriately resolved by the functional team, it becomes a barrier to the program's objectives. The program manager must then intervene to ensure that the appropriate focus and resources are applied to the issue to reach closure.

Figure 13.1 illustrates the four competency areas in which the program manager must become proficient.

One aspect from the summary view of the core competency model is that project management is but one element of the program management discipline. This underscores the discussion in Chapter 1 that showed the distinction between program management and project management. One can visualize from Figure 13.1 that program management is a broader role than project management. For this reason, project management certification is helpful for the practicing project manager but is insufficient for the program management role. In addition to strong project management competencies, the successful program manager needs to gain proficiency in business, market, and leadership. Without this added proficiency, the transition from a project management role to a program management

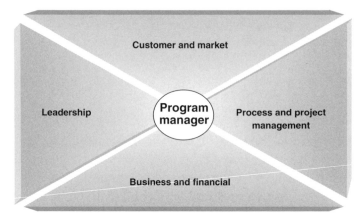

Figure 13.1   Program management competency areas.

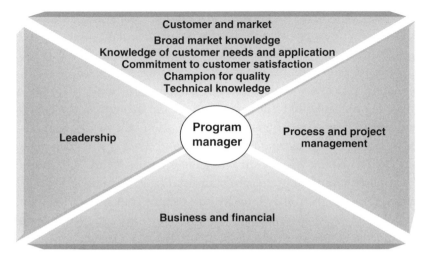

Figure 13.2   Customer and market core competencies.

role is many times frustrating for the individual, the organization, and the customer.

CUSTOMER AND MARKET CORE COMPETENCIES

Figure 13.2 illustrates the areas in which the program manager should gain proficiency within the customer and market core competency.

Customer and market competency means that the program manager has a full understanding of the market or organization that the product, service, or infrastructure capability is being deployed into and knows how it will be used by the customer and end user. This is important because the program manager and his or her team can closely align the product, service, or infrastructure capability with the customer's needs in a more efficient way. It will also enhance the potential for customer satisfaction and the successful achievement of the business results intended. Specifically, there are several technical competencies that the program manager should become adept at, including knowledge of their company's products, services, and infrastructure capabilities and key applications. Generally, the program manager should possess a working knowledge of the key technologies that a firm's products, services, or infrastructure capabilities are based upon, as well as the future technological trends within their industry.

As a part of a firm's development approach today, it is common to have key customers participate in the development process. Therefore, the program manager should be knowledgeable and sufficiently skilled to incorporate the valuable input of the customer into the design and development process—what some call customer intimacy. The program manager should be able to demonstrate a commitment to the customer, demonstrate knowledge of customer application and needs, and be able to apply market and customer validation and assessment tools (see box titled, "Do You Know What Your Customers Really Want?"). In addition, as the quality champion, the program manager should possess a bias for action; be able to think globally; and assure that quality, reliability, manufacturability, serviceability, and regulatory compliance targets are achieved.

Do You Know What Your Customers Really Want?

Program engineers in Accuracy Inc. were proud that they practiced the concept of customer intimacy, and as they put it, "it showed in their practice." They saw customer intimacy as the ability of their product development teams to recognize, internalize, and build the customer's needs into their customized interconnecting cable products. To check how much they were "customer intimate," they were advised to design and administer a survey that would ask 36 customers how satisfied they were with Accuracy's performance. The survey identified 15 dimensions of customer intimacy, which included, for example, the quality of the joint product definition process, quality of the rapid prototyping development process, ability to manage product changes, adherence to program milestones, and adherence to the final delivery date. For ratings, the survey used the following scale: 1-poor, 2-fair, 3-good, 4-very good, and 5-great.

On 2 of the 15 dimensions, Accuracy Inc. received very good ratings or higher for adherence to the final delivery date and program milestones. The remaining 13 dimensions revealed average ratings below 3, of which, most dimensions were rated about 2, or fair. As he looked at these ratings, the director of product development wondered if his group was really "customer intimate." What do you think?

Customer and Market Enablers

Key organizational enablers that support the program manager in the customer and market competency area include the following items: (1) Adequate funding provided by the company for thorough market research to identify such things as market segmentation, customer demography, market trends, and competitive positioning; (2) the development of good voice of the customer programs on the front and back ends of the program to help identify customer needs and customer satisfaction respectively; and (3) getting the customer involved in the product, service, or infrastructure development and validation processes to ensure the program output satisfies customer and market needs.

BUSINESS AND FINANCIAL CORE COMPETENCIES

Business competencies include the ability to develop a comprehensive program business case that supports the company's objectives and strategies; the ability to manage the program within the business aspects of the company; and the ability to understand and analyze the related financial measures pertaining to the product, service, or infrastructure capability under development. As we detailed in Chapter 12, managing the business aspects of the program is a primary role of the program manager. The areas in which the program manager should gain proficiency to effectively manage the business elements of a program are shown in Figure 13.3.

First and foremost, the program manager needs to have a working knowledge and level of proficiency in business fundamentals. As we've stated multiple times, program management is about the business. Business fundamentals include capabilities in financial analysis and accounting, international management, economics, law and ethics,

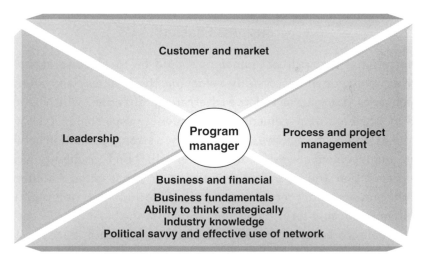

Figure 13.3   Business and financial core competencies.

resource management, negotiation and communication, and management of intellectual property.

The program manager is required to think strategically to align the program, and the projects within the program, to the strategic objectives of the organization. A part of strategic thinking involves a basic understanding of the industry in which the business operates. Industry trends, knowledge about competitors, and supply chain implications are a fundamental part of keeping a program viable from a business perspective. The program manager should possess a strong business sense and integrative capability to successfully manage the broad array of strategic, business, and financial attributes related to his or her specific program.

It is important that the program manager possess both a keen understanding of the organization and is politically savvy to build strong relationships. Company politics are a natural part of any organization, and the program manager should understand that politics is a behavioral aspect of program management that he or she must contend with to succeed. The most effective method for playing the political game is to leverage the powerful members of one's network who can help achieve the program objectives.[2] The key is to not be naïve and understand that not every program stakeholder sees great value in the program. A program manager must be sensitive to the interests of the most powerful stakeholders, and at the same time, demonstrate good judgement by acting with integrity.

Business Enablers

Organizational and cultural enablers should be in place to assist the program manager in successfully achieving the business and financial competencies. Long-term vision and strategic direction, which is identified by business management, should have a proper balance between short- and long-term objectives. Additionally, a strong correlation between senior management's words and actions will build credibility across the organization and, specifically, with the program team. An effective portfolio management process helps to prioritize and select programs that best align to the long-term vision and strategy of the business. This establishes a balance between resource capacity and demand. Lastly, robust and stable financial systems need to be in place to accurately measure program progress and results.

LEADERSHIP CORE COMPETENCIES

Leadership competencies are how we describe the "people skills" that program managers need to be successful in leading a team of people and managing a set of stakeholders. Leadership competencies involve those that are needed to effectively lead multiple cross-disciplined project teams that are a part of the program. Specifically, we are referring to organizational leadership competencies. The ability to lead the cross-project and cross-discipline teams is a key role of the program manager, as discussed in Chapter 12. The program manager needs to have the capability to build, coalesce, and champion the team to achieve product, service, and infrastructure solutions that will satisfy the company's customers. The areas in which the program manager should gain leadership proficiency to effectively lead the program team are shown in Figure 13.4.

Strong leadership begins with the ability to create and continually champion the program vision. If a program manager expects the program team to follow him or her through all phases of the PLC, he or she needs to explain how the team's hard work and dedication contribute to the broader vision of the program and company. People need to understand how their work contributes to the success of the business.

The second aspect of the leadership core competency is team building. The program manager should be able to build a team that will meet the program objectives; create an environment of trust and respect;

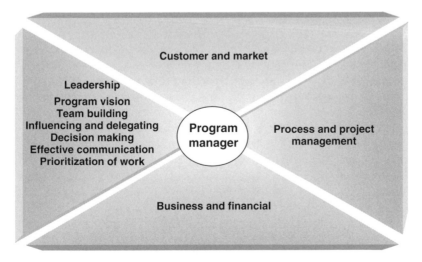

Figure 13.4   Leadership core competencies.

demonstrate value in diversity of people, styles, and opinions; acknowledge and recognize team member accomplishments; resolve conflicts and differences; and act as a coach and mentor to junior members of the team. Today's business models have created additional team building challenges for program managers, as discussed in the example titled, "Virtual, Discontinuous, and Condensed Team Building."

The ability of the program manager to achieve the firm's objectives lies with his or her expertise, background, and competencies to lead and accept accountability for the entire team's effort and resulting outcome. To effectively lead the team, the program manager needs to be respected by senior management and the members of the team. This requires personal characteristics that include credibility, integrity, and responsibility.

Virtual, Discontinuous, and Condensed Team Building

True program leaders forge the vision for the team, evangelically communicate it, and inspire and motivate team members to follow them. All this is typically associated with leading a single program at a time in a colocated manner. But business models of today have brought new leadership challenges for program managers. Two of which are discussed below: leadership of virtual programs and simultaneous leadership of multiple programs.

The first challenge, leading virtual program teams, raises the following question: Is it possible to build true leadership in virtual teams when members are geographically, culturally, organizationally, and time-zone dispersed? Patrick Little, a senior IT program manager for a leading research hospital answered that question this way:

"Leading a virtual team is possible, but it takes additional effort from all members of the team. Our virtual program team has an assigned program leader—myself—but leadership roles are shared within the team. Leadership responsibilities include motivating, information- or opinion-seeking, mediation, facilitating communication, removing barriers, lubricating interfaces, and making each conflict functional so it could be used to improve the quality of our decisions. All these aspects are important for building team spirit. However, doing this in a dispersed team structure requires a lot of travel on my part. Frankly, I am on the road all the time, constantly trying to tie all the pieces of the program together. This is very difficult and, at times, fails to work because I don't get to see some team members face-to-face for months."

Obviously, leadership in successful virtual teams is possible with true leaders and face-to-face meetings.

The second challenge, leading multiple simultaneous programs, is equally difficult. In a classic situation of leading a single program at a time, program managers can use any activity in their agenda throughout the program life to communicate their messages within the same program team. Similarly, they can focus on leading a single team without needing to split their time and effort between multiple program teams. Because program managers managing multiple programs need to build multiple teams concurrently, their time for each team is limited. Consequently, they need to apply a condensed, fast team-building method. Similarly, they are expected to operate in a discontinuous manner—lead one team, discontinue, and lead another team.

As business models continue to change, so do program management practices and competencies. And as managing programs become more complex, we find that program management core competencies expand to adjust to the complexity to continually provide improved business results.

In contemporary program management, members of the program team seldom report directly to the program manager. This requires that the program manager become proficient in influencing the actions of the team. Influencing competencies are also required to positively affect the actions and decisions on the part of the senior management team, key partners, suppliers, and support organizations. Successful program managers also are adept at the skill of conflict management and resolution. Conflict can, of course, occur at all levels of the program and between the program team and other groups within the firm. Anticipating these types of problems and being able to properly and objectively address and finesse them is one of the most important abilities of an excellent program manager.

Just as the program manager must be empowered by the senior leaders of the business, the program manager must be willing to empower the project managers on the program. This means being willing to delegate authority and responsibility. To effectively manage horizontally, or across the project teams, the program manager cannot spend an inordinate amount of time or effort working with any one project team (see Chapter 1). He or she must set clear expectations of each project team and then allow the project managers to execute and deliver on the expectations.

The program manager's ability to set and balance program priorities is one of the key indicators of success. The first step in setting priorities is checking to make sure the program priorities are correct. It is one thing for a program manager to set the program priorities, but it is most important to validate the priorities with the primary sponsors of the program.

The most critical aspect of the priority validation process involves testing the assumptions driving the program priorities with the program sponsors. If the assumptions behind the priorities are incorrect, it's quite possible the priorities themselves will be incorrect. Once the priorities are validated, the program manager should manage accordingly. For example, if cost containment is the highest priority of the program, then the program manager must emphasize the need to stay within the financial constraints of each project and the program as a whole. If technological leadership is the highest priority, the program manager will need to keep the project teams focused on the technical aspects of the program and provide the necessary resources to ensure technological success.

The success of the program manager's ability to make timely decisions will be considerably influenced by their degree of experience, knowledge acquired, and their ability to think of the program from a holistic perspective. Making good, timely decisions is a critical factor in achieving time-to-money goals. There can be thousands of decisions, large and small, that a program manager will encounter during the course of a program. To prevent even a small number of these decisions from being barriers to progress, a program manager needs to be proficient in collecting all necessary facts, analyzing the pertinent data, and then driving to a decision. Nothing can be more frustrating and paralyzing to a program team than waiting for a decision that gates progress.

Program decisions need to be aligned with the strategic direction of the program and the strategic objectives of the business; therefore, the decisions should be approached from a holistic perspective.[3] Losing sight of the strategic reasons for a program, while making the series of large and small decisions during the course of a program, is a primary reason why some programs become misaligned with the strategic objectives of the business. Likewise, decisions made on the program can also have an effect on the profitability of the business. Many times, decisions that a program manager makes during the program have the most important consequences after a product, service, or infrastructure capability is delivered. For example, a decision to add a new feature to a product may affect the gross margin in a couple ways. The added feature may allow the company to increase the price of each product sold, therefore, positively affecting the gross margin. However, the additional feature may change the physical configuration of the product. This may increase the manufacturing cost of each product, therefore, increasing the cost of goods sold and negatively affecting the gross margin. The ability of the program manager to make program decisions with all business consequences in mind is a critical skill within the leadership core competency.

Communication is another critical leadership competency in which the program manager should foster open communication across all functional elements and throughout all levels of the organization. This requires the ability to speak multiple disciplinary languages—business language when communicating with senior management, user language when communicating with the customer, technology language when communicating with technologists, and so on. Effective communication competencies also mean that the program manager should be able to actively listen and provide clarity in difficult situations, many times serving as the translator in multidisciplinary discussions.

A senior program manager we talked with told us the story of why the second program he managed was a major failure (his first program was very successful).

> Because of the success I had with my first program, I thought I knew all the answers on the second program and stopped listening to what was being said to me by the program team. I listened to what I wanted to hear versus what they were really telling me. I wasn't looking for problems, even though they were being communicated to me by my team. Eventually, the problems became insurmountable and I was removed from the program manager position.

Finally, effective program managers are continually prioritizing their work and the work of their team to achieve the greatest return from the actions of the team.[4] There is an enormous opportunity for the program team to spend too much time on work that is of little value, which is due to the complexity and scale of many programs. The ability of the program manager to focus the work of the project teams on the highest priority needs of the program, for example the next set of deliverables due, is crucial for maintaining a high level of team productivity. The effective program manager is continually sharing the program vision and the next set of targets to hit to keep the work of the project teams focused on the priorities of the program.

Leadership Enablers

Several enablers should be in place to assist the program manager in successfully developing and utilizing the leadership core competencies.

A stable organizational structure with clearly defined roles and responsibilities establishes the foundation for good leadership. The program manager must be empowered by senior management, with an established balance between responsibility, authority, and accountability to effectively lead the program team. A well-defined decision structure and program governance plan should be in place to facilitate time-to-decision performance. Finally, a consistently visible and active senior management team is needed. Senior managers must understand and be willing to perform their role on the program on a consistent and continual basis.

PROCESS AND PROJECT MANAGEMENT CORE COMPETENCIES

The program manager should be well versed in core program processes and project management competencies and be able to abstract the competences for use at the program management level. The program manager needs to be competent in program processes to handle program-level issues. Also, both the program manager and the functional project managers on the PCT must possess operational competencies, including project management methods and tools, to effectively manage the tactical elements of the program. The areas in which the program manager should gain process and project management proficiency are shown in Figure 13.5.

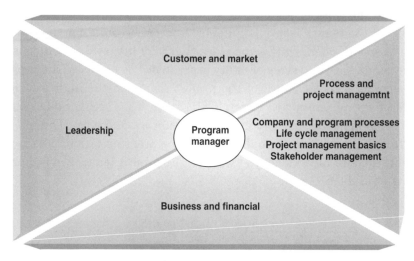

Figure 13.5 Process and project management core competencies.

Another important aspect of this core discipline set is the program manager becoming proficient at possessing a solid working knowledge of the specific processes and practices of the company. If he or she knows how things get done, the policies and procedures that must be adhered to, and understands who must be involved and approve various aspects of the program, the manager will increase the likelihood of successfully completing a program. If it is a product-based company, for example, the program manager must be thoroughly familiar with the firm's new product design, development, and market launch processes to ensure that the development team's efforts adhere to management's requirements and expectations as to how products must be designed and built. The program manager who develops this competency will increase his or her probability of gaining team member's confidence and trust. They will become confident that their leader knows how to get things done in a timely and successful manner.

Project management basics are best described and represented in the Project Management Institute's (PMI) PMBOk. This reference guide presents nine knowledge areas for the effective management of projects:[5]

- Integration management
- Scope management
- Time management
- Cost management
- Quality management
- Human resources management
- Communications Management
- Risk management
- Procurement management

The PMBOK provides key information and understanding that both project managers and program managers should master to possess the highest probability of success in achieving the objectives and expectations of their customers and management. When it comes to project management competencies, the program manager's role is similar, with the key added dimension of integrating and coordinating the work being done on two or more projects simultaneously within a program.

In addition to the knowledge areas called out in the PMBOK, stakeholder management is a critical competency that needs to be called out separately. The program manager can have many stakeholders that he or she has to manage both internally and externally (see the box titled, "Even a Salad Vendor May Be a Stakeholder"). In some cases, stakeholder management can become a full-time job in and of itself, if not performed effectively. Good stakeholder management involves understanding who the program stakeholders are and their needs, the level of influence each stakeholder has on the program, and their allegiance and attitude toward the program (never assume all stakeholders want the program to succeed). From this information, the program manager can determine which stakeholders he or she needs to manage and how to manage them (see Chapter 8). Effective stakeholder management helps the program manager gain cooperation from the highly influential stakeholders, cut through competing stakeholder agendas, and confront those who may be inhibiting program progress.

Even a Salad Vendor May Be a Stakeholder

This is a bizarre example of what can happen if a comprehensive stakeholder analysis is not conducted on a program. The program was a fast-track transfer of manufacturing technology from Europe to a Middle East country. Just after the beginning of the program, the program manager got a call from a vendor claiming that all roads leading to his office were blocked by groups of violent people. As a result, some important computer equipment couldn't be delivered. A quick check proved this call correct. The people were local butterhead salad farmers, who were unhappy that the foreign contractor was not buying salad from them, rather, importing it from Europe. The siege went on for days and the delivery delays were impacting the program schedule. Finally, the program manager figured out his mistake—one cannot ignore the relationship with local communities that have a big stake in the program. The salad farmers, in this case, were part of the local community, therefore, a stakeholder in the program. Once the oversight was discovered, the contractor began buying local salad and program deliveries proceeded without delays.

Project Management Enablers

Key company and industry enablers should be in place to assist the program manager in successfully performing the process and project management competencies. The firm should encourage the connection of their program and project managers to professional organizations to enhance and broaden their learning experience. It is helpful for career development if senior management assists in funding personal skill and knowledge development. It is also helpful if a firm views program management as a functional discipline and provides and encourages ongoing development of personnel in this skill category. Additionally, some firms require basic skill certification in project management as an entry-level requirement for employment as a project manager.

COMPETENCY-MATURITY MIX

Each company should adapt the mix of competency groups in the described model and tailor it according to its needs and level of program complexity. Specifically, in the descending order of complexity, companies do breakthrough programs (yielding products, services, and infrastructure capabilities that are new to the world) platform programs (yielding multiple configurations); and derivative programs (yielding products, services, or infrastructure capabilities derived from the platform products). Each of these program types is of different complexity and requires program managers of different maturity and competency levels.

For example, as demonstrated in Figure 13.6, junior program managers typically have higher levels of technical competency and lower levels of business and leadership. This is mainly because junior program managers are commonly promoted from a technical function or the project management ranks to lead a program. They have not yet had time or

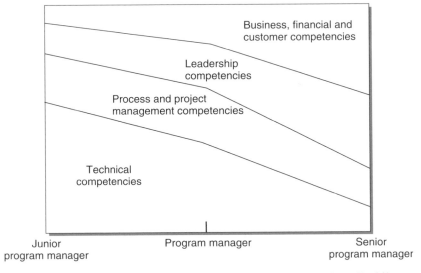

Figure 13.6   The program management competency—Maturity Mix.

experience leading larger, more business significant programs to hone their leadership competencies.

In contrast, senior program managers who lead larger and more significant breakthrough and platform programs have typically not been involved in solving specific technical problems for a period of time; therefore, they are unable to maintain their technical competency. Rather, long exposure to business, financial, and customer affairs help the managers polish these broad-based competencies. Similarly, managers with years of experience, who lead cross-discipline program teams, have the opportunity to polish their business, financial, and leadership competencies.

Generally, direct technical skills become less important as the program manager takes on larger, more complex programs. He or she then delegates management of the technical aspects of the program to the project managers but may intervene on a technical issue, if it becomes a barrier to the success of the program. The need for business, financial, market, and leadership competencies grows as program managers face more complex and strategically important programs.

Process and project management competencies are somewhat different. Junior program managers have modest competencies of this type because they normally have project management experience within the company prior to taking on program management assignments. As their maturity and experience grows in the program management discipline, process and project management competencies become less significant, and more focus is applied to the business management and team leadership aspects of program management.

ALIGNING COMPETENCIES TO BUSINESS STRATEGY

Understanding and managing the program management competencies and mix of maturities is important but insufficient by itself. A company utilizing the program management discipline to develop products, services, and infrastructure capabilities should also align program management competencies to the business strategy for maximum effectiveness. As discussed, companies select a business strategy as a means of competing with their rivals in the market place. Although each type of business strategy has the same goal to create competitive advantage, ways to accomplish the goal differ. For example, one company may choose to build competitive advantage by using a strategy to be the low-cost leader within an industry, while another company within the same industry may choose a strategy of technical differentiation.

The business strategy of a company should then drive the particular competencies that are necessary. For example, a program manager within a company employing a low-cost strategy will need to develop competencies that focus on cost first and foremost. This requires customization of the program management competency model to emphasize cost-based skills and abilities within the four competency categories. The box, "Customizing Competencies in the Projector Industry," provides an in-depth example of how program management competencies are tailored to align to a company's business strategy.

Customizing Competencies in the Projector Industry

To understand the effect of business strategy, let's use as an example of Porter's model of the strategies (Figure 13.7) for three companies producing liquid crystal display (LCD) projectors.

The core of differentiation strategies in Figure 13.7 (high-differentiation/ high-cost quadrant) is an ability to offer customers something different from a company's competitors. This may include fast time to market, which we used as an example in Figure 13.7, high quality, innovative technology, special features, superior service, and so on. When LCD projector companies strive for product superiority, they pursue these strategies and build in whatever features customers are willing to pay for. That enables them to charge a premium price to cover the extra costs for differentiating features.

Companies focusing on low-cost strategies aim at establishing a sustainable cost advantage over rivals (low-cost/low-differentiation quadrant). The intent is to use the low-cost advantage as a source of underpricing rivals and taking away market share from them. Another option is to earn a higher profit rate by selling at the going market price. This is pursued with a good, basic product that has few frills and continuous quest for cost reduction, without giving up quality and crucial features.

Best-cost companies combine upscale features with low cost (low-cost/high-differentiation quadrant). This should lead to superior value by meeting or exceeding customer expectations on product features and surpassing their expectations on price. At the same time, the aim is to become the low-cost provider of a product that has good to excellent features and use that cost advantage to underprice rivals with comparable features. Because such a company has the lowest cost compared with similarly positioned rivals, the strategy is called best-cost strategy.

The blank quadrant in Figure 13.7, (high cost/low differentiation) is not a viable option in competitive battles for survival and prosperity.

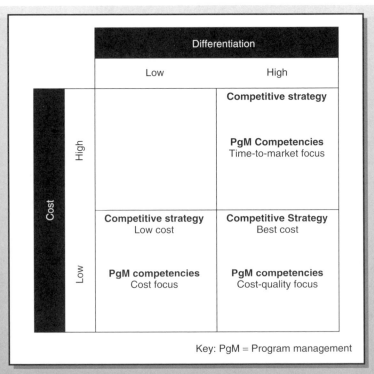

Source: Porter, M.E. Competitive Strategies: Techniques for Analyzing Industries and Competitors, 1st edition. New York, Ny: Free Press, 1998

Figure 13.7   Customizing competencies by strategy.

Now, let us use the competency model to see how the business strategy shapes the competencies within each group. We are going to use examples of three different companies called Sirius, Park, and Plan and show how the strategy impacts the changing importance of individual competencies within the process and project management competency. Sirius's business strategy is one of differentiation. The strategy targets innovation and time-to-market speed as competitive advantages. Within it, a significant role belongs to product development programs that roll out new advanced LCD projector chips faster and faster. This is when program management processes come into play, focusing on overlap across and within phase activities to shorten time-to-market on risk management and interface integration of new technology. Project management emphasizes competencies related to time management.

Park's business strategy is different from Sirius. Instead of differentiation and time-to-market emphasis, Park has set out to become the low-cost leader in the industry. To develop the ability and become the leader in the industry, Park has had to emphasize competencies to streamline program management process and continuously lower the program development cost goals. Part of that effort has been perfecting project cost management competencies for managing cost cutting within the program's projects. Nurturing these competencies support Park's low-cost advantages.

Sirius's and Park's competitive advantages exploit their schedule- and cost-focused program and project management competencies, respectively. In contrast, Plan relies on a best-cost strategy. The goal is to have the best cost relative to competitors whose LCD projectors are of comparable quality. Accordingly, their competencies concentrate on a program management process that emphasizes high quality and low development cost. Project management competencies aim at accomplishing cost and quality goals through excellent cost and quality management.

Consequently, in the case of Sirius, Park, and Plan, each company's perspective and focus is different, as follows: schedule-focus (Sirius), cost-focus (Park), and cost-quality focus (Plan).

MANAGEMENT OF MULTIPLE PROGRAMS

We would be remiss if we did not address the subject of a program manager managing multiple, simultaneous programs. This is relatively common when a company has a number of smaller programs occurring at the same time. It is not unusual for a program manager to manage two to four smaller programs concurrently. It is our point of view that managing multiple, smaller programs simultaneously is not a unique competency but rather a matter of experience and proficiency gained in the program management core competencies. As program managers become more proficient in their core competencies, their skills become scalable, which allows them to manage larger, more complex programs and multiple programs.

STAFFING THE PROGRAM MANAGEMENT POSITIONS

Staffing of program managers within the organization is a key exercise of capacity loading for the number and complexity of programs that the firm plans to manage over a given period of time. The importance of doing an excellent job of staffing is obvious—the ability to select the best candidate for the role will enhance the organization's ability to succeed in their development objectives. Once the staffing needs have been determined, the organization can fill these positions either from external new hires or by selecting individuals within the firm who possess the potential to grow into fully qualified program managers. The latter approach assumes that at least a portion of the firm's existing program managers are appropriately trained and sufficiently experienced to provide coaching and mentoring to any new internal candidates.

Selection of individuals should be measured against established job descriptions that characterize the specific role needed for the firm; the key responsibilities; and the required skills, abilities, and qualifications of the position. The job description will serve as the basis for establishing the screening, selection, and hiring criteria. Assuming the firm has multiple levels of program management experience identified, job descriptions should be tailored to fit each position. These positions may range from an entry-level program manager who manages smaller, less complex development efforts to a senior program manager who manages large, complex, and multisite development programs.

PROGRAM MANAGER PERFORMANCE EVALUATION

Once the program manager job description has been approved and implemented, it can serve as the basis for individual performance evaluation. The job description, the program management competency model, and expectations by management will serve as the basis for determining and documenting the accomplishments and overall performance of the individual. Performance evaluation, if done correctly, can provide recognition for good performance, increase personal motivation, and identify course corrections needed for performance misses. The performance evaluation, coupled with training and development, become powerful tools for driving continual improvement in program management practices within an organization.

PROGRAM MANAGER PROFESSIONAL DEVELOPMENT

The knowledge, skills, and abilities described as part of the program management competency model are most useful for growing and developing the firm's program managers, once they are hired. This is important because future gains are made through a model of continual improvement in performance. Business competition has proven that if you are at status quo, others will pass you by. Systematic training and development will increase the probability of staying ahead of the competition. Of course, much of what the program managers will learn will come from the on-the-job training while running their specific programs. However, to further broaden their capabilities and achieve the potential of the individual, it is helpful if senior management provides encouragement for continual improvement in their discipline.

Many program managers are in their disciplines because they enjoy the role and want to make it their career. Most highly effective program managers are self-motivated and demonstrate a desire to self-assess, take on a philosophy of continued improvement in their growth and responsibility, and are persistent in attainment of their personal objectives. In doing so, they may pursue growth and advancement opportunities within their company, which is important for the long-term viability of the firm. In many firms, the role of program management is viewed as a career path to business management. This is because much of the program management role embraces and overlaps the competencies associated with business and general management.

As stated earlier, it is rare that program managers come into the role fully proficient in all core-competency areas (see box titled, "How Many Skills are Required?"). The program management competency model has been used by some companies as the guideline for continual career development of their program managers. Ongoing dialog between the individual program manager and his or her direct manager are focused on understanding the program manager's growth and career aspirations and balancing that with management's short- and long-term performance expectations. The results from this exercise serves as the basis for an individual development plan over a given period of time (annually in most cases) for the program manager. It is a process of targeting where the individual currently is in their performance and capability, where they want to be in their career and performance at some time in the future, and then laying out a plan of development on how to get there. The specifics of the plan may include some or all of the following:

- *Internal training courses*: Some firms offer their own structured training modules for the program manager. These may be web-based and available to the entire population of the company regardless of location.
- *External training seminars and courses*: Many professional organizations and consulting firms offer a variety of courses. However, good courses focused on program management are few and difficult to find.

- *College-degree programs*: There are currently no degree programs offered in program management. However, some Masters of Business Administration degrees are valuable for providing the broad core competencies needed by a program manager.
- *Industry certifications*: At the time of this writing, there are no program management certifications currently offered outside of the U.S. defense industry, and project management certifications are insufficient for program management. However, some companies have instituted their own internal certification.
- *Specific assignments for on-the-job training and experience*: This is an effective technique for broadening the experience and capability of a program manager.
- *Rotational assignments through various organizations within the firm*: If the firm views program management as a career path to broader and larger roles in the firm (such as general management), this is a useful way to evaluate the potential of the individual.
- *Mentoring relationships*: Newer program managers can gain much knowledge and capability enhancement by mentoring from an experienced program manager.

How Many Skills are Required?

The depth of skills a program manager needs depends upon a number of factors: the size and complexity of a company's programs, the level of maturity of its program managers, the industry the company is in, and the culture of the company. For example, one world-class company that we have consulted for requires a range of 11 to 28 different skills for its program managers, depending upon the maturity of the program manager and the types of programs he or she is required to manage. The company uses the following three-level program management career path: junior program manager, program manager, and senior program manager. Skills for each maturity and experience level are shown below:

Junior program manager	Program manager	Senior program manager
Leadership Skills	*Leadership Skills*	*Leadership Skills*
• Conduct effective meetings	• Conduct effective meetings	• Conduct effective meetings
• Conflict management	• Conflict management	• Conflict management
• Team development	• Team development	• Team development
• Team leadership	• Team leadership	• Team leadership

Junior program manager	Program manager	Senior program manager
Leadership Skills • Decision making • Effective-problem solving	*Leadership Skills* • Decision making • Effective-problem solving	*Leadership Skills* • Decision making • Effective-problem solving • Coaching • Influencing • Facilitation
Process and Program Management Fundamentals • Differentiation between programs and project mgmt fundamentals • Characteristics of a good program manager • Program and project differentiation	**Process and Program Management Fundamentals** • Differentiation between programs and project mgmt fundamentals • Characteristics of a good program manager • Program and project differentiation • Company planning • Program life cycle • Procurement • Program and project tools • Supplier management • Program/project tools	**Process and Program Management Fundamentals** • Differentiation between programs project mgmt fundamentals • Characteristics of a good program manager • Program and project differentiation • Company planning • Program life cycle • Procurement • Program and project tools • Supplier management • Program/project tools
Business and Financial • Business and financial skills • Alignment to corporate strategy	**Business and Financial** • Business and financial skills • Alignment to corporate strategy • Negotiating to yes **Managing Complex Programs** • Geodispersed teams • Matrixed teams • Virtual teams • Managing without line authority	**Business and Financial** • Business and financial skills • Alignment to corporate strategy • Negotiating to yes **Managing Complex Programs** • Geodispersed teams • Matrixed teams • Virtual teams • Managing without line authority **Managing Multiple Programs**

Junior program manager	Program manager	Senior program manager
		• Managing multiple programs • Resource allocation across programs • Prioritization of programs

The junior program manager needs 11 core skills, while the senior program manager needs 28. As one can see, the more senior the program manager, the more depth in skills required.

If the program management model is embraced by a company, then growing and nurturing it's program management personnel not only enhances performance but also provides greater individual advancement opportunities and the ability to grow talent within the firm for the future. According to Steven Wheelwright and Kim Clark, authors of "Revolutionizing Product Development", it is essential that managers give attention to issues of career paths, education and training and, in general, the development of human resources in the organization.[6] If a company does not provide an established career path for the program manager, then a program manager's best chance for advancement within his or her discipline may be pursuing more senior-level positions in other companies or industries.

ORGANIZATIONAL CONSIDERATIONS

As mentioned earlier in this chapter, there are certain management philosophic and cultural considerations that bear on the ability to advance the program management discipline within the firm, as follows:

Strategic importance of program management to the firm: Firms that see the strategic importance and value that the program manager brings to the business are more motivated to invest in the growth of the function for the future. These firms are willing to fund personal skill and knowledge development for program management personnel.

Viewing program management as a discipline: Program management needs to be recognized as a professional discipline and managed as a function within the company. Organizations are much more inclined to recognize the need to fund development of competencies within the formal functional disciplines of the company. Having program management recognized as a function and a professional discipline, which is lead by a senior manager or director of program management increases the probability that senior management will fund its skill and development needs.

Formalized professional development planning: A formal approach to professional development contributes to an environment of continual improvement of the firm's employees. If the company formally looks at individual performance jointly between the employee and their manager and finds ways to improve performance for the future, it will naturally lead one to see the value of improving the competencies and capabilities of the employee.

Training philosophy: Larger firms are more capable of providing internal training and development resources and fund external opportunities for their employees than smaller firms. However, it also depends on the philosophy and management view about the value of training and development. Some senior managers are much more inclined to view training and development as an important weapon in their arsenal to stay competitive. These managers foster a philosophy of continual professional development of their employees and increased capability of their organization.

One executive who has experience implementing program management in several organizations is David Churchill, vice president and general manager of the Network and Digital Solutions Business Unit of Agilent Technologies. Churchill describes effective management philosophy in support of program management this way:

> Organizations that recognize the strategic value of program management will do the following: treat program management as a critical talent and skill set and establish it as a functional discipline like engineering and marketing; elevate the program management function in stature and place it at the senior level in order to provide program managers the necessary level of influence across an organization; and empower program managers as leaders within the organization with sufficient authority to implement and achieve the intended business objectives.

SUMMARY

The program management competency model[7] has been designed to provide the necessary knowledge, skills, and organizational enablers to systematically support the recruiting, staffing, professional development, and career planning of the program manager. The model specifies the areas in which the program manager should gain proficiencies to help the organization achieve expected business results. These areas are as follows:

- Customer and market
- Business and financial
- Leadership
- Process and project management

Each company should tailor the mix of competencies to its needs. Customization is necessary due to variations in program complexity, business strategy, and specific characteristics of the industry in which the enterprise operates. Finally, this chapter discusses the connection between the program management competency model and program staffing, performance evaluation, and professional development.

The Principles of Program Management

▼ A broad set of skills and capabilities are needed to fulfill the demanding role of the program manager

▼ A balance between technical competencies and behavioral and human-oriented competencies is required

▼ The four competency areas are: (1) business and financial, (2) customer and market; (3) leadership; and (4) process and project management

▼ The mix of the four competency areas are tailored per a program manager's maturity

▼ A program manager seeks to constantly learn and broaden your knowledge and experience

---

Program Management in Practice

They Are Business Leaders at Spotlight Corporation
Authors Peerasit Patanakul and Dragan Z. Milosevic

Prologue

This industry example focuses on program management competencies as practiced in a technology-driven organization within a high-velocity environment and on the new product development programs, which fuel the organization's growth engine. The organization is Spotlight Corporation, a leader in the LCD projector industry.

The leaders of Spotlight Corporation see program management competencies as an essential element of their performance and continued growth, as well as a key factor in whether a program will succeed in achieving the desired business goals.

Competencies for program managers at Spotlight Corporation are driven by and adapted to the business strategy. This principle is widely used in the business world, as we emphasized in Chapter 13. In particular, companies such as Spotlight Corporation make sure that the competency-group mix for program managers is in tune with the company needs. This also includes adjusting the various skills within a particular competency area, as needed.

The management of multiple programs is an additional competency of experienced program managers introduced in this example. It involves developing the skills necessary handle the interdependency and multitasking responsibilities associated with managing multiple concurrent programs. Lastly, this example introduces the competency-metric concept, which is used to gauge a program manager's competency level and determine if it is sufficient for leading a company's most strategically important programs.

Spotlight Corporation

Having arrived at work a little early, Dave Moskhill reflected on the business of the company he had recently joined, Spotlight Corporation. As the newest member of Spotlight's executive management team, Moskhill spent the last couple of weeks getting familiar with his new company. Although he collected a lot of information about Spotlight, many aspects of the business were still unfamiliar to him.

For Spotlight, product development is the engine of its growth. To maintain a leadership position in the market, the company emphasizes leveraging its core competencies in advanced research and engineering design to develop new markets, while maintaining customer focus and improving efficiency and effectiveness in product development processes. Rapidly changing technology, customer demands, and aggressive competitors require a clear and well-developed strategy.

Like many U.S. corporations, Spotlight recently outsourced its manufacturing to China. On average, Spotlight implements 40–50 product development programs per year. The programs include all types, from derivative to breakthrough, and range from 1 million to more than 5 million dollars in budget and 9 to 24 months in duration.

Moskhill scheduled a meeting with Brian Hall, director of the PMO, to discuss the way things work at Spotlight. Their discussion, focusing on Spotlight's program management function, follows.

Program Managers at Spotlight

Moskhill had a particular interest in program managers and the competencies they need to be real business leaders. He opened the discussion by asking Hall how many program managers there are, who they report to, and what they do.

According to Hall, there are 12 program managers in the PMO that report to him. On average, each of them leads two to three programs at a time—some big, some small. Some of the programs are derivative with added features and others are new product development programs. Because Spotlight's goal is to consistently make its products physically smaller, it requires new technologies. New product development programs are fast paced and involve high market uncertainty and high technological and organizational complexity.

"You need true leaders with significant competencies and a wide range of experience to lead programs like those," Moskhill said. Hall agreed, and added, "In our program management office, we have three levels of program managers—entry level with very little experience, highly experienced veterans who've seen it all, and those in between."

Hall added that Spotlight uses a competency metric to help determine what level of maturity their program managers are at. "We have a comprehensive list of competencies that we think our program managers need, and we rate their competency level against that list," he said.

Hall explained how Spotlight's competency list was developed. "When we first thought about the competency metric, we sat down, studied models in the technical press, and benchmarked some companies in our industry. We put the list together, but when we used it to gauge program managers, some mid-level program managers only scored as a level 1, so we knew the model had some serious flaws. We went back to the drawing board, testing different models of the competency metric. Finally, it dawned on us that the competency metric we developed corresponded to models from the technical press and benchmarked companies. What we really needed was a model for our own business strategy and our program managers executing that strategy. We picked our five most successful program managers, evaluated their competencies, averaged them out, and proclaimed them a temporary competency benchmark. Then, we refined the metric for almost a year until we polished it to reflect our company and the competencies of our program managers. We take pride in the fact that it is not just a competency list, but competencies that help Spotlight's program managers improve program results."

Hall put the competency metric sheet on the desk and offered its rationale (see Table 13.1). The metric had six groups of competencies on the list, as follows: technical, administrative and process, interpersonal, intrapersonal, business and strategic, and multiple program management. Spotlight program managers need to have multiple program management competencies because

Table 13.1   Spotlight Competency Metric

Competencies

Technical
- Knowledge of product applications
- Knowledge of technology and trends
- Knowledge of program products
- Knowledge/competencies of technology tools and techniques
- Ability to solve technical problems

Administrative/process
- Monitoring/control
- Risk management
- Planning/scheduling
- Resource management

- Company's program management process

Interpersonal
- Leadership
- Communication
- Team management
- Problem solving
- Conflict management

Intrapersonal
- Organized and disciplined
- Responsible
- Proactive and ambitious
- Mature and self-controlled
- Flexible

Business/strategic
- Business sense

- Customer concern
- Integrative capability
- Strategic thinking
- Profit/cost consciousness

Multiple program management
- Experience of managing multiple programs
- Interdependency management
- Multitasking
- Simultaneous team management
- Interprogram process

they lead two to three programs simultaneously. These competencies will help them coordinate multiple concurrent programs.

Technical Competencies

Hall explained to Moskhill that technical competencies constitute the knowledge, skill, and experience of a program manager related to the technical facets of the program product. At Spotlight, the program managers are not required to have extensive knowledge of product technology, to know how to use technical tools and techniques related to product development, or to have the sophisticated ability to solve technical problems. "According to our experience, our program managers need to know not only the latest product technologies in the market but also understand when we are talking about the importance of system architecture being defined and how critical that is to the program," Hall added. "In general, they understand the technological concepts of products and their application, including the knowledge of technology and dominant trends."

"That concept is easy for me to grasp because it is similar to what we used in my previous company," Moskhill said. "However, the technical aspects of products are very important. How do you know that the technical issues will be taken care of?" he asked.

Hall emphasized that it is not just that the technical aspects are very important, but in some programs, there are also a lot of uncertainties. If a program is highly complex and involves many technological uncertainties, an experienced technical leader will be assigned to the program, and the team will be staffed with technical people. The program manager focuses on the big picture and the business aspects of the program.

Moskhill agreed and emphasized his point that because program managers lead two to three programs simultaneously, it is almost impossible for them to have technical knowledge of all of them. "In general, they should have the knowledge of dominant technology trends," he said.

Administrative and Process Competencies Are a Given

Moskhill turned his attention to the administrative and process competencies. Common sense told him that if program managers do not have the knowledge, skills, and experience in planning, scheduling, organizing, monitoring, and control, they shouldn't be program managers. But he wondered to what extent those competencies helped program managers enhance program success.

Hall gave Moskhill good information about this issue. He emphasized that, besides planning and scheduling, Spotlight's program managers should be proficient in monitoring and control. "This is very important, since the business environment changes rapidly, and those changes often impact their programs directly," Hall said. "In other words, if the changes impact the direction of their program, program managers should be able to recognize those changes, talk to executives about setting new goals, and control the new direction of the program."

Hall continued that risk management is also important because most of the programs involve high levels of technological uncertainty. "Program managers know that they have to identify risks in their programs, estimate the probability of those risks occurring, identify the severity of risks, and propose the counter measure or strategies to prevent or mitigate them," he said. "I have to admit that we are not very good at risk management yet," he added, "but I believe we will be."

Moskhill was glad to hear that there was a desire to improve risk management and decided to emphasize it. "Good," he said. "Risk management is very important in our environment, and I understand that it is not easy. Maybe we have to provide more training on effective-risk management practices." Moskhill went on to inquire if program managers have the ability to negotiate and allocate resources.

"Yes," Hall replied. "The program managers know how to identify and estimate resources, negotiate resources with the functional groups, allocate them, and monitor and control resources. I also expect our program managers to understand how things work here." He added that program managers understand the program management processes, including policies, procedures, and tools that are used to manage programs. In other words, even though they have other competencies, they have to learn the company's lingo, forms, process, and get to know people. This will help them manage programs effectively and, in fact, will help them properly employ their program management competencies. "All in all, I want our program managers to have a solid foundation of program management that they can tie into any type of program" Hall said. "I call all of these administrative and process competencies, and they consistently help to enhance Spotlight's program managers' results."

Moskhill agreed and the discussion turned toward people competencies.

Soft Competencies Do Matter

Spotlight learned that only having process competencies is not enough. Their program managers are expected to have interpersonal and intrapersonal competencies as well. The company considers these soft competencies, and Hall was a strong advocate of this approach. Interpersonal competencies of leadership, communication, team building, problem solving, and conflict management were the competencies listed in Table 13.1 he and Moskhill were looking at.

"Let me give you the details, Dave," suggested Hall, "Our program managers are proficient in setting direction, delegating authority, and influencing the team with fairness. They also have political competency to be able to set the priority of program activities to be in line with management and the goals of Spotlight. In addition, they should be able to influence people and have credibility in the eyes of all the program stakeholders, including senior management. They should also know when to involve me or senior management in program activities," he added.

This was a hot button issue for Moskhill, and he jumped in to comment. "I agree. Based on my experience, program managers should have good communication competencies both verbally and in writing," he said. "I think a successful program manager should be a good listener and ask the right questions. I see that you have communication competency on the list. But I want to emphasize that our program managers need to know how to articulate and handle any kind of information, whether it is technical, legal, administrative, or interpersonal in nature. Now, what about team management?" Moskhill asked.

Hall commented that team building is important, especially at Spotlight where some team members are from Europe and Asia. "There is an expectation that program managers put a team together that is committed and mutually accountable," he said. "Then, they should be able to keep the team motivated in a group setting. They are expected to be good at problem solving and conflict management. As we know, these issues increase the level of complexity when dealing with distributed teams," Hall added.

Moskhill asked about intrapersonal competencies, and Hall explained that he expects his program managers to be organized, thoughtful, and methodical. "By being organized and disciplined, program managers are able to perform their job better in a high-velocity environment," Hall said. "They should also be proactive and ambitious. Program managers should be action-oriented and self-motivated, so that they can anticipate issues and develop a plan to account for them." He added that he believes program managers are responsible for getting people motivated. "If they cannot get themselves motivated on the program issues, their teams won't get motivated. The other characteristics that they need, as I listed in the table, are being mature and self-controlled. These will help them have emotional stability, patience, poise, and tolerance toward uncertainty. Our program managers need these characteristics, especially at Spotlight where things change quickly," he said.

Moskhill added that program managers should be spontaneous, adaptive to the working situation, and open to change. "Sometimes program managers have to lead a nontraditional program that requires them to stay up until 2:00 or 3:00 in the morning to have a teleconference with a team overseas," he said. "Therefore, being flexible is important."

Hall agreed, adding that besides the intrapersonal competencies on the list, being entrepreneurial, creative, visionary, and competitive are also important.

Business and Strategic Competencies Are a Must

Moskhill noticed that business and strategic competencies were on the list, and Hall commented that program managers need to understand business and strategic aspects of their programs. "These will help them make the right trade-off decisions on the program. We can tell from Spotlight's experience that having business sense assists them in formulating any program issues in a business context; recognizing fine variations among schedule, budget, and performance needs; and making benefit/cost trade-offs," he said. "In addition, they pay attention to customers in order to understand and respond to their concerns. We believe that these competencies have helped improve our programs over time."

Hall added that the competencies are integrative capabilities because they help a program manager make decisions in the systems context. "They have to take into account the big picture, the revenue associated with the products, and customer involvement, while making decisions," he said.

Moskhill stared at the list, saying that he was somewhat confused because he thought that the strategic thinking competency was one of top management. "At Spotlight, people believe that it is very important for our program managers to have this competency as well—maybe not as extensive as what top management has," Hall clarified. "We expect them to understand and adapt to the strategic direction of Spotlight and recognize our competitive components. These will help them manage their programs more effectively. Last but not least, our program managers should always take profit and cost into account when they manage program details. In a nutshell, we are doing business. Therefore, in our experience, our program managers should understand the business aspects. This has made our organization more competitive and helped improve program results," he added.

Multiple Program Management Competencies

Not having seen this group of competencies before, Moskhill asked Hall to explain more about it. Hall explained that the competencies they had previously discussed are important for managing each individual program. But when it comes to coordinating multiple programs simultaneously, multiple program management competencies are needed.

Hall added "To be good in multiple program management, program managers should have at least two years experience in managing multiple programs for Spotlight. The point is that they need time to establish their credibility and network inside the company. Program managers should also be competent in interdependency management to be able to manage interdependencies and interactions among programs related to shared milestones, resources, and technology," he said. "They should see the big picture and not get lost in details. There are an incredible amount of details in any one program, and program managers have to be able to step back and focus on the right things."

Hall continued that program managers need to move out of the mindset that they need to know the task level of details and move up to managing the deliverable level across the board for all disciplines. "I think this will help them in problem solving too," he said. "They have to understand how to solve a problem to the benefit of all the programs they are working on."

Multitasking is also important at Spotlight, according to Hall. He stated that good program managers are able to estimate their own resource capacity (for example, number of work hours per week or month) to set priorities and switch contexts and multitask among different programs. "Multitasking is a significant challenge when managing more than one program because each program often has unique characteristics," he said. "Balancing elements of time, cost, and performance across the metrics of two or three programs is always challenging."

"Sure, they have to get refocused when they move from issues of one program to the next," Moskhill agreed. "But does being good at multitasking mean losing time when switching from one program's issues to another program's issues?"

Hall nodded. "Yes. Our program managers once told me that they, on average, lose 20 to 30 minutes during context switching. This can be a big loss of their time if they change gears, that is, programs, many times a day," he said. He added that another issue needing emphasis is that program managers should know how to lead several teams simultaneously. "They should be able to select and use different management styles specifically for each team. Since they have limited time to spend with each team, program managers need to be able to organize the team and have it become effective in a timely manner," he added. "In addition, they have to be expert in communications. The communication has to be concise," he added. "It's hard to find time to have a lot of face-to-face communication with many people."

Competency Metric in Action

Finally, Moskhill and Hall came to the point of how to gauge these competences and determine the competency level of program managers. "The competency metric is used in many ways," Hall said. "First, it is used to measure the competency levels of our program managers. Once that information is available, Spotlight senior management will know what kinds of programs can be assigned to them. This is a part of the program manager assignment process."

Hall continued that, among other things, it is important that the competencies for each Spotlight program align with the skill level of the program manager assigned. "In addition, we consider how important the programs are to Spotlight and issues limiting the assignments, for example, the time availability of our program managers," he said. "We have quite an elaborate model for program manager assignment, but I don't think that we will have time to discuss it now. The idea is that we want to assign important strategic programs to program managers who have sufficient competencies to lead those programs. We also use this metric to determine what kind of training our program managers need for professional development. Additionally, the metric helps us identify qualified candidates for promotion when new positions for program managers are created," he said.

Moskhill concluded the discussion by saying, "Thanks for a very thorough description of our program management competencies. It is clear to me from our discussion that these competencies are a pivotal factor in Spotlight's competitiveness and our ability to consistently execute our product strategy."

REFERENCES

1. Crawford, L., "Competence Development," in P. W. G. Morris and J. K. Pinto. *The Wiley Guide to Managing Projects*. Hoboken, NJ: John Wiley & Sons, 2004.
2. Pinto, Jeffery K. *Power and Politics in Project Management*. Newtown Square, PA: Project Management Institute Publishing, 1998.

3. Cohen, Dennis J., Robert Graham, and Robert J. Graham. *The Project Manager's MBA*. San Francisco, CA: Jossey-Bass Publishing, 2001.

4. Maxwell, John C. *The 21 Irrefutable Laws of Leadership*. Nashville, TN: Thomas Nelson, Inc Publisher, 1991.

5. Project Management Institute. *A Guide to Project Management Body of Knowledge*. Drexell Hill, PA: Project Management Institute, 2004.

6. Wheelwright, Stephen C. and Kim B. Clark. *Revolutionizing Product Development: Quantum Leaps in Speed, Efficiency, and Quality*. New York, NY: Free Press Publishing, 1992.

7. Martinelli, Russ and Jim Waddell. "The Program Management Competency Model". Project Management World Today (July-August 2004).

Part V

Organizing for Program Management

Part V focuses on the organizational aspects of program management. In Chapter 14, the process of transitioning from project to program management is explained. It covers three major steps: understand the transition, execute the transition, and continuously improve the results of the transition. In the first step, understand the transition, we present business and operational factors that have driven senior management teams to consider moving to a program management model for developing their products, services, and infrastructure capabilities.

Once the need for transition is understood, the second step—executing the change—a comprehensive plan for implementing the transition is presented. We present the SECURE (Scope, Engineer, Confirm, Ultra-plan, Realize, and Enhance) process to execute the transition. In each stage we define boundaries for the transition; design an organizational structure, systems, and culture; test pilot the designed program management model; plan for implementation and political and cultural changes accompanying the transition; implement the transition; and terminate the transition.

The final step involved in transitioning to a program management model is continuous improvement. The PMO, the topic of Chapter 15, provides the infrastructure for continuous improvement. The PMO is also a pillar for managing and controlling multiple development programs. It is often responsible for managing scarce resources across multiple programs, monitoring program status, and taking actions to make sure that deliverables are accomplished in an acceptable and timely manner.

The PMO's job is also to ensure that senior management is given appropriate and consistent information for decision making.

Finally, we provide an example from practice of an IT organization that illustrates that there are various avenues to build a successful PMO. The example also points out the value of the PMO for providing improved business results for the company.

Chapter 14

Transitioning to Program Management

Organizational transformation is always a challenging endeavor and transitioning to a program management-based organization to developing products, services, or infrastructure capabilities is no different. Implementation of the program management model will change the culture and politics of the organization because it affects all levels of management. Additionally, it changes the rules of engagement, the decision-making hierarchy, roles and responsibilities, core competencies required of some functions, and the political landscape of the organization.

Our experiences in leading organizations through this transformation have revealed several key factors that come in to play and must be appropriately managed. These factors include the following:

- There must be a compelling reason to change.
- Senior management must drive the transformation and have a clear vision and end state in mind.
- Changes to the organizational structure may be required, which may change the culture of the organization.
- People need to understand their new roles and responsibilities and the new power equation.
- New core competencies may be required within the organization.
- Behavior change on the part of individuals affected has to be modeled, monitored, and managed.

This chapter will focus on the management of these factors through a disciplined three-step transition. First, we will address understanding the transition. Some of the drivers that motivate companies to change the way they develop and deliver their products, services, and infrastructure capabilities. During this process, we will characterize the continuum between projects and programs and identify the key factors that determine when projects should be redefined and restructured as programs. Next, we look at the critical elements that constitute a firm foundation from which a strong and sustainable program management model can be built, and, then, present a detailed stage-by-stage transition using the SECURE process. This chapter, therefore, will help senior managers attain the following objectives:

- Understand the factors that indicate a need for transition to program management
- Explain the major steps for implementing program management in an organization
- Describe the fundamental elements that must be in place to sustain the program management model
- Present the case for doing a cultural risk analysis and political plan
- Understand common challenges to expect and suggested methods for resolving them

THE TRANSITIONING PROCESS

There are three major steps in transitioning to program management, each including several substeps (see Figure 14.1):

- Understand the transition
- Execute the transition
- Continuously improve

Understanding the transition to program management tells us in broad terms what factors push for the transition, and whether there's

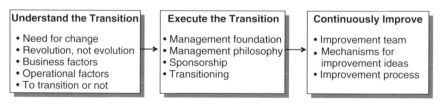

Figure 14.1   The steps for transition to program management.

a compelling case to transition or not. If the case is good, it drives the next step—execution of the transition process. The deployment of program management practices on real-world programs will reveal its glitches and generate new learning, resulting in a need for continuous improvement—the third step. Details about each step follows.

UNDERSTAND THE TRANSITION

The goal of the transition is to achieve a fully functioning program management discipline with as little organizational pain as possible. To achieve the goal, you need to fully understand the whole game, the associated risks, and roadblocks surrounding the transitioning. This can be achieved by performing the following steps:

- Understand the need for change
- Understand that it is mostly revolution, not evolution
- Understand business factors driving change
- Understand operational factors driving change
- Decide whether or not you need to transition

Understand the Need for Change

How development efforts are managed within a company is the result of the firm's management philosophy and the business model they practice. Many companies have found that the traditional project management approach is sufficient to meet their development needs and are quite satisfied with the results. Other companies have either chosen, or are seeking, a more strategic approach to optimize their business results. When a firm decides to undertake a major organizational transition, such as moving from a functional to a program management-based organization, there are many key factors that must be considered. We will look at these factors in detail later in this chapter, but let us first explore why companies make major organizational changes.

Many companies over their life span face times of severe difficulty in meeting the performance expectations of their shareholders and other stakeholders. These performance problems are many times the result of a major upheaval or shift in the business environment within which the firm operates—such as a shift in the primary markets, competitor actions, or macro- and micro-economic change. These changes usually result in some form of adverse or negative impact to the firm's ability to maintain market share, generate sales, sustain profits, and experience growth in earnings. These adverse conditions result in actions on the part of a senior management team to search for opportunities to remedy the performance problems and make necessary changes to get the enterprise back on track to deliver planned and expected results.

This then becomes the crux of the discussion for whether a company needs to consider moving to a program management-based organization. If a disconnect exists between the company's ability to convert business strategy to work output, or if they have identified that they lack an integrated systems approach for developing and delivering their new products, services, or infrastructure capabilities, then it may be the appropriate time for senior management to consider transitioning to a program management-based organization.

Understand That It is Mostly Revolution, Not Evolution

To obtain the broad-based benefits of applying program management to the business model of the firm, several fundamental changes may be required. Our experience has been that the probability that a firm can successfully convert a functional or project-based organization to a program management model by evolving into the model over time is low. In other words, can the firm increase the responsibility of the project manager and the capability of the project team to deliver upon the strategic and business objectives by providing training for these individuals over time? Our belief is that the answer is no and this is not likely to happen. The successful transition to a program management-based organization is unfortunately not evolutionary but mostly revolutionary in nature.

The senior management team of the firm must consciously decide that they will make the fundamental structural and cultural changes necessary to move the organization to the program management model and then lead the organization through successful implementation of the changes. There are a number of key factors that typically drive a senior management team to decide if they are ready to implement the program management model. The key factors are broken down into two major categories of business factors and operational factors.

Understand Business Factors Driving Change

Firms that have their development efforts either functionally aligned or operationally and tactically focused are most likely missing the opportunity to leverage their efforts directly toward the core business objectives of the firm. This direct link between work output and business objectives provides the potential to drive improved business results. There are a number of business factors that senior management should assess to determine if this opportunity for improved business results exists in their firm. These include the degree of linkage between development output and strategic objectives, the need for time-to-money improvement, and the level of business expertise needed, as follows:

Degree of linkage between execution output and strategic objectives: This factor involves the degree of linkage between the deliverables and final output of a product, service, or infrastructure development effort to a specific set of strategic business objectives of the firm. If the current link is low or nonexistent, then consideration of transitioning to the program management model is an opportunity for the firm to pursue (see Figure 14.2). As we state throughout this book, a primary advantage of the program management model is the ability to directly close the gap between the firm's strategic objectives and the successful execution of plans and actions to achieve those objectives.

The need for time-to-money improvement: The level of importance the firm places on the need for time to money improvements is the second factor for senior management to assess. It is a well-known fact that in today's highly competitive world, time to money is a critical factor in gaining

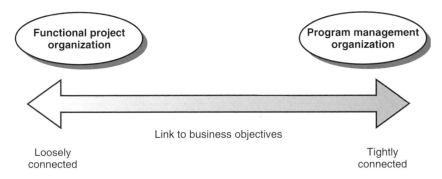

Figure 14.2   Link to business objectives.

an advantage over one's competitors. Gaining time to money advantage means decreasing the development cycle time, and many companies have achieved some gains through the concurrent development approach. However, greater improvements in time to money may be deemed necessary by senior management. Best-in-class companies accomplish greater time to money gains through an integrated development approach driven by a program manager, as described in Chapter 4. In the integrated approach, the functional teams work shoulder to shoulder through the PLC, from conception to end of life. Achieving time to money advantage requires tight management of the cross-discipline interfaces, shared risks, and open communication channels between the interdependent project teams. This requires strong leadership from a program manager to orchestrate, coordinate, and direct the work of various cross-discipline project teams.

Level of business expertise needed: The degree of general management and business expertise required of the team leader to successfully manage the product, service, or infrastructure development deliverables is another business factor to consider. A direct indicator as to whether the firm's management is using their development teams to drive strategic business objectives is the degree of knowledge, training, and skills required of their team leaders. For complex, multi-discipline development efforts, the team leader must be able to fulfill the role of the generalist who coordinates the efforts of the functional specialists toward achievement of a set of business results (see Figure 14.3). The program manager is a generalist who manages the overall program, and must possess the knowledge and capability based on broad-based cross-discipline experience and understanding to be successful in this role.

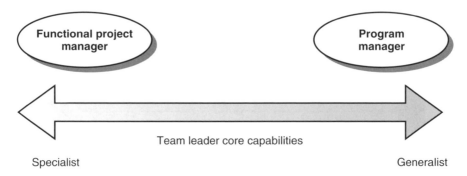

Figure 14.3   Team leader capabilities.

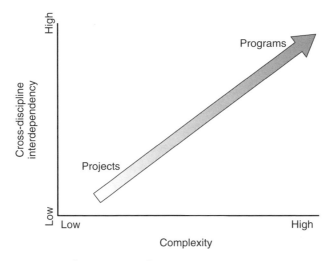

Figure 14.4   The project and program management continuum.

Understand Operational Factors Driving Change

A continuum may be the best way of visualizing and characterizing the distinction between project and program management-based organizations, with complexity and degree of cross-discipline interdependency being the distinguishing factors along the continuum (see Figure 14.4). The further a development effort moves toward increased complexity and cross-discipline interdependency, the greater the need for a program management approach for managing the development effort. Those who have been involved in more complex efforts appreciate that the challenges facing the leaders and members of these teams surpass the typical or traditional project team requirements. There is a point in which the increased complexity of a project makes it ineffective to manage as a single entity. At that point, the best option is to break the project into multiple interdependent projects that deliver one or more elements of the total solution and reorganize the development effort into a program.

Complexity: To better understand complexity as a determinant for transitioning to a program management-based organization, we need to review what we mean by complexity (also see details about this topic in Chapter 4). Complexity manifests itself in several areas as follows: product, service, or infrastructure designs have become more complex as features and integrated capabilities increase; the process to develop and manufacture the solutions has become more complex; the ability to integrate technology and end-user needs has become a very complex undertaking; and the current global, multinational approach to development, where teams are geographically dispersed across multiple sites and time zones, has added yet another level of complexity to the problem. Thus, complexity of a development effort occurs from elements related to technical, structural, and business complexity factors.

Technical complexity is driven by a number of factors. It may be directly related to the technical aspects of the product, service, or infrastructure under development, or from the knowledge and capability of the existing resources of the firm to successfully address the technical complexities. Examples of variation in technical complexity are illustrated in Table 14.1.

Structural-complexity factors are those that involve the organizational elements of the development effort. Like technical complexity, structural complexity may also be exacerbated by the lack of skills, knowledge, or capability of the firm's existing personnel to address or have experience in addressing structural complexities. Examples of variation in structural complexity are shown in Table 14.2.

Business-complexity factors are those that involve the external business environment in which the firm operates. Examples of variation in business complexity are listed in Table 14.3.

Complexity factors will vary for each company, but the primary point that is being made is that greater technical, structural, and business complexity drives a greater need for transition to a program management-based structure.

Cross-discipline interdependency: The size and number of cross-discipline groups involved in a development effort is also a consideration in the

Table 14.1   Examples of Technical Complexity

Technical Complexity	
Low complexity	*High complexity*
• Feature upgrade to an existing product	• New product architecture and platform design
• Development of a single module of a system	• Development of a full system
• Use of existing and developed technologies	• Use of new and undeveloped technologies

Table 14.2   Examples of structural complexity.

Structural Complexity	
Low complexity	*High complexity*
• Team is colocated	• Team is a virtual team
• Single site development	• Multisite development
• Single geography development	• Multigeography development
• Single cultural team	• Multicultural team
• Single company development	• Multicompany development

Table 14.3   Examples of business complexity

Business Complexity	
Low complexity	*High complexity*
• Selling into traditional and mature markets	• Selling into new and emerging markets
• Receptive customers and/or stakeholders	• Unreceptive customers and/or stakeholders
• Flexible time-to-money requirements	• Aggressive time-to-money requirements
• Existing end-user usage models	• New end-user usage models

decision to transition to a program management-based organization. Cross-discipline interdependency is directly related to the complexity and scope of the development effort.

For example, a small development effort addressing a software feature enhancement to an existing product in the marketplace may require only three cross-discipline teams to implement. Conversely, a requirement to develop a product with new hardware and software architectures, along with a platform design utilizing a new user interface and electro-mechanical package, may require a large team with ten or more disciplines involved.

Figure 14.5 shows just how explosive the growth in the number of interdependencies can be as more cross-discipline elements of a solution are required. By looking at the chart, we can see that the small software feature enhancement project requires three cross-discipline teams to implement; the technical complexity is low and the potential number of interdependencies between disciplines is less than ten. By contrast, on a new product development project that requires ten cross-discipline teams,

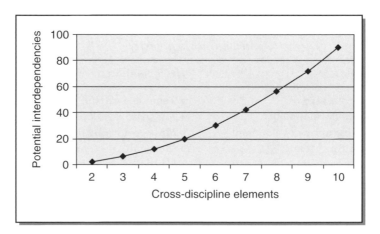

Figure 14.5   Interdependencies versus cross-discipline elements.

the potential number of interdependencies between disciples increases to nearly 100! This requires a strong development model that facilitates the required horizontal cross-discipline collaboration between teams.

So then, one might ask, *When a project becomes large enough in scope, should it just be considered a program?* The answer is *no*. It is not a matter of size or scope alone. Rather, there comes a point in which interdependencies between functional teams become large enough in number that the best option is to reorganize the development effort into a program consisting of multiple interdependent projects. The focus of the program then shifts from development of technical discipline-specific solutions to development of a cross-disciple solution provided by the interdependent relationship between functional project teams. The horizontal nature of this cross-discipline solution is a key indicator that an organization needs to move to a program management-based organization to become more effective and efficient in its product, service, or infrastructure development efforts.

Skilled personnel: The availability of trained and experienced personnel in the firm that are capable of fulfilling the role required is the next factor for senior management to consider. If management is convinced that the program management model is the right model to pursue, then, before they proceed, they need to assess the current availability of skilled talent within the organization. Candidates with the requisite skills can be derived in the following ways:

- Hire qualified candidates from the external marketplace
- Grow the talent internally through training and development
- An appropriate combination of external and internal acquisition of talent

Hiring qualified candidates externally is by far the quickest approach for obtaining the much needed talent and capability. Program management is not a new discipline and experienced and qualified program managers exist in many industries today. A focused search for external talent should be fruitful. There is, of course, the traditional risk all firms face in hiring externally—properly assessing a candidate's skills for the role and his or her ability to successfully meld into the firm's culture.

Growing program management talent internally has the advantage of utilizing known company personnel for this new role. Internal candidates may lack the total breadth of skills necessary to perform the program management role (see Chapter 13), but they may have the talent and potential to grow into the role. Management can pick high achievers that have demonstrated they are leaders in the firm and are well respected by their peers and team members. This may contribute to more rapid team building than by hiring a program manager externally. The key disadvantage to this approach is that it may take considerable time and cost to achieve the necessary skill level and experience for the internally grown candidates.

In our experience, the third approach, a combination of external hiring and internal development offers the best opportunity for successful long-term growth of the program management capability within an organization. This approach provides an appropriate percentage of the total resource acquired externally, while also growing the firm's current talent pool with the remainder of the positions needed. The externally hired resources can initially help institute the new model and provide mentoring and assistance to those internally attempting to learn the new role. Additionally, this adds motivation to the internal candidates, as it encourages a "promote from within" philosophy with the existing staff. Melding the two sources of program management candidates enables the chance for a more balanced model that fits the culture of the firm.

To Transition or Not to Transition

Now that we understand some important factors involved in the transition decision, we can essentially consider the following two options: stay

Table 14.4   Decision table for transitioning to program management.

Factor	Project Management	Program Management
Understand business factors driving change		
Degree of linkage between execution output and strategic objectives	Loosely Connected	Tightly Connected
The need for time-to-money improvement	Lower	Higher
Level of business expertise	Specialist	Generalist
Understand operational factors driving change		
Technical complexity	Lower	Higher
Structural complexity	Lower	Higher
Business complexity	Lower	Higher
Cross-discipline dependency	Lower	Higher

with a functionally oriented, project approach or transition to a program management-based model. Each fits one situation better than another. To decide whether to transition or not, we devised a simple tool in Table 14.4

In Table 14.4 we listed which of the two approaches may be a better fit in each of the seven factors. (We also assessed and explained earlier how and why the factors favor each of the two approaches.) For example, for factor one in the table, we assessed and explained that if the degree of linkages between execution output and strategic objectives is required to be tightly connected, a company should favor program management. And, so goes assessment for the remaining six factors.

Table 14.4 can be used as a guide in the decision to transition to a program management-based approach. First, tailor the factors for your case. Determine if each factor favors project or program management by placing an X beside your choice. Repeat the process as needed.

If the majority of X's are on the project management side, stay your current course and do not transition. If the majority of X's end up on the side of program management, you have the case to transition. A word of caution may be of help here. This is a simple but practical tool to make your transition decision. Like any other tool, it is only as good as the information contained in it. Therefore, prepare excellent quality information for your case and assess the factors professionally.

EXECUTE THE TRANSITION

Before transitioning to the program management model, a firm foundation has to be established from which to build upon. This foundation consists of a strategically focused management philosophy and strong senior management sponsorship.

Management Philosophy

Senior management's philosophy as to whether they choose to view development efforts strategically and linked to the success of the business, or rather to view the efforts as tactical and operational in nature, is critically important for a successful transition to a program management-based organization. For a program manager and his or her program team to perform at their highest potential, senior management of the firm must make it apparent that they believe the team's efforts are directly tied to the success of the core business and are integral to the strategic success of the enterprise. With this view in place, it elevates the importance and resulting influence of the program managers within the firm and enables them to coalesce action to get things done throughout the enterprise.

Additionally, senior management should position program management as a core business function within the organization. Too often, program managers are embedded in other functions within an organization, such as engineering or marketing. This is counterproductive because program managers are generalists who are effective in managing across functional lines. By embedding them within one specific function, an inherent functional bias is established based upon the direct reporting relationship between the program manager and his or her functional manager. Senior management must be careful to preserve the functional neutrality of the program management function. Firms that view program management as a true functional discipline that is equal in influence to the other functions of the organization experience increased program management talent and capability; ensure the long-term viability of the discipline; and, potentially, develop future GMs and business leaders for the enterprise.

Senior Management Sponsorship

Firms that have been most successful in implementation of the program management model have had strong sponsorship from the senior management team of the organization. Implementation of the model works best when it is managed from a top-down approach, rather than a bottom-up approach. When senior management buys into the approach and drives the implementation, the rest of the organization begins to fall in line. It is difficult to derive the necessary and significant changes in an organization when it is attempted through middle management, as barriers across the functional organizations quite often become too difficult to overcome.

One senior executive told us how his first attempt at implementing program management within an organization was a complete failure. As he told us, "The transition failed because it didn't have executive support, and, worse yet, the executives weren't even open to learning about the value of program management." This was an organization that was too hard to move without top management supporting and driving the transition and necessary change in culture. According to the senior executive, their philosophy was "if a person was not a technologist, he or she wouldn't be able to lead a development team." The problem with this philosophy is that technologists get rewarded for developing technologies, and product development managers get rewarded for *integrating* system technologies into successful products—an obvious mismatch of skills and responsibilities within this organization.

The senior sponsor must be an influential change agent. This means someone with sufficient power and influence in the organization who can ensure that the organizational structure, behavior, and culture move in a new direction. The change agent must have a clear vision of what the change is and why is it important for the success of the organization. This value proposition should be sufficient for the senior executive to champion the change.

Transitioning to the Program Management Model with the SECURE Process

Once a strong program management foundation is established, transition to a program management-based organization can begin. Many day-to-day tasks have to be managed during the transition, but several structural and cultural elements are worthy of focus for a successful transition.

To illustrate this, we chose to demonstrate the transitioning through the SECURE process (Figure 14.6). Other processes can be used just as effectively, provided that they are comprehensive and systematic.

One important point must be made. It is highly recommended that a qualified organizational development expert within the business or an experienced consultant be hired to lead the transition process and team. They have the requisite skills to scale the barriers that will be encountered, and they represent a neutral position within the company.

To proclaim that the old way is dead and shift overnight to the new way of program management, may lead to chaos in an organization. Therefore, major changes call for planning. To create a program management-based organization, a good road map is needed. The SECURE process is such a road map. It guides the transition from the current business model to the program management model by a stage-by-stage process. We shall describe each stage in short, and then we will delve into the details.

In stage 1 (scope), a situation analysis is conducted to clarify why the transition to program management is needed and explain it in a business case. Next, a scope statement is developed detailing the "wills" and "won'ts" of the transition. Finally, the team vision, values, and objectives for program management are established. Stage 2 (engineer) includes developing a program management design or blueprint that shows all details regarding changes in structure, systems, and culture that are needed to build a program management function. In stage 3 (confirm), a pilot test of the program management model is conducted. The purpose is to see if the program management design performs as expected in a contained environment. In small-scale transitions this stage is sometimes skipped.

During stage 4 (ultraplan), a realistic action plan for implementing the program management design is developed. The reason we called it "ultraplan," rather than "plan" is simple—organizational transition requires planning to an extreme degree. Part of planning is the development of a political plan and cultural analysis. Stage 5 (realize) focuses on implementing the program management design to turn the design into reality. This is the most difficult stage when the real changes happen and people respond, resist, or fight back. In stage six (enhance), the transition project team activities come to a close and focus shifts to continuous improvement, in which the emphasis is on line organizations continuing to improve the new program management environment.

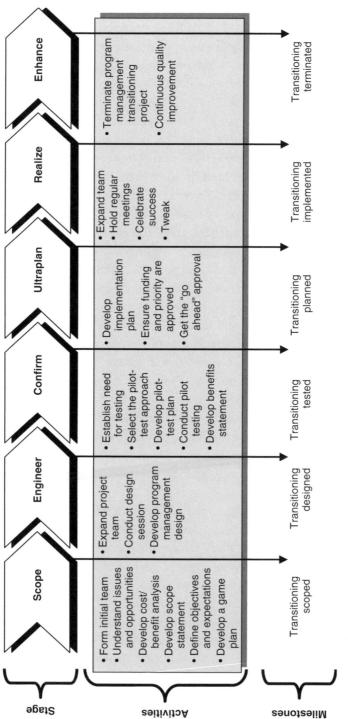

Figure 14.6   The program management transitioning with the SECURE process.

Stage 1: Scope

If you think thoroughly about the program management transition you want to undertake, it will significantly increase your chances for success. Think through every step of the changeover to program management. But first, justify the need for program management and determine what it will include. The scope stage has six steps:

1. Form initial team
2. Understand the issues and opportunities
3. Develop a cost/benefit analysis
4. Develop the scope statement
5. Define objectives and expectations
6. Develop a game plan

First, form an initial team to pursue the transition implementation plan. The transition effort is most effective when it is managed as a project in and of itself and is properly planned and executed. In doing so, a formal implementation team can be established to develop and execute the whole SECURE process. As mentioned earlier, the team should be led by a qualified organizational development expert or experienced consultant who is in charge of forming the transition team.

The transition team will work to understand the issues and opportunities, involve knowledgeable personnel to conduct the gap analysis between the current situation and where the firm wants to eventually be, and identify the problems and make recommendations for improvement. For this, it is best to conduct a situation analysis, observe the work flows, identify the problems, and perform root-cause analysis. Don't forget to analyze customer relations, market conditions, and organizational culture. This is a basis for the subsequent steps.

What is the business case for implementing the program management model? What does your cost/benefit analysis say? Your case will demonstrate what would be the benefits of having program management and the costs of sticking with the old way. If the case creates "massive discomfort with the status quo," you are destined for change.

The scope statement should clearly spell out what will be done and what won't be done during the transition process. It includes details on functional scope; organizational boundaries; interface points; possible benefits; political and cultural issues as to the transition project;

root causes of the current problems; risks of inaction and action; and assumptions, constraints, and limitations.

Defining the purpose of the program management function means defining the vision, objectives, and values then communicating them to all appropriate personnel. Vision is the end state that you want the program management discipline to fulfill. Values set the ground rules for expected behaviors and cultural shifts needed to make program management successful.

To get approval from senior management to proceed to the second stage, a preliminary transition plan needs to be developed. It should include scope of work; the transition team; the transition process, schedule, and resources needed; and, most importantly, a risk assessment.

Stage 2: Engineer

The next stage in the SECURE program management transition process is engineer. Based upon the outcomes of the scope stage, design of a program management function can now be completed. The engineer stage has three steps:

1. Expand the project team
2. Conduct design sessions
3. Develop the program management design

In the first step, the initial team formed in stage one is expanded to include all those who will be affected during the implementation of the program management transition. Hence, the team will include business executive sponsors, a project director, a core team, business-unit champions, and subject experts. The implementation groups will be added once the transition enters the realize stage.

Conducting design sessions is a bit of an art. The challenge of having all key players present (and with very diverse ideas about how to design the program management function) calls for firm direction and superb coordination skills on part of those conducting the sessions.

When refined, the design should include the elements necessary for building the program management function—structure, systems, and culture components (see Figure 14.7).

- *Structure components*: program management process, organization, information, and other technology

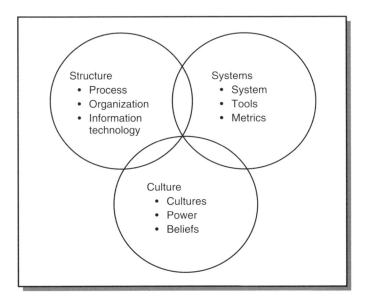

Figure 14.7   Elements of program management design.

- *Systems components*: program management systems, tools, and metrics
- *Culture components*: program management culture, power, and on-the-job behaviors

We will now address each element, some that are including referring to in other parts of this book or not deemed critical to our scope of the book. These elements are as follows:

Program management structure: Structure consists of process, organization, and information technology elements. This process is to be designed as described in Chapters 5 through 8. The options for organizational structure are described in the following section.

Cross-project and cross-discipline program structure: New products, services, and infrastructure capabilities are designed and developed through the combined efforts of many personnel across an organization. These employees work together to address the tremendous number of activities and tasks necessary to achieve a complex development effort. This combined effort requires effective communication, coordination, ability to resolve issues and barriers, and effective decision making both within the team and by senior management. An organizational structure that supports and facilitates this collaborative approach is necessary.

Traditional organizational structures range from the purely functional structure to the purely project-oriented structure. Both extremes create some significant limitations from the perspective of the program management model. The purely project-oriented approach minimizes the critical importance played by functional managers in providing for the long-term viability of the functional capabilities by staying current on the latest technology and maintaining the highest-skilled and best-trained resources to support the enterprise. However, a purely functional organization minimizes the importance of cross-discipline knowledge and a broader view of the development effort. Functional organizations tend to stifle cross-discipline thinking, which is paramount to the program management model. Authority rests only with the functional managers and tends to be inwardly focused. As Christopher Meyer (author of "Fast Cycle Time") states, "By far, the most serious problem with the functional structure is that it serves its members better than its customers."[1]

The most effective organizational structures for the program management model is a compromise of the two extremes—the matrix structure and the program structure. With both approaches, formal responsibility and authority for the development effort resides wholly with the program manager.

The matrix structure: Figure 14.8 illustrates one form of the matrix structure that is used in various companies but is quite commonly used by electronic- and consumer-product companies. There are, of course, many variations of the matrix structure.

In a matrix structure, program team members continue to report directly into their functional organization (depicted as solid lines) and are loaned to the program manager (depicted as dotted lines). The program manager is responsible for integrating the cross-discipline contributions. The functional managers are responsible for overseeing the core capabilities, process, and tools within the function.[2]

Like all structures, matrix management has its strengths and weaknesses. The strengths of the matrix structure are considerable and include the following:[3]

- The program manager has access to the entire pool of experts within the functional organizations.

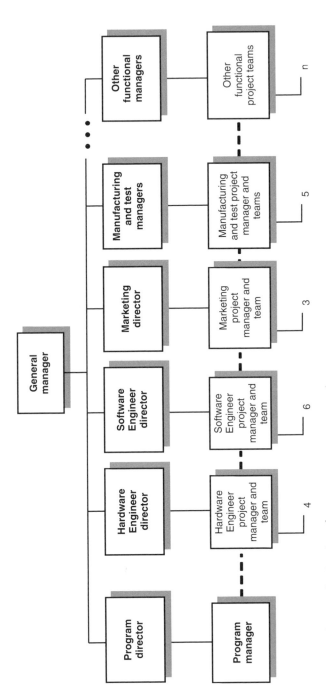

Figure 14.8 The matrix structure for program management.

- Duplication of functional departments is eliminated, with a single organization for each function.
- Resources can be shared intelligently across programs to ensure effective use of critical and scarce skills.
- Integration of the functional disciplines is enhanced, which serves to reduce power struggles and improve collaborative teamwork.

However, the matrix structure also has its weaknesses that must be handled. These weaknesses create tension between functional managers and the program manager, it can lead to competition among program managers for scarce and critical resources, and program team members need to learn how to work for multiple bosses.

The second organizational structure that we will discuss is the program organization structure shown in Figure 14.9. This structure comes in many forms and can be found in the aerospace, defense, and automotive industries.

In the program organization structure, the project teams report directly to the program manager. Whether the functional managers report directly to the program manager is specific to the organization. All resources reporting to the program manager are dedicated to and work full-time on his or her program. Generally, cross-program functions such as marketing, finance, and human resources support all programs, therefore, are not dedicated to a single program.

Strengths of the program organization structure include the following:

- The program manager is in direct control of all resources on the program.

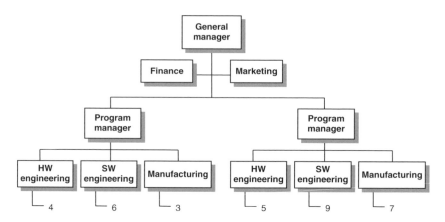

Figure 14.9   The program organization structure.

- Resources are completely focused on a single program, which leads to improved productivity and cycle time.
- A high degree of cross-project integration occurs.
- A high level of program identification and team cohesiveness emerges.

Like the matrix structure, the program organization structure also has its weaknesses. It tends to be more expensive due to the duplication of functions and middle management, divisiveness between a program and the parent organization and other programs can develop, the program manager may not have access to the most qualified expertise in the enterprise, and a dilemma can exist on what to do with personnel once the program ends.

An important element in both organizational structures is that program management is a stand-alone function. This is important to the program management model because the positive impact of program management is considerably diluted when it is contained or assigned within any one specific functional organization.

The last piece of program management structure is information technology. Information technology is the tool used by program management teams to communicate and store information. While collaborative technology tools vary widely in complexity and information richness and offer different grades of efficiency, the crucial point is to devise standard communication guidelines for the use of the tools. In one company, for example, the rule is to use mostly e-mail for daily communication, although it is an information-poor medium that has its own problems. This team member observed, "Sometimes e-mail has a one-day lag. A team member from Romania is going to send us e-mail to the United States and we are going to read it the next morning. By the time we send back the reply, we have wasted a day."

Help for this can come from exchanging instant messages, as in the case of another program management team. "We use instant messaging very heavily. It is installed on all of our laptops and desktops. Obviously during the day we do not expect all team members to be online but a lot of times they are." If more in-depth information is needed, companies use phone conference calls, such as weekly teleconferences. When even richer information is to be communicated, teams use teleconferences and net meetings. But before even thinking about specific collaborative technologies, companies first resolve the issue of the infrastructure. As one program manager pointed out,

"I've made sure that everyone on the team has the right office setup at home, so everyone has DSL or cable-modem access [and] mobile headsets so they don't have to have a phone cradled to their ear for three hours, if they're working on a long call, and we'll pay for such an office."

Product and project management software are also elements of information technology design and may be a success factor in programs. For example, in a software development program management function, the most frequently used product management software is for development (for example, Visual Studio); source control system (for example, CVS); enterprise-based configuration management system; defect/issue tracking; test-case tracking spreadsheets (Excel); and document storage solution for sharing documents. Each piece of the information technology has a specific function; the overall purpose is to ensure efficient support to the success of programs. This approach to collaborative product management and project management technology tools—insisting on standardized, integrated, simple, and relatively few of the tools—not only provides a high-communication potential but also helps with all components of program success. The results are more punctual schedules, more satisfied customers, better cost effectiveness, and higher-quality accomplishments. If this approach is not offered to program management teams, programs end up struggling to find the right information technology and how to use it, introducing variability into the program management process. This may lead to subpar program performance. Therefore, information technology should be part of program management design.

Program management systems: These include systems, tools, and metrics. Systems, for example, may include the following:

- Changes in policies and procedures
- Well-defined roles for all positions that are important to the program management-based organization, including senior management; functional management; core team project managers; individual team contributors; and, of course, the program manager (see box titled, "Roles for Effective Program Management")
- Skills to perform the program management role
- Program management best practices (see box titled, "Program Teamwork")
- Setting performance goals for the program teams

- Rewarding performance
- Employee development plans

Roles for Effective Program Management

Well-defined roles for the management team involved with a program is critically important to clarify program governance and decision making, lines of communication, and program responsibilities.

Senior Management: The role of senior management is to provide strategic guidance for the program and to establish and maintain the processes and procedures by which programs are managed within their organization. Senior management owns the portfolio of programs for the organization; selects and empowers the program manager to form a team and organize a program; assists the program team in setting the key objectives, targets, and constraints for the program; makes key approval and direction setting decisions; and provides the necessary leadership and guidance on issues and barriers outside the control of the program team.

Program Manager: As detailed in Chapter 12, the program manager has two primary roles. First, the program manager must manage that portion of the business represented by his or her program as a proxy for the GM of the business. Second, the program manager must lead the cross-discipline project teams toward an integrated product, service, or infrastructure solution that helps to achieve the business objectives of the enterprise. There are many other responsibilities that the program manager is accountable for, but these two are by far the most critical. A senior vice president from a high-technology product development company was quoted in regard to the criticality of the program management role and stated: "No enterprise manager can manage all the aspects of his or her business. It is too complex and too fast moving. Program management is the engine that drives the success of product development."

Project Manager: The functional project manager serves on the PCT as the expert of the function he or she represents and manages the individual contributors on the project team. The project manager drives the creation of the project plan and integrates that plan into the overall program plan. He or she then drives the implementation of the project plan, manages the work commitments within the project, and ensures completion and closure of all project deliverables. The project manager also works closely with the program manager and other PCT members to seamlessly integrate all functional elements of the final product, service, or infrastructure capability under development. The functional project manager is also the key conduit for communication on the program between the program team and the functional organization he or she represents.

Functional Manager: The functional manager is responsible for the hiring, training, and development of the functional personnel. He or she is also responsible for the maintenance of the skills, capabilities, best practices, and tools to sustain the long-term functional expertise. The functional manager makes the resource and work commitments for functional activities in support of the program and assigns a qualified project manager to represent the function on the program. The functional manager must also notify and negotiate with the program manager if any previously committed resources need to be reallocated.

Program management culture: The third element of the program management design (see Figure 14.7), culture, consists of organizational culture, power, and beliefs. No matter what approach is taken to handle the culture element of program management, a conscious effort should be invested to design it, not to let it evolve.

Program Teamwork

Teamwork is a key attribute of the program management-based organization. In fact, it is one of the pillars that support its power. According to McGrath, Anthony, and Shapiro, "Team-based organization is one of the most essential elements of product development, yet, few companies have implemented a consistently effective approach to it. Some do not even have a clearly stated way to organize each development effort and leave this to the individual team to figure out how to organize."[4] Many times, teams can't work together effectively due to organizational barriers that inhibit effective collaboration and decision making. When this is the case, fundamental changes to the organizational structure may be required.

The matrix and program structures both facilitate effective team-based collaboration. Teamwork occurs at multiple levels within the organization. First, each of the project teams within the program work collaboratively to create the deliverables for its element of the product, service, or infrastructure capability. The collaborative teamwork of the PCT consists of the second level of teamwork. As presented in Chapter 5, the PCT is the leadership body of the program whose responsibility it is to ensure the project elements integrate into the whole solution. This is accomplished through effective cross-project and cross-discipline teamwork.

The next two levels involve teamwork between the program manager and the senior management of the organization. First, effective teamwork between the program manager and the functional managers is critical to ensure the project teams are staffed with the right team members who have the right skills for the tasks at hand. It is important that the program manager and functional managers work together to keep the vision for the program and functional organization in alignment. Lastly, effective teamwork between the program manager, the executive sponsor, and his or her staff is important for program success. The program manager and senior managers also work together to keep the program in alignment with the strategic objectives and broader business environment. Additionally, this teamwork is crucial to make key decisions when needed and remove barriers to progress to keep the program moving through the development process.

Stage 3: Confirm

The third stage in the SECURE process, confirm, tests the program management design elements. During this stage, the transition team will see if the design performs as expected. Why would one test the program management design? A lot of time, money, and resources are involved in the transition, and expectations probably run high, at least on part of the transition team and primary sponsors. So, testing the implementation within a controlled environment to detect errors and correct them is wise and inexpensive insurance that the full implementation will succeed.

The confirm stage has five major steps:

1. Establish the need for testing
2. Select the pilot test approach
3. Develop pilot test plan
4. Conduct pilot testing
5. Develop benefits statement

There are several reasons for conducting tests of this kind. If executives are reluctant to allocate resources without "hard" data substantiating the expected benefits, or the program management design is radically different from the current model, it is wise to conduct a pilot test to gather data and gain learning (see box titled, "Pilot, Learn, Improve"). We recommend to pilot test a small number of programs, collect and analyze results, report preliminary results to the team, and change the design as needed.

Pilot, Learn, Improve
Individuals within an organization are only able to comprehend and act upon a finite amount of change to the way in which they perform their jobs.

Therefore, transition to a program management-based organization must begin slowly with a fairly narrow scope to build confidence, credibility, and momentum. It is recommended that pilot programs be utilized initially to gauge the amount of change that can be sustained within the organization and to capture early learnings that will be valuable going forward. Broad-based communication up, down, and throughout the organization is critical during this time frame to ensure that progress, changes, and issues are well understood. Rumors and misinformation may be frequent and numerous, which will require active management by the team.

By taking an iterative process to implementation, it allows the transition team, management, and the broader organization to do the following:

- Start slowly
- Learn by doing
- Build confidence and credibility
- Achieve early successes and demonstrate tangible results
- Seek feedback and involvement of personnel across the organization on a regular basis
- Adjust the plan as needed when things change
- Create new benchmarks and raise the bar for driving continual improvement
- Ensure the internalization of new values and behaviors

Stage 4: Ultraplan

This is stage 4 of the SECURE process. Developing the program management design is easy compared to implementing it. Without a meticulously, well-thought-out plan, the program management model will never become a reality. Like many projects, transition implementation always takes more time than anticipated simply because we have a tendency to look only at the technical, hard aspects of the transition. "Soft" aspects such as politics of change and resistance to change tend to be neglected. In addition, keep in mind that planning is generally no more than preparation for the future.

The ultraplan stage has three major steps, as follows:

1. Develop the implementation plan
2. Ensure that appropriate funding and priority is set and approved
3. Get the go ahead approval.

Getting the right set of people together who have diverse ideas about how to go about transitioning to program management and then developing a transition plan can be challenging. It will test one's ability to lead people and synchronize their ideas and activities.

When completed, the transition plan should have all elements necessary for building a program management function. For example, these elements are as follows:

- An activity plan that clearly specifies tasks, deliverables, who does what, schedule, resources, and costs
- Comprehensive training plan for executive management, functional management, program and project managers, and individual contributors
- Plan for reviewing transition change progress and issues
- Cultural transformation plan
- Political plan

The first three bullets from the implementation plan are well known to practitioners undertaking change initiatives. What is less known are the cultural and political plans. The latter is described in Chapter 15. As for the cultural transformation plan, it is a key factor that a successful transition to program management hinges upon and, yet, very little is known about it. Therefore, we will discuss it in detail in the following section.

The cultural transformation plan: What is organizational culture? In academic books, culture is usually defined as a set of values and beliefs that are shared by members of the organization.[5] For example, believing that programs exist to support organizational goals and strategy. Practitioners define culture as "how we do things around here," such as using a matrix organization to manage programs. Some others argue that culture is collective programming of mind.[6] For example, a company's program team members are programmed by on-the-job training and management mandate to use their company's PLC.

The program culture should be designed to express a set of clearly articulated, performance-oriented values and on-the-job behaviors that are built into program practices. For example, one such value is "being proactive," in which team members practice a behavior of periodic progress reporting that includes predicting the business results, schedule, and budget at completion of their tasks. The intention is that program team members have a sense of identity with the cultural values and accept the need to invest both materially and emotionally in their program. This should make them more engaged, committed, enthusiastic, and willing to support one another in accomplishing the program goals. As a result, they should work harder and be more effective, thereby, increasing success.

No matter what organizational culture definition you use, two things are important. First, the organizational culture is used to direct general behaviors of company employees; for example, the use of consultative decision making. Second, the organizational and program culture are deeply ingrained in program members' minds and behavior. So why is this important knowledge in developing a program management transition plan? Roger Lundberg, director of Chrysler development and vehicle engineering operations for DaimlerChrysler provides this excellent explanation: "There is no single right organizational solution. Each company implementing program management will have to align it with their company's culture." If the program management design does not align with current company culture, there will be human resistance to the transition. Resistance may not result in cultural difficulties, but to put it brutally, it may kill the transition to program management if the resistance and cultural change are not properly managed. One way to manage cultural change is to develop the cultural transformation plan.

There are four steps to the development of the cultural transformation plan:[7]

1. Understand your current culture
2. Evaluate the strategic importance of the program management design
3. Assess cultural fit of the program management design
4. Develop the cultural risk plan

Understand your current culture: This step involves studying the current project management culture, which again is just a manifestation of your organizational culture when it comes to "how we do things around here." For example, let us assume that the project management culture in a fictitious company uses the same framework from Figure 14.7. Then, study all the elements. To demonstrate, we will give examples of one element of each—structure, systems, and culture.

- *Structure*: We have been using a well-developed matrix organization for all projects, including smaller and larger, routine and innovative, and tactical and strategic ones.
- *Systems*: We use project coordinators to lead projects. They do not have the decision-making power for important project decisions but collect and bring information to functional managers to make the decisions.
- *Culture*: Some of our project coordinators apply a reactionary, others a proactive approach in how they coordinate projects. The former behavior means reacting to daily fires becomes the dominant mode of operation.

Again, this kind of the analysis is done for every major element of the framework.

Evaluate the strategic importance of program management design: In this step, the importance of each major element of the program management design is assessed. The design is that which was developed in the engineer stage and refined in the confirm stage of the SECURE process. We again follow the framework from Figure 14.7 and evaluate only one element of each—structure, systems, and culture.

- *Structure:* According to the program management design, a matrix organization will be used. This structure is crucial to ensure program management is based on effective resource allocation for cross-discipline program teams. The matrix structure is of high-strategic importance for program management implementation.

- *Systems:* According to the program management design, an empowered program manager who is responsible for all major program decisions and results will be in charge of the program (see box titled, "Empowering the Program Manager"). The concept of the empowered program manager is the cornerstone of program management and, logically, has high-strategic importance in the organizational strategy.

- *Culture:* According to the program management design, the program managers will not use a reactionary approach to managing programs but a highly proactive approach. The strategic importance of this element is medium.

Empowering the Program Manager

Empowerment of the program manager is a critical element of the program management model. Empowerment is provided by senior management and involves giving the program manager the responsibility, authority, and accountability for the success of the program. It enables the program manager to negotiate and acquire resources, make key decisions concerning the operation of the program without permission from senior management, make tough trade-off choices to balance constraints, and take all appropriate actions necessary to achieve the business objectives that are specific to the product, service, or infrastructure development effort they are managing.

According to Roger Lundberg, director of Chrysler development system and vehicle engineering operations for DaimlerChrysler, "One of the most critical aspects of implementing a program management system is to ensure people are adequately empowered to make key trade-off decisions." However, Lundberg also explains that empowerment is not granted to a program manager—it is earned by the program manager. "At Chrysler, a program manager has to establish cross-organizational respect prior to becoming a program manager through work in one or more specialist functions and then demonstrate leadership competence as a program manager in order to fully earn empowerment."

In Figure 14.10, we provide a way to organize and visualize the strategic evaluation of each element of design. Degrees of strategic importance are mapped on the y-axis as low, medium, and high. Assessments from this step are entered along the y-axis of the cultural assessment matrix.

Assess cultural fit of program management design: The purpose of this step is to evaluate how well each element of the program management design fits the current culture. Of course, the higher the fit, the lower the cultural risk and the easier it will be to implement the element. In Figure 14.10, assessment of the following is entered on the x-axis:

- *Structure:* According to the program management design, a matrix organization is used. Our current culture has been using a matrix structure for all projects for a long time. Obviously, there is experience with the matrix, which supports the change to a program management matrix. Based on this analysis, the new structure is supported by the current culture, and we assess cultural fit as high.

- *Systems:* According to the program management design, an empowered program manager is responsible for all major program decisions

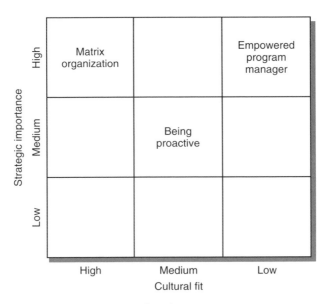

Figure 14.10   Cultural assessment matrix.

and results of the program. The current culture relies on project coordinators without power to make major project decisions. Obviously, the gap between the current culture and the program management's way of "how we do things around here" is large. Therefore, we assess the cultural fit as low.

- *Culture:* According to the program management design, the program managers don't use a reactionary approach to managing programs but a highly proactive approach. The current culture uses both reactionary and proactive, apparently somewhat in conflict with the new way of managing programs. Hence, we assess the cultural fit as medium.

Develop the cultural risk plan: In this last step, we look at the strategic importance and cultural fit, which define the cultural risk, of each element of program management design. Again, we show only one element of each—structure, systems, and culture.

- *Structure:* The implementation of the organization matrix is assessed to have high-strategic importance and high-cultural fit. Do we need a cultural risk plan for this element? Not really, because the high-cultural fit means the cultural risk is very small, simply, because our fictitious company's current culture is used to the matrix structure. In risk management, this is called the ***absorb* strategy**, which means accept the risk.

- *Systems:* The concept of the empowered program manager is of high-strategic importance and shows low-cultural fit. This means high-cultural risk and a high probability of cultural resistance because it is so new to the organization. Several risk management strategies are possible to manage around the risk. We will show two of them. The first is termed the ***transfer* strategy**, which means that the high-cultural risk is transferred to a third party. For example, the company may outsource program management, as NASA does in some programs. If this is culturally unacceptable, the company may use the ***avoid* strategy**, in which case they would not use the empowered program managers. Instead, they would first have to develop a comprehensive training program for all prospective program managers and senior managers. The company should also hire several seasoned program managers and pilot test several programs prior to full implementation of the program management model to better manage the required changes of behavior and decision making. The idea is to slowly overcome cultural hurdles and implement the concept of the empowered program manager.

- *Culture:* Proactive program manager behavior shows medium strategic importance and medium cultural fit; therefore, this has a medium cultural risk. One way to mitigate this risk is to have more experienced program managers who exhibit the correct behavior, serve as mentors to others, and still manage in a reactive manner (see box titled, Misunderstanding of Lines of Authority").

A successful cultural transformation plan is one that reflects the organization's current culture, evaluates the strategic importance of the program management design, assesses cultural fit of the program management elements, and includes a cultural risk management plan.

Misunderstanding of Lines of Authority
During transition, the lines of authority are not always clearly defined. This can lead to program managers either overstepping the bounds of authority or not

accepting enough responsibility. This balance of authority tends to become better defined over time as program teams and managers get more familiar with their new roles and responsibilities. Setting boundary conditions for the program manager in the program strike zone (see Chapter 11) and clearly articulating program priorities are two ways in which senior management can establish a better understanding of authority boundaries.

Stage 5: Realize

Using the approved transformation plan as a road map, the transformation can now be executed. In this stage, the organization begins to operate under the program management model.

The realize stage has three major steps:

1. Expand team
2. Hold regular meetings
3. Celebrate success

The first step, expand team, means that implementation groups should be added to the transformation team. The groups are assigned accountability for specific implementation actions such as develop materials, train people, change policies, develop systems, or restructure the organization. The transformation team will need to meet regularly to monitor implementation, evaluate it, face the implementation challenges, solve them, and share learning.

To not expect challenges when transitioning to a program management-based organization can be foolish, as challenges most likely will be many. From our experience and research, the most common challenges one can expect are resistance to change, tension between program management and functional management, and a misunderstanding on the part of program managers about their boundaries of authority. More details about the challenges are in boxes titled, "Resistance to Change" and "Tension Between Program and Functional Management."

Resistance to Change
Resistance to change is not unique to the transition to a program management-based organization, as resistance occurs any time a significant change to an

organization occurs. One can expect resistance from both middle management and some individual contributors. A key contributor to this resistance is that any change is a change from an employee's current status quo, and most times change from status quo breeds resistance due to fear of the unknown. A company may have operated under a business model that generated good to excellent results for years; therefore, people within the organization build their comfort zones around the existing business model. However, if the current business model has run its course of effectiveness and is no longer enabling a firm to keep pace with environmental changes, the business and the people it employs must either adapt to the changing business environment or prepare for a potentially painful demise of the enterprise. Resistance is likely to occur more frequently and severely in the early days of the transition.

Don't underestimate the power of celebrating success. Celebrate both small and big wins. The point is that social events (for example, team lunch) are necessary to show the team members that you appreciate their hard work. As improvement becomes visible, which is a big win, you need formal celebrations and rewards! Remember, celebrations keep executive commitment and focus strong.

Tension Between Program and Functional Management

A shift in the balance of power from functional managers to program managers commonly leads to tension between the two groups. Many organizations transition from a functional structure in which the functional managers possess resource, budget, and decision-making power to the program management model in which budget and decision-making power shifts to the program manager. Any shift in organizational influence can cause people to react unpredictably. Christopher Meyer calls this "the golden rule of organizations"—those who have the gold (in this case, the power) make the rules.[1] For the program management-based organization to work successfully, this power must be appropriately redistributed between the functional managers and the program managers by senior management. The key distinguishing point is that the program manager needs to be formally empowered by senior management to be responsible for all operational and financial aspects of their approved program.

Stage 6: Enhance

As the organization transforms into a program management-based entity, it becomes self-reliant and no longer needs the support from the transition team. It is at that point when the transition ends and the never-ending game of continual enhancement of the reengineered operation begins.

Terminating the program management transition process involves tying up loose ends such as performing a final account of the transformation, developing a final transition report for historical reference, having a final celebration of success, and disbanding the transformation team.

CONTINUOUSLY IMPROVE

Once the organization begins full operation under the program management model, continuous improvement in practices, methods, tools, and infrastructure begins. Without such improvement, the program management discipline will gradually deteriorate, losing the ability to support the business strategy of the organization.[8] Avoiding such a predicament and instead developing a continuously improved program management discipline can be achieved through the following steps:

- Form a program management improvement team
- Identify mechanisms for improvement
- Follow improvement process

Form an Improvement Team

The program management improvement team is usually part of the team responsible for designing and managing the program management function, typically through a PMO which is the subject of the next chapter. This team has the responsibility for simplifying, improving, and managing the implementation of the program management practices. When forming a team, we recommend that the majority of the program management team members come from the program management ranks.

Identify Mechanisms for Improvement Ideas

Ideally, there should be a continuous stream of suggestions and ideas to improve program management. To secure such a stream, one can require that program teams perform post-program retrospective reviews. If the reviews identify needed changes to the program management practices, a change request can be submitted to the improvement team. Additionally, change requests may come at any time from anyone involved in programs. Note that requests are not the only way to collect program management improvement ideas, other methods include surveys, small-focus groups, and one-on-one discussions with practicing program managers.

Follow Improvement Process

This process defines steps in acting upon change and deviation requests, as well as an escalation procedure, in case the requests are turned down and those proposing them want management to evaluate them. Change requests are suggestions for making changes to program management, most frequently coming from program teams. Quickly collecting and responding to them is of vital importance. Requests are also significant to deviate from program management. If appropriate, the deviation requests should be allowed to make sure the program management is flexible. These are requests to deviate one time only from some aspect of the program management model 1 for example, a change to the standard metric used. Because they are submitted while the program is in progress, it matters that management responds as soon as possible. Later, the requests can be evaluated to determine if program management should be adjusted.

Approved changes to the program management model should be implemented as quickly as possible, but too much change should not be introduced at any one time into the system. Additionally, to minimize risk and resistance to change, various changes may be best implemented through a pilot process prior to full deployment across all programs.

SUMMARY

The process for transitioning to program management involves three major steps: *understand the transition, execute the transition, and continuously improve.* In the first major step, business and operational factors drive senior management teams to consider moving to a program management model for developing their products, services, and infrastructure capabilities. Business factors include the level of linkage between development output and the strategic objectives of a firm; the need for time-to-money improvements to gain or sustain competitive advantage; and the need to have team leaders with the requisite level of business expertise. Operational factors include the level of technical, structural, and business complexity of a development effort; the number of cross-discipline interdependencies that have to be managed; and the ability to attract and maintain skilled program managers.

We introduced the SECURE process for planning and executing the transition. This is only one model that can be employed. In the scope stage boundaries of the transition are defined. The engineer stage focuses on designing an organizational structure, systems, and culture that supports the program management discipline. The confirm stage is used for pilot testing the designed program management model. Emphasis in the ultraplan stage is placed on planning for full implementation and political and cultural changes that accompany the transition. The realize stage involves execution of the transition plan. The final stage, enhance, is used to terminate the transition and move into continuous improvement of a company's program management practices, methods, tools, and infrastructure.

Some of the most common challenges include resistance to change, tension between program management and functional management, and a misunderstanding about new lines of authority.

The Principles of Program Management

▼ When an increase in project complexity and cross-discipline interdependencies becomes difficult to manage as a single project, redefine and reorganize the project into a program of multiple interdependent projects.

▼ Senior managers must fully support and drive the transition to a program management-based development model.

▼ A systematic process for transitioning to a program management business model is needed.

▼ Cultural changes such as rules of engagement, the decision-making hierarchy, roles and responsibilities, and core competencies of some functions will change when implementing program management and be the change agent.

REFERENCES

1. Meyer, Chris. *Fast Cycle Time: How to Align Purpose, Strategy, and Structure for Speed*. New York, NY: Free Press Publishers, 1993.

2. Gray, Clifford F. and Erik W. Larson. *Project Management: The Managerial Process*. New York, NY: McGraw-Hill Publishing, 2005.

3. Larson, E. and D. Gobeli, "Organizing for product development projects," *Journal of Product Innovation Management* Vol. 5, No. 3 (1988), 180–190.

4. McGrath, Michael E., Michael T. Anthony, and Amram R. Shapiro. *Product Development: Success Through Product and Cycle-time Excellence*. Stoneham, MA: Butterworth-Heinemann Publishers, 1992.

5. Schein, Edgar H. *The Corporate Culture Survival Guide*. San Francisco, CA: Jossey-Bass, 1999.

6. Hofstede, Geert. *Culture's Consequences*, 2nd edition. London, UK: Sage Publications, 1984.

7. Schwartz, H. and S. M. Davis. "Matching Corporate Culture and Business Strategy," *Organizational Dynamics*, American Management Association (1981).

8. Juran, Joseph M. "Managing for World-Class Quality," *PM Network*, 6 (1992), pp. 5–8.

Chapter 15

The Program Management Office

As an organization grows in its implementation and maturity of the program management discipline, a natural progression that many firms follow is the formation of a program management office, or PMO. The PMO addresses two of the most common problems that arise as the use of program management increases within an organization. First is the realization of the need for consistency in the definition, planning, and execution of all programs within a business unit. Without consistency across the portfolio of programs, business results are not predictable or repeatable on a recurring basis. The second realization that arises is the need for a single program management point of contact within an organization. A single point of contact for program management provides improved communication, decision making, and program oversight.

The PMO provides leadership and infrastructure for managing and controlling multiple programs. It represents a compilation of program management infrastructure, support, tools, and best practices that have been melded together to improve business results and drive continual gains in a firm's management of its product, service, or infrastructure development efforts.

Roger Lundberg is a former PMO director of the Jeep division at DaimlerChrysler and currently the director of Chrysler development system and vehicle engineering operations. According to Lundberg,

> "Successful implementation of program management within an organization requires a certain amount of infrastructure. Key elements of that infrastructure includes a good portfolio management process, a well-defined and executed development process, a good requirements

system, and good processes such as resource management, change management, decision making, budgeting, scheduling, and financial target setting."

An effective PMO provides the right level of program management infrastructure and aligns it with the overall organizational culture and structure.

This chapter focuses on the role the PMO plays in establishing program management as a functional discipline within an organization. The description of the PMO, its primary functions, and operational elements are discussed. Additionally, PMO implementation guidelines and a variety of organizational factors that are worthy of consideration when a management team is evaluating the formation of a PMO are presented, as well as how a PMO contributes to improved business results. An example from an information technology organization is included at the end of this chapter to demonstrate the use of a PMO. Senior managers and practicing program managers will find the information contained in this chapter helpful in understanding the following:

- Why companies have implemented PMOs
- What constitutes a PMO
- The functions and operational elements of an enterprise PMO
- PMO implementation guidelines

THE PROGRAM MANAGEMENT OFFICE

So what is a program management office and how does it differ from the project management office?

The PMO is a centralized body within an organization that is responsible for instilling structured leadership, methodology, and infrastructure across all programs to make the best use of a company's time, money, and human resources (see the box titled, "PMO: A New Idea?"). The objective of the PMO is to promote and drive consistent, repeatable program management practices within an organization. It represents a compilation of product, service, or infrastructure development best practices that has been melded into an integrated approach that is targeted at improving business results. More specifically, it is a central function utilized for simultaneous management of multiple development programs.

PMO: A New Idea?

The first program office was formed and organized in 1957, then called the special project office (SPO), within the United States Department of the Navy. The SPO was established to manage the development of an underwater ballistic missile-launch system. Indeed, the structure of the missile-launch system program mirrors the program management structures utilized today—a series of interrelated projects (launcher, missile, guidance, installation, navigation, operations, and test) collectively and coherently managed as a program. So, to answer the question from our title, the PMO is not a new idea.

However, there is a fundamental difference between that PMO from 1957 and today's offices. The office from 1957, and the wave of offices that followed, had a common attribute of supporting one program only, and, certainly, those programs were large.

The new wave of PMOs support multiple programs. This is, of course, an issue of productivity, resource, and cost efficiency. Additionally, these multiple programs are smaller than the likes of the development of an underwater ballistic missile-launch system. So, from this perspective, today's PMO *is* new in that it supports multiple programs.

The program management office differs from the more common, typical project management office in at least the following two ways:[2]

- First, the PMO is focused on consistency of methods, tools, and practices across all programs within an organization, as well as consistency of methods, tools, and practices across all projects that make up a program. Therefore, the PMO encompasses both program and project management practices.
- Second, the PMO is focused on business and strategic success. It's responsible for defining and measuring program metrics that measure business goals and ensuring that each program supports and helps to achieve the strategic objectives of the company.

PMOs are used across a wide array of organizations and industries for coordinating various types of programs. It can manifest itself in a wide spectrum of implementations ranging from an informal community of practice to a comprehensive, enterprise-level PMO. The appropriate implementation of the PMO is dependent upon the objectives, complexity, and culture of the specific firm. Normally, a formal PMO is more useful and cost effective for larger firms and is especially beneficial for companies with geographically distributed sites and teams. Large, decentralized firms; single-site organizations; or those with a limited number of product, service, or infrastructure development programs may not warrant the need for a comprehensive and formal PMO. For these organizations, it may be more appropriate to create an informal PMO in the form of a community of practice.

It is also important to point out what a PMO is not. It is *not* a cure-all for poor business strategy, and it will not substitute for comprehensive portfolio planning. Likewise, it is not a strategic planning function for a business unit or a company. Even though it can be formalized within the organization, its structure and use does not have to be rigid or inflexible. It often is not as effective in a decentralized organization, and it is not a panacea. It won't solve all the problems that exist within an organization.

AIMING FOR BETTER BUSINESS RESULTS

The impetus for the creation of a PMO most often originates out of the need for continual improvement in achievement of business objectives caused by major change within the organization (see box titled, "Gaining Control Through the PMO"). This could be fueled by such things as a firm's rapid growth, mergers and acquisitions, or relocation of operations to multiple sites. Changes such as these may yield a greater number of programs; larger programs; increased complexity due to design and team structure; and codevelopment of new products, services, or infrastructure capabilities involving multiple geographical sites. These changes, although strategically important to the firm, may create challenges to achievement of business objectives because of inconsistency in program performance, poor communication between program managers and key stakeholders, variability in program measurement techniques, and the lack of central coordination of all programs within an organization.

These challenges become even greater as growth occurs internationally, adding language, cultural, and time-zone barriers. If these challenges and barriers are not properly addressed, they may result in a firm's inability to manage and control development efforts, leading to adverse impacts on business results due to missed schedules, cost overruns, customer dissatisfaction, and competitor gains.

Gaining Control Through the PMO

As mentioned in an earlier chapter, Tektronix, Inc. is a global test and measurement equipment company within the high-tech industry. It installed a

worldwide PMO to oversee all product development and program management activities.[1] Tektronix grew the firm's product and technical capabilities
through several company acquisitions. Many of these were internationally
based and each acquired business entities that had their own unique development processes and tools, many of which were only partially compatible
with the parent company. The firm's normal acquisition integration process
addressed most of the compatibility issues. However, several communication,
coordination, and control challenges remained. The resulting inefficiencies
tended to adversely impact the company's ability to consistently achieve
its business objectives. The solution was to consolidate responsibility for all
worldwide product development programs and provide management and
control through a PMO centrally located at the company's headquarters in
Beaverton, OR. Standardization of metrics, tools, and processes contributed
to improved communication and business results.

The Tektronix PMO continues to evolve and develop as improvements and
best practices are identified. The PMO is credited with having improved team
performance and contributed to Tektronix's ability to achieve their business
and operational objectives. As an added benefit, the company has derived a
cohesive worldwide team of trained and experienced product development
program managers.

PMO VARIATIONS

While there are many variations of support offices, we discuss three
types of program management offices that are commonly found within
businesses. They vary with respect to scope of their charter, functions that
they perform, amount of influence they possess, and the organizational
level at which they are established. Figure 15.1 demonstrates the three
most common types of program offices. The first type, the project control
office, resides at the project level of the organization; the functional
program office is found at the functional department level; and the
enterprise PMO is established at the business-unit level. One, two, or all
three types of program offices may exist within a company.

The Project Control Office

The project control office is normally established to provide administrative
and tracking support for the project teams. Work is focused on maintaining project procedures, schedule maintenance, earned-value tracking, tool
usage and support, and project metrics and report generation.

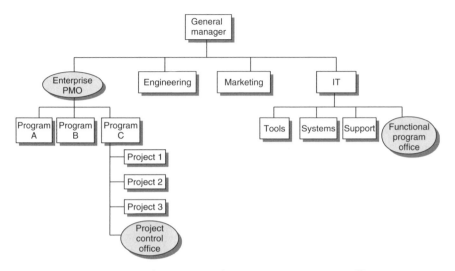

Figure 15.1 Three types of support management offices.

The Functional Program Office

The functional program office is set up to support the program managers within a department but does not operate as a true function within the company. Rather, it is more administrative in nature. The operation of the functional program office is focused on maintenance of program schedules, program and project-data tracking, and development of program indicators and reports. The functional program office also maintains a central repository of program and project information pertaining to the particular function in which it operates, such as the IT department, as shown in Figure 15.1.

The Enterprise PMO

According to some experts, establishment of an enterprise PMO is one measure of maturity of an organization, in which program management is recognized as a true function within the company.[3] The fully functional enterprise PMO is the center for program management competencies and practices within a company. It should be established at a level within the organization comparable to other critical functions. Placing the PMO high in the organization hierarchy is critical for two reasons. First, the PMO manager needs to be part of the senior management team of the organization to properly align programs to business strategy. Second, the PMO manager must have sufficient political and decision-making influence to broker tensions between other functions—such as the natural tension that exists between marketing and engineering. The enterprise PMO is normally led by a director of program management who is responsible for establishing commonality of methodologies, process, tools, and program management practices across the company.

Companies that have much of their product, service, or infrastructure development in a distributed or virtual environment that spans multiple sites and geographies derive a tremendous benefit from the leadership, monitoring, and control offered by an enterprise PMO. Remote development efforts that are located far from the main offices of the firm often have difficulty effectively linking and staying consistent with requirements that are established by the main body of the corporation.

An effective enterprise PMO can have significant influence throughout the company. Figure 15.2 illustrates how an enterprise PMO can be used to influence critical aspects of a business's operation such as building a stronger program link to strategy; building program management maturity; forging strong relationships between the program managers and the functional and senior managers; and establishing commonality in program management practices, processes, metrics, and tools.

The remaining sections of this chapter pertain to the various aspects of the enterprise type of PMO.

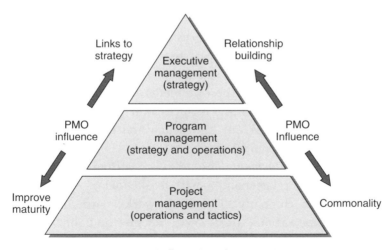

Figure 15.2   Influencing the organization.

FUNCTIONS OF AN ENTERPRISE PMO

PMOs function in a variety of ways depending upon the structure, size, culture, and needs of the organization. The challenge is to create an enterprise PMO that delivers a balance between program efficiency and creativity and promotes an environment of continual learning and improvement. It is also important that creativity is not stifled by an overburden of process bureaucracy.

Common PMO functions that support an environment with a healthy balance of efficiency and creativity include the following:

- *Commonality of methods, tools, and metrics*: To gain efficiencies through consistent program management practices, all programs within an organization need to use the same methods and tools for defining, planning, and executing programs. Likewise, commonality in data collected, utilized metrics, and information reported facilitates improved communication between program teams and senior management (see the box titled, "Monte Carlo in Italy").
- *Alignment to strategy*: Because the enterprise PMO is highly placed within an organization, it can assist the program manager and ensure that his or her program stays in alignment with, and continues to support, the strategic objectives of the business (see Chapter 2).
- *Promotion and improvement of the program management discipline*: The PMO is the company champion for program management and promotes the discipline as a true organizational function on par with other functions such as engineering, marketing, and manufacturing. The PMO is also the program management competency center within an organization. Best practices are collected, documented, and implemented through a central repository. Additionally, program management *next* practices are developed, piloted, and implemented by the PMO in response to new problems and opportunities that arise over time. Finally, the PMO ensures that program managers have access to the most current knowledge and training to continually increase their core competencies.
- *Relationship management*: The PMO is continually focused on managing program relationships with stakeholders both across and outside of the company. The PMO works with all stakeholders to ensure expectations are known and met and that the program management function is viewed as a valued partner in product, service, or infrastructure development efforts.

Monte Carlo in Italy

When KET, a large technology company from the United States acquired Eros, a small Italian company, the purpose was purely business. Eros had a specialized, state-of-the-art line of products that was catered to a small niche market, consisting of several international companies. There was no real competition, and executives in KET felt that Eros's product line would nicely complement KET's product line and further strengthen their position as an international leader.

David Hackham, director of the KET's enterprise PMO, had two roles in the acquisition effort. First, Hackham was put in charge of managing the acquisition integration, which was organized as a program. Second, he was to introduce KET's program management processes to Eros. Similar to how he handled other KET acquisitions, he first audited Eros's existing program management process, hoping to find some good elements he could retain. Just like in earlier acquisitions in which he found such elements, he was fortunate again to find his program management nugget in Eros. It was a Monte Carlo analysis, a risk tool, which Italian engineers simplified in Excel but added amazingly attractive and telling graphs.

Hackham saw many examples in Eros's programs, where Monte Carlo analysis was consistently used in meaningful ways. Because he was responsible for standardization of KET's program management tools and did not have a good risk tool, he decided to adopt the Eros's Monte Carlo tool. Soon, thereafter, he provided training in Monte Carlo for more than 30 of KET's program managers.

This example demonstrates that good program management tools can be found in all corners of a company, even in newly acquired subsidiaries.

FUNDAMENTAL ELEMENTS OF THE ENTERPRISE PMO

As stated previously, PMOs are structured in a variety of ways, but regardless of structure, the PMO is the central function for the program management discipline within a business. It provides infrastructure, consistency, and coherency to facilitate program definition, planning, and delivery. In its fully developed state, it should contain all elements necessary to ensure that development programs are managed as effectively and efficiently as possible. If the elements are centrally managed, coordinated, and integrated into a systemized approach, it provides unified coordination and control, while enhancing communication across all participating organizations, functions, physical sites, and time zones. Figure 15.3 illustrates an effective PMO system that consists of eight key elements.[4]

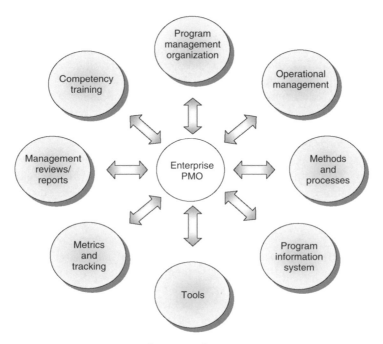

Figure 15.3   Elements of an enterprise PMO.

Program Management Organization

An enterprise PMO needs be established as a true function within an organization and have a well-defined structure with roles and responsibilities communicated across the firm. This is critical for larger organizations that have numerous program managers. We discussed the need for empowerment of program management by senior managers in Chapter 14. This implies that in establishing a PMO for the firm, senior management is empowering it and supporting it to operate as chartered. One of the important roles of the PMO director is to develop and administer a career ladder for the growth and advancement of program managers within the firm. The career ladder should include detailed job descriptions that outline responsibilities and experience levels that range from entry-level to senior-level positions. These can be used for hiring, setting performance expectations, promotion, and other professional needs.

Operational Management

Many PMOs provide operational and administrative assistance to the program managers in the organization. This may involve maintenance of the program master schedules, tracking of budget spending versus plan, development and delivery of program indicators and reports to support program decision making, change management documentation and communication, facilitation of program mapping workshops, and risk tracking and reporting. The PMO also provides consultation and mentoring for sensitive issues and problems, program start-up and closure activities, as well as a structured process for the escalation of issues and problems. This is achieved by an array of management and organizational tools such as the program strike zone, dashboards, and program reviews (see Chapter 11.) These provide an escalation procedure up through senior management for decision making and resolution of the issues that are outside the responsibility and control of the program manager.

Methods and Processes

A primary objective of the PMO is to promote consistent, repeatable program management practices that result in efficient use of business resources. This is accomplished in large part through the development and use of common methodologies and processes. For example, all programs should be managed via a common PLC as well as consistent scheduling, risk management, change management and requirements management methods and processes (see Chapter 8). These should be common across all sites and geographies to improve coordination and communication. Process consistency provides improved predictability of performance on all programs across the organization, aids in the adoption of a common development language, drives more effective decision making, develops economies of scale with respect to infrastructure and tool deployment, and increases flexibility of resources from program to program.

Program Information System

The PMO is the focal point for program management information and knowledge management within an organization. Centralized storage and organization of program information improves accessibility of the information. Program information includes historical data on program performance to business results, risks, issues and problems encountered, and best-practices employed on previous programs.

This central repository also allows the opportunity for wide distribution of post-program review information collected to enhance communication of key lessons learned.

The program knowledge base includes a repository of the common methodologies and tools to be employed on all programs including templates, checklists, and instructional material. Additionally, the PMO drives research, development, and innovation of new methodologies, processes, tools, and technologies.

Tools

The PMO is responsible for evaluating, selecting, deploying, and many times maintaining the tools that are implemented by the program teams (see Chapters 10 and 11). This also includes training and mentoring of the effective use of the tools. Like program methodologies and procedures, the PMO looks to drive commonality of tool deployment and use across all programs within an organization. This increases economies of scale to support, use and maintain the suite of tools, and increases productivity by allowing increased capabilities in tool usage across the organization.

Metrics and Tracking

The PMO is responsible for developing and maintaining metrics that accurately measure program progress and reflect the achievement of the business objectives intended (see Chapter 9). This includes providing the information as needed and publishing the results to the appropriate audiences. The PMO also tracks, collects, and analyzes the data necessary for the development of the metrics through common program tracking tools. New tools are consistently being developed to more adequately monitor and track progress of programs within the simultaneous, multiprogram environment. The role of the PMO is to consistently research new tools, evaluate the use of tools within the organization, and implement the tools that provide improved data collection and information dissemination.

Management Reviews and Reports

The PMO may be responsible to orchestrate and run the formal program reviews for senior management. This includes defining the format of the meetings, agendas, and attendees. The PMO provides an organized and structured approach for management's visibility of programs under way in the firm. The structure of management reviews, information to be presented, and who attends the meetings should be negotiated between the PMO director and other senior managers. The PMO is then responsible for scheduling and leading the reviews for the senior management staff and capturing action items from the meeting for follow up. The focus of these reviews is to determine the status of the individual programs toward achieving the specific business objectives.

Competency Training

The program management discipline will only be as effective as the program manager's skills, capabilities, and experience. Skills training and development can and should be made available to all program managers and part of the PMO charter. Much of this material can be maintained on websites and be available to all personnel. It can also include online modules as well as face-to-face group sessions and forums addressing specific topics. For multi-site firms, training and best practices for managing in a distributed or virtual team environment can be provided. Training and career development of the program managers in the organization is driven and supported by a program management competency model, as discussed in Chapter 13.

The PMO creates a learning environment through the alignment of all program managers and their teams under the same leadership, philosophies, and practices. Program managers master their skills as they communicate and coordinate their programs with other program managers in this structure and, therefore, are continually observing and learning from one another.

IMPLEMENTING AN ENTERPRISE PMO

When implementing an enterprise PMO within a company, much thought has to go into *how* the PMO will be implemented, as a fully functional PMO will have implications on current roles, responsibilities, decision making, and political elements of the firm (see box titled, "PMO: It Is a Power Game Too!"). Therefore, how the PMO is established will have a large impact on its overall success or failure as an organizational entity.

PMO: It Is a Power Game Too!

The potential power of a PMO raises questions in the minds of some stakeholders. This sets in motion resistance, conscious or unconscious, that is aimed at killing the PMO before it gets established. Let's look at some power traps: players, resistance, PMO viability, and overcoming barriers.

Power players: First, you face **highlanders,** top executives who have their way of thinking that ordinary people can it always figure out. If a PMO runs against some of their firm principles, they can obliterate the PMO. **Buddies** are people who sit on the same level as you, the champion of the PMO. They may resist because of a belief that PMO may erode their power. **Diagonalists** are managers with a diagonal relationship to you, below your level. The PMO may also reduce their power, so they may resist the PMO in an attempt to maintain their power. **Workerbees** are program practitioners who unless they understand *"what is in it for me?"* may sabotage the effectiveness of the PMO.

Resistance: Why would people resist a PMO? The first reason is *ignorance*, meaning that they do not understand the concept and value of a PMO. *Poor justification* of the PMO idea to the stakeholders is a sure way to generate detractors motivated to resist. If a *cultural fit* does not exist, then resistance to the PMO will be high. The PMO existence should be both in line with the corporate culture and technically justified. Pick the right political moment, when people are not occupied by thoughts of potential mergers, layoffs, internal power battles, and so on—all of which may cause PMO demise.

PMO viability: If program management works just fine the way it is or if a company exists in a stable, slow-changing environment, a PMO may not be needed. The point is that the PMO must be a viable idea and provide value to the organization.

Overcoming barriers: The above mentioned players who may resist the PMO concept are not intentionally acting badly. Quite the contrary, they are people with good intentions who just do not see the need for a PMO. To overcome this barrier, one needs a strategy for these power-related issues. *Take time* to inform, indoctrinate, and involve people. Building a PMO is a big organizational change that will take time. *Show benefits* of a PMO to each major stakeholder group. *Seek political help* by finding a high-profile and high-powered sponsor, building a support coalition, and getting your allies involved in implementing the PMO.

Businesses have had the most success when they have used a project approach to implementation. An iterative methodology, with a multiple-phase framework has shown to be a pragmatic way to introduce the PMO into the organization.

Figure 15.4   Evaluate, plan, and act methodology.

Iterative Cycles of Learning and Improving

The iterative methodology consists of multiple cycles—evaluate, plan, act—in which the current state is evaluated; the next state is identified and planned; and changes in methods, tools, and practices are piloted. The cycle is than repeated until the final end state is achieved. Figure 15.4 illustrates the evaluate, plan, and act methodology for organizational transformation, which is a form of Deming's well-known plan-do-study-act methodology.[5] The following section describes the evaluate, plan, and act cycles:

Evaluate: During this cycle, an organization audit is conducted by the PMO implementation team to determine the "as is" state, in which current methods, tools, culture, and practices are reviewed. A gap analysis is then conducted, in which the current state is evaluated against the next state of maturity. Objectives for the next cycle of transition are then identified.

Plan: During this cycle, changes in methods, tools, and practices that are targeted for the next phase of implementation are planned based upon the gap analysis, which is completed in the evaluate phase. Templates and other documents are developed. Training is also developed and delivered to increase capabilities and educate the organization about changing practices.

Act: During this cycle, the new methods, tools, and practices are piloted to prove their feasibility and test the capability of the organization to move to the next stage of implementation. Key learnings are captured from the pilots and incorporated in the next iteration of implementation.

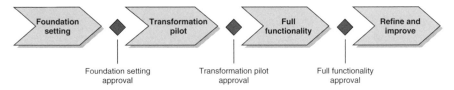

Figure 15.5   PMO implementation phases.

Phase Implementation

The use of a multiphase implementation framework, such as the one shown in Figure 15.5, has been quite successful for firms implementing an enterprise PMO. The primary reason for the success is that the framework brings a structured and methodical approach to introducing change into the organization. The framework shown in Figure 15.5 consists of four phases, as follows: foundation setting, transition pilot, full functionality, refine and improve.

Phase 1: Foundation setting: This is the phase in which the PMO concept is derived, defined, and sold to the senior management team of the business. Assessments in program management methods, tools, communication, and decision-making practices are conducted, and the business case for change is developed and presented. The primary activities that are conducted in this phase are as follows:

- Define the PMO charter and objectives
- Assess current program management methods, tools, capabilities, and practices
- Define end-state methods, tools, capabilities, and practices
- Create an organizational transformation plan
- Identify and recruit a senior sponsor to champion the need for and value of the PMO (see box titled, "Politics and the PMO")
- Identify the PMO director and core team
- Develop a communication strategy

Politics and the PMO

Do you want to build a PMO? Then think twice before acting. It is critically important to understand the elements of a PMO and how to implement it in a business, but it is not enough to be fully successful. The reality is that a PMO will change the power and influence balance within an organization, and some may not like it! Therefore going in, one needs to have political acumen and political strategy. The basic elements of a political strategy include the following:

Determine the lay of the land: Get to know who really has the power in the organization. Who will benefit, and who will lose from developing a PMO? A thorough stakeholder analysis is required (see box titled, "Knowing Your Stakeholders"). The point is to know your supporters and detractors when building a PMO. The stakeholder analysis should also identify a powerful sponsor from the senior management rank. Implementing a PMO without a powerful sponsor is like trying to push a one-ton boulder up a hill.

Show people how they will benefit from the PMO: Some will benefit from the PMO and support it; others will lose and resist. Don't ignore the second group. You are better off showing them how they will benefit from a PMO, which may neutralize their resistance. Get to know who may resist and show them how they will benefit.

Assess the friction to change: Over time businesses develop processes for how things are accomplished in the organization. They are used, refined, passed from generation to generation, and become the organizational truth. This truth is the friction one encounters when moving an organization from one way of doing business to another. The older and more successful the organization, the more difficult it becomes to change. To implement a PMO, one needs to assess the amount of friction that will be encountered and adjust the implementation plan, communication plan, and timeline accordingly.

Learn from the past or repeat It: Those who do not remember their past are condemned to repeat it. Go out and talk to people and learn what major change initiatives failed in the past. Then, determine what you can do differently to avoid the same fate when implementing a PMO within your organization.

Knowing Your Stakeholders
Knowing how to build a PMO, understanding the political game around the PMO, and having political acumen, is not enough. You must have a proper political plan to play the game in favor of a PMO. A prerequisite for building an effective political plan is a thorough understanding of the stakeholders involved. This requires a good stakeholder analysis. A stakeholder analysis will

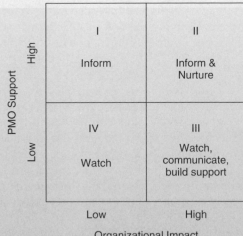

Figure 15.6   Stakeholder influence.

tell you who the stakeholders are, what motivates them, what their concerns are, who supports or does not support the PMO, which stakeholders have high organizational influence, and who could disrupt the PMO implementation.[6]

Armed with the analysis, you can devise a strategy for each of the four groups of stakeholders. See Figure 15.6.

Group 1: High support, low impact: They support the PMO but don't have much power. To continue their support, keep them informed.

Group 2: High support, high impact: These are your trump cards. Keep them informed about everything going on around the PMO and nurture them at all times. You may ask for their influence to help gain support from stakeholders with lower levels of support or opposition.

Group 3: Low support, high impact: They dislike the idea of the PMO and have power to kill the idea. They cannot be ignored! Build steady communications with them and try to figure out how the PMO can support their interest. Then, use that to increase their support. Also, try to enlist support from Group 2 stakeholders to increase the support of these stakeholders.

Group 4: Low support, low impact: They oppose the PMO but are not influential. Keep them informed and watch that they do not encourage more influential stakeholders to oppose the PMO.

Phase 2: Transformation pilot: This phase is when the evaluate, plan, act implementation process is applied. Multiple iteration cycles are conducted to move the organization from its current state to a fully functional enterprise PMO. The primary activities that are conducted during this phase of implementation include the following:

- Repeated assessment of current and next state capabilities
- Repeated development and piloting of new methods, tools, and practices
- Training developed and conducted to educate and increase organizational capability
- Coaching and mentoring of pilot teams
- Development of new reporting structures
- Establishment of centralized communication channels and messaging

Phase 3: Full functionality: During this phase, an enterprise PMO is established as a true function within the organization. Pilots are completed; new methods, tools, communication channels, and decision-making processes are established; and the PMO continues to mature in its practices. The primary activities that take place in this phase of implementation include the following:

- The PMO is fully resourced
- A central repository of program and project information is established
- Roles and responsibilities are defined and documented
- Program and project manager job descriptions are developed and career paths are established
- Consistency in methods, tools, and practices is established across all programs
- The PMO orchestrates and runs the formal program review and PLC decision-checkpoint meetings
- A long-term training curriculum is established
- Continual coaching and mentoring of program teams is conducted

Phase 4: Refine and improve: Once the PMO becomes fully functional and is established as an independent function within an organization, the final phase of implementation begins. The PMO must be committed to continual improvement and maturity to remain a viable and valuable function within the organization. Long-term improvement involves capturing and innovating new best practices, evaluating and implementing more effective tools, and establishing and evolving a central knowledge base for program and project management. Also, as companies continue to become more globally dispersed, the PMO must become more effective in establishing the links to the dispersed organizations and closing the communication and operational gaps that may exist.

IMPLEMENTATION CONSIDERATIONS

If the decision is made by senior management to implement a PMO, one consideration is how quickly they adopt its capabilities. Russell Archibald states,

> It is not recommended that an organization attempts to establish a PMO overnight to the full range of responsibilities and authority. Rather, a logical evolutionary plan is suggested for adopting the portions as they make sense and the organization's skill level and experience can successfully absorb the new entity.[7]

Another consideration is the firm's management philosophy regarding centralization versus decentralization of various functions and activities. Clearly, most organizations would agree on the benefit of centralizing processes, policies, tools, methods, and training. Controversy normally arises when the business unit managers are reluctant to relinquish the planning and control of individual programs that are directly tied to the manager future success to a central group or function.

The question as to whether all program managers should report to one person can also be controversial. This may need to be decided and resolved at the executive level. As stated earlier in the chapter, both the matrix and program organization structures work well with individual program managers who report directly to the PMO director, while being accountable to the senior management of the organization for program results.

The overall success of the PMO needs to be assessed and periodically evaluated by senior management. In the short term, senior management needs to determine if management is getting sufficient and reliable information that is needed to manage the business in a timely and effective manner. From a longer-term perspective, senior management should ensure that operationally the PMO develops some quantitative measures of performance. For example, the PMO should a composite measure use to gauge time to market, profitability, quality in the field, and so on, for all programs. These metrics can be evaluated over time to see whether continued improvement is occurring.

IMPROVING BUSINESS RESULTS

One executive we interviewed described the problem within his organization this way. "Prior to establishing a central PMO, we had no way of measuring the overall success of our programs as they were dispersed in various groups across the organization with little or no consistency of information, measures, and metrics."[8] This type of environment contributes to reactionary management and creates numerous surprises to senior management as things go out of control. A well managed and appropriately chartered PMO should significantly reduce surprises to senior management and should reflect a philosophy of "raising the bar" of performance through continuously improving business results.

The primary objective of a PMO is to improve business results by a focus on execution. The PMO should be held accountable by senior management for the execution of and results from the development programs. The PMO improves business results by achieving the following:

- *Predictability and repeatability*: A PMO establishes common and consistent implementation and use of key business processes, tools, and measures across the organization. Consistency provides improved predictability of performance on all programs and aids in the adoption of a common language. Productivity gains will also be achieved as program teams become more experienced in process execution and as cross-site development efforts utilize common methods. Commonality helps to achieve productivity gains by decreasing the learning curves for new program team members and increasing execution expertise on new development programs.
- *Concurrency*: The PMO provides the means to coordinate and control numerous interdependent programs that are being executed simultaneously through a common organization and infrastructure. The PMO establishes a single focal point within the organization for program-related information exchange and direction. This keeps consistency in information flow and content, which leads to improved

decision-making capabilities within the organizations. Additionally, by focusing the execution of the portfolio of programs within a central organization, the PMO is instrumental in ensuring the programs remain strategically focused by communicating the business benefit and value of each program.

- *Speed*: By establishing common and best methodologies and practices, consistent metrics and measures, and focused decision making, the PMO lays the foundation for increased skill development and productivity of the program teams. Increased skill and productivity ultimately leads to improved development cycle time and consistently decreasing time-to-money results. Products, services, or infrastructure capabilities are released from development into the operational environment quicker, resulting in improved business results such as earlier revenue and cost-reduction gains for the business.

David Churchill, vice president and general manager of the network and digital solutions business unit of Agilent Technologies, summarized the business value of the PMO this way:

> The consequences of the failure to successfully execute new product development are numerous, including missed revenue and profit objectives, lost market opportunities, and unhappy customers. Many organizations, when faced with these consequences, will attempt to establish improved processes. Process improvement, however, is not enough as it contains no content or time element. What is needed in addition is the ability to successfully execute to an approved plan in a coordinated manner across all functional disciplines bounded by time and performance metrics, which are used to track the necessary deliverables for achievement of the business objectives. The functional discipline of program management is designed to accomplish this responsibility, and the best organizational entity to manage these activities and ensure effective communication of development progress is the program management office.

SUMMARY

The PMO is an innovative approach for providing leadership and infrastructure for managing and controlling multiple development programs. The PMO provides the ability for an organization to successfully manage and coordinate scarce resources, consistently track development program progress, ensure that deliverables are completed correctly and on time, and that senior management is receiving relevant and consistent information for decision making. PMOs are currently used in many organizations for managing various types of programs and are generally more cost effective for larger firms due to economies of scale. Smaller organizations that either cannot afford or justify the formal PMO will benefit from an informal approach that provides some of the elements of the formal PMO. Because of the PMO's significant potential to positively impact business results, it is important for the management team of the company to understand the benefits and use of the PMO to improve business results.

The Principles of Program Management

▼ The PMO as a centralized body within an organization that is responsible for instilling structured leadership, methodology, and infrastructure across all programs.
▼ The PMO provides the means to coordinate and control numerous interdependent programs that are being executed simultaneously.
▼ The PMO promotes and drives consistent, repeatable program management practices with a company.
▼ The PMO is the focal point for program management information and knowledge management.
▼ The enterprise PMO is comparable to other critical functions within a business.
▼ The PMO establishes and maintains a true career path for program and project managers.

Program Management in Practice

The PMO
Author: Sabin Srivannaboon and Dragan Z. Milosevic

Prologue

The topic that we explore in this industry example is the establishment of a PMO in an IT group of a company in a niche, cell phone market. The corporation has seen a major decline in sales and hopes to cut costs by terminating second-rate programs to save money.

The PMO is a mature concept that has been recently redesigned and massively applied. For this reason, there are many PMO implementation options for a company to evaluate, as presented earlier in this chapter. The CIO for Trust Corporation has put a team together to explore the various options and has recently hired an experienced PMO director to lead the implementation.

The example begins with an assessment of the PMO feasibility by looking at both the tangible and intangible returns. To determine the contributions, focus is directed on the purpose of the PMO. Some examples offered from Trust Corp. are the following: The PMO is seen as an important means to create a common and standard language; the PMO forces the program manager to adhere to company policies and procedures to increase program success; and the PMO improves business results through improved and common processes.

When looking to implement a PMO, one of the central topics that business's need to comprehend is the PMO scope. For Trust Corp., the IT PMO scope includes process standardization, personal development, consulting, program reviews, and program information system. (Other areas of implementation scope are discussed earlier in this chapter). With scope determined, the discussion turns to how the PMO should be organized, or more precisely, where within the organization should the PMO be located and who should the program managers report to?

Perhaps the most fundamental aspect of the PMO that it is a business entity. Unless the PMO has very limited scope, the purpose is to help the company make money and convert strategy into successful programs.

Background

Trust Corp. is a specialty instrument company in the U.S. cell phone industry. Six years ago, Trust Corp. introduced a breakthrough product to the market, leaving the competition in the dust. In the high-technology industry, product advantages over the competition are short term, unless followed by the continuous stream of equally good products. Because Trust Corp. has not had another successful new product introduction, the company has seen a decline of business in the last five years.

While Trust Corp. was developing and introducing a few lackluster products, the competition caught up, the market slumped, and Trust Corp. sales dropped significantly. The company's annual sales were $360 million in 1998 but only $130 million in 2004. As a consequence, management began focusing on cost cutting, especially in Trust Corp.'s IT group. The general feeling was that too much money was being wasted on poorly scoped programs that resulted in scope creep, delays, and late cancellations. There was also a feeling that the establishment of a PMO would help put IT's ducks in a row, terminating poor programs earlier and saving money through efficiency improvements.

Saul Cognito, the chief information officer, put together a team led by Barry Senders, a longtime program manager, to study the feasibility of creating a

PMO. After the team spent some time working on the problem, it suggested going forward with the establishment of the PMO and hiring an experienced program management officer from outside the corporation. Peter Deerling, a former program management officer in the IT group of a much larger and more successful company named Stellar Corp., was hired.

Cognito's plan was to first organize a series of meetings between Deerling and major stakeholders to further promote the PMO as a vehicle to increase program success. His second goal was to get input from the stakeholders. A summary of one such meeting, attended by Cognito, Senders, and Deerling follows.

The Real Purpose of the PMO

The first thing Cognito planned was to collect data based on Deerling's experience with the PMO at his previous company. He wanted to know whether it helped the success of program management and, if so, how. He also wanted Senders to hear Deerlings story because Senders was an influential stakeholder who would have a major role in shaping the PMO to be program manager-friendly. He was also a longtime program manager at Trust Corp. and was respected enough to be viewed as the voice of program managers. In addition, Senders helped Cognito develop a plan to establish a PMO before Deerling came aboard. Therefore, Cognito let his ally, Senders, ask the questions.

Senders had already heard a high-level story about Deerling's PMO while interviewing him for this job. It was interesting, but he wanted more details.

Senders: Can you tell me about the history of the PMO at your previous company, like how it started and was developed?

Deerling: I believe that to truly understand a PMO, one has to fathom the organizational context. Therefore, I'll begin with the background of Stellar Corp. The company is one of the leaders in North America in specialty manufacturing and has more than 10,000 employees. It started about 50 years ago, and at the time I left, it was operating five major manufacturing plants and one parts manufacturing plant in North America. The company has several business units, which are supported by the IT group, including verticals like the infrastructure division and the support competency center.

The PMO was established in the company about seven years ago. Overall, it was seen as an important means to create a common and standard language, where program teams could utilize program management tools, processes, and methodologies. Prior to that, we did not have any standardized ways of using tools or sharing the program management methodologies. Some program managers followed the policies and procedures much more closely, and programs became more successful as a result. However, some program managers didn't follow them closely enough and did things their own way. Sometimes, they were successful, but many times programs required rework late in the development cycle because something was missing. We needed the PMO to standardize our practices.

Cognito: I feel that I need to interject because that answer seems commonplace. If I understand it right, one of the PMO's purposes is having program managers adhere to the policies and procedures in order to increase program success. Don't take this wrong, but we want to learn what aspects of the PMO bring success and what is just company folklore.

Deerling: The purpose was to increase program success. Beyond that, the PMO is viewed as a business function whose job is to develop, implement, and continuously improve program management processes and tools. That solely is infrastructure to improve program performance to help the company achieve its business objectives. That was made clear by the senior management in the PMO founding charter—show that improved practices and processes improve business results.

Cognito: Okay. But how did you make it happen?

Deerling: I'm going to try to approach this from a different angle to help you understand. To achieve its business objectives, the PMO includes working with the community of program managers on how to develop processes (and tools to help enable the process), how to provide training, mentoring, and coaching, and how to implement those things, all with the simple purpose of making our business better. In addition, the PMO is a governing approval body to evaluate and determine if programs should go forward from one program phase to another. In other words, programs need to get approval from PMO committees or officers at every major milestone to be able to move forward. We got very good at terminating poorly defined, planned, and executed programs.

The Tangible Contribution of the PMO

Senders: I am afraid that my thought process needs more details to fully fathom the concept. For example, what were the specific benefits of running the PMO? What success did the PMO bring? What was the PMO's return on investment?

Deerling: I want to try to avoid theory and explain it all by means of the company bottom line. I think that the PMO can really make a difference with its contributions. For example, the PMO develops and installs the standardized process. It makes programs shorter, cheaper, and faster—meaning increased productivity—which translates into lower cost, better sales, and higher profits. But management wanted to know the return on PMO investment. So, we had to take a macroview of the operation and spend a huge amount of time trying to provide our executives with an accurate return on investment value for the PMO. I mean, something like how many dollars we get back for every dollar spent on the PMO. That is the language of senior management—the language of money. We proved our case by making money for Stellar Corp. We proved that our PMO increased program success in dollars. But as much as we understood what senior management wanted from us, we figured out it was a one-sided approach. There are so many intangible benefits you simply cannot quantify.

The Intangible Contribution of the Program Management Office

Deerling: Let me give you another example. At times, we needed to make sure that we did use case analysis for programs correctly, and we were struggling with end-user surveys for determining the wants and needs of the infrastructure users. In my opinion, they were full of leading questions. "Would you like to have this capability that can do this and that for you?" Those kinds of surveys gave us misleading data. Wrong decisions were made based on those answers. "Oh, they like this capability," we heard many times. But when push came to shove, you could not really implement it because, for whatever reasons, customers we had surveyed did not want to use it. The PMO helped drive common end-user surveys and use case development. Stellar Corp.'s end-user surveys now help make better decisions about the infrastructure feature set, and its use. But for figuring the PMO return on investment, we needed to determine how much money was saved because of the PMOs help with end-user surveys. We were not able to determine that. Sometimes, things do not look black and white, but rather gray. That is the world of the intangible.

Senders: Aren't there too many intangibles that our executives wouldn't like?

Deerling: Lots of the PMO benefits I see come from tangible dollars, like I explained, but many of them are intangible—like having a common process, having a common standard language, and having a common understanding of what program management is. We utilized that in a program business case, for example, so that we were all on the same page, thinking about the same thing. We utilized that to identify risks before they became problems, so we were much more proactive. One of the real big benefits was improved communication so that we could do more upfront planning and avoid more crisis or reactive management later downstream. All those things create intangible benefits that were difficult to quantify for the PMO return on investment. Also, what looks more quantifiable, for example, gains in time-to-market speed that can be attributed to improvements created by the PMO, is not as tangible as it looks. There is no doubt that they can be roughly estimated, but many question their accuracy or see it as splitting hairs.

The Scope of the PMO

Cognito: What does the PMO do now and what will it do in the future?

Deerling: I've written down the PMO's duties. Essentially, the PMO does what it was originally charted for. I can summarize what the PMO does.

Process standardization: This means that the PMO organizes the company's program management practices by standardizing the program management elements:

- Processes
- Procedures
- Metrics
- Tools
- Vocabulary

Personal development: This includes organizing the career development aspects for the program manager:

- Competency set development
- Training
- Mentoring
- Coaching

Consulting: This means that the PMO organizes delivering program management consulting services to programs to increase the effectiveness of our program management practices.

Program reviews: The PMO is a program governing body that organizes, reviews, and approves:

- Program phase/gate decision-checkpoints
- Major program milestones
- Program audits

Program information: The PMO organizes and standardizes data and information to support:

- Program reporting to senior management and other key stakeholders
- Information technology system

I want to emphasize two points. One is to pay attention to the word 'organize.' This means that Stellar Corp's. PMO only organized all these tasks. For this, the PMO borrowed human resources from functional groups because the PMO did not possess its own resources. In fact, the PMO had only two administrative assistants and me. For example, to organize the development, deployment, and improvement of a company's standardized program management process, the PMO would borrow program managers from functional groups and form a team. The team members were doers—developing, deploying, and improving the process.

The second point was that exactly this approach and the chosen scope of the PMO were also contributors to the success of Stellar Corp.

Senders: You did not mention how you decided what the PMO would do and what the PMO would not do. Can you tell me about that?

Deerling: Stellar Corp. struggled a lot with that choice. They benchmarked PMOs in several companies and saw different versions of the PMO application. Some PMOs only cared about standardization. Another managed all programs in the company. Yet, another PMO in a big consulting company was virtual in nature because they had sites in different cities, and because their business units (profit/loss centers) were strongly independent minded, so they established a PMO on the web.

The more PMOs Stellar Corp. benchmarked, the more they believed in what they saw. The major factors influencing the decision about the scope of the PMO are the company's business strategy and organizational culture. For example, the company with the virtual PMO pursued a strategy of independent-business units, so what they wanted of the PMO was to develop a common, standardized process and make it easily available to all business units. That's all. The organizational culture was such that the PMO could not even check or ask business units if the process was used. If they tried, the answer would be along the lines of "if we need further help, we'll call you."

So, in my previous company, we did the same. We asked ourselves two questions:

- What does our business strategy need the PMO to do?
- What kind of PMO will our organizational culture tolerate?

Answers to these questions pretty much shaped our decisions about the scope of the PMO. I think these answers about our PMO scope of activities helped us to develop more successful programs.

Organizational Location of the PMO

Senders: What organizational structure, location, and size made Stellar's PMO successful?

Deerling: The Stellar Corp. IT department had about 20 full-time and 20–30 part-time program managers. Their jobs were to manage programs in their verticals (functional groups), supporting all business units as assigned per the program charter. Formally, the program managers are not located in the PMO but in verticals. Depending on their experience and knowledge, some of them are very powerful, exerting full control over the program and making all major decisions. Others, typically less experienced, act more like program coordinators, collecting and bringing information to their vertical manager to make all major decisions.

How is the PMO structured and who does it report to? The PMO is structured in a simple way—the head of PMO or, as we called it, the program management officer, is responsible for the work of the PMO. That, of course, was my job. It was very lean, as the organizational culture dictated. Also, part of our business strategy is to be cost leaders in the industry. So, we had to have a cost-effective PMO. Let me show you the organizational chart [Figure 15.6].

Cognito: Okay, let Barry and I read the chart, and if we make any mistakes, Peter, jump in.

Deerling: Agreed.

Cognito: A program manager reports to his or her division, that is, a vertical manager. Then, the division manager reports to the CIO, who, in turn, reports to a VP of the company. A program manager has that direct line relationship with their division manager but, at the same time,

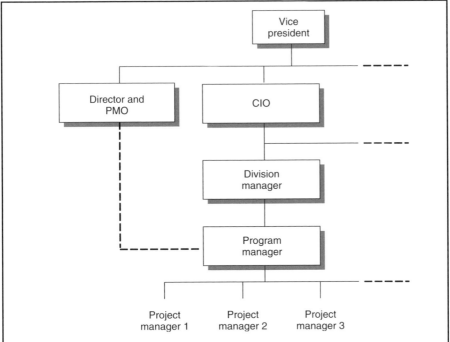

Figure 15.7   Partial Trust Corp. organization chart.

there is an informal dotted line relationship to the PMO because the PMO helps develop, deploy, and continuously improve all the processes and tools that they use. So, Peter what you show here on the chart is the organizational structure that helped Stellar do—stellar programs, if you will. A lean PMO that does not own any program managers. Verticals own them—pretty simple!

Deerling: Exactly.

Senders: Were there any problems in terms of establishing the PMO?

Deerling: Resistance to change would definitely come to the top of the list because we never had this role and function before. So there was some tough resistance for some time, especially from the divisional managers who were worried about the division of power. First, some people doubted if the benefits of having the PMO for the organization were significant. So communication and support from the CIO and vice president who served as the sponsor was key.

Second, I think we were struggling at first with how to establish the standardized process. My job at the PMO was to ask program managers to make sure that they understood our business strategy and how to use our procedures to align their programs with the business strategy. At first, I did not know how to ask those questions, and such a standardized procedure did not exist. That caused a lot of dissatisfaction among senior managers and program managers. But a few of us put our heads together and developed the standardized procedure. After that, I saw no resistance.

Cognito: Okay, I have another meeting in five minutes. We will meet again. In the meantime, I want you to understand that you guys have to have a plan for any anticipated and unanticipated resistance. When it comes time to spend money on building the PMO, we want to look prepared for success, and I want all players in this company to know we are prepared. That is, Peter, why I hired you and that is why, Barry, I chose you to help Peter.

REFERENCES

1. Waddell, Jim. "The Good, the Bad, and the Benefits of a PMO," *PMI panel discussion*. Portland, OR. (2003).
2. Block, T. R. and J. D. Frame. "Today's Project Office: Gauging Attitudes," *PM Network*, 15 (2001), pp. 50–53.
3. Archibald, Russell D. *Managing High Technology Programs and Projects.* Hoboken, NJ: John Wiley & Sons, 2003, pp. 149.
4. Martinelli, Russ and Jim Waddell. "Achieving Common Leadership and Infrastructure Through the Program Management Office," *Project and Profits Magazine*, 4 (2004): pp. 75–80.
5. Evans, James R. and William M. Lindsay. *The Management and Control of Quality*, 6th edition. St. Paul, MN: South-Western College Publishing, 2004.
6. Englund, Randall L., Robert Graham, Paul C. Dinsmore. *Creating the Project Office: A Manager's Guide to Leading Organizational Change.* San Francisco, CA: Jossey-Bass, 2003.
7. Archibald, Russell D. *Managing High Technology Programs and Projects.* Hoboken, NJ: John Wiley & Sons, 2003: pp. 152.
8. Waddell, Jim. "The Good, the Bad, and the Benefits of a PMO," *PMI panel discussion*. Portland, OR. (2003).

Part VI

Industry Case Examples

During the course of this book, we have explored five stages of program management. First, we explained what program management is and then what and how it helps the organization accomplish their business strategies. We also addressed the need to align organizational business strategies with program operations and to illustrate how program management aids the company in delivering the winning value proposition in the form of the whole product.

We then detailed the process of defining, planning, and executing a single program and explained how program management metrics and tools are enforcers for successful execution of programs. Next we turned attention toward the program manager in which we explored the roles, responsibilities, and core competencies needed to successfully manager a program. Finally, we showed how a transition to program management can be executed in an enterprise and how a PMO can serve as both a business function and operational support organization for the program management discipline.

We now have one stage remaining. Part VI demonstrates through comprehensive case examples how program management is applied in four real-world examples. In the four cases that follow, we look at multiple elements of program management as they interact in more complex examples and contexts than the focused industry examples contained in the previous parts of the book.

In designing this section, we looked at multiple program characteristics such as industry, program size and nature, and program phases. The examples are from different industries, specifically, high-tech manufacturing, software product development, space exploration, and IT.

While one company adapts a hardware product for a new market, another develops a new software product; still another organization is attempting to identify terrestrial planets in the universe; while the last looks at the design and deployment of an IT solution in a hospital. The size of the programs varies from 1 million to more than 400 million U.S. dollars.

American Shogun, Appendix A, shows how program management is made simple in a high-tech company when the fundamental principles of the discipline are followed. The teams' dedication to coordinated collaboration between projects, focus on business goals and the bottom line, understanding of cross-project dependencies, and effectively utilizing horizontal- and vertical-management techniques are things we have observed in well-managed programs. Sure, nearly every program experiences at least one crisis, and Shogun is no different, but it is interesting to follow its resolution on a well-run program.

ConSoul Software, Appendix B, is a story about a good company that is also overly ambitious. Under senior management pressure, the program team commits to an unrealistic product feature set and schedule. Two months before the planned customer delivery date, senior management decides to drop two features to recover the schedule slip and hit the announced program launch date. Unfortunately, that is not a simple task at the current phase of the program. Additionally, the program team disagrees with management and prefers to finish the program with the planned 10-feature scope and the related delay because of prelaunch announcements to customers. While sparks fly, management and the program manager meet to resolve the issue.

Planet Orbit, Appendix C, offers a story of the possibility of extraterrestrial life. However, conflict between the scientists and organizational management is highlighted. Senior management ignores the cosmic glory of "to boldly go where no one has gone before," and focuses on earthly issues of schedule delay and cross-site organization. After waiting for more than 10 years for funding of the program, the scientists are not willing to see it stopped and are ready for a fight.

In General Public Hospital, Appendix D, the program team wants to make what it considered to be a big change. However, the program hits a stalemate at the baseline gate review, in which senior management is considering the cancellation of the program. Creative thinking on the part of the program manager breaks the stalemate and brings the possibility of approval to progress to the next phase of the PLC.

Appendix A

American Shogun

Bjoern Bierl and Andrea Hayes-Martinelli

GETTING STARTED

It was in late May 2002 when Jan Vesely, sales manager for Southeast Asia and the Pacific region at International Instruments, Inc., received a call from RisingSun, one of its key accounts in Japan. "They told us that they were interested in our 1001-series monitors if we were able to provide audio capability—a feature that our competitor already had implemented in their product," Vesely said. "Additionally, RisingSun wanted us to deliver the product in 11 months, which was an aggressive time-to-market goal. Since RisingSun was one of our most important customers, we jumped into action." International Instruments, Inc. was a global market leader in the field of monitoring systems, and the 1001 series was their main-product line of monitors addressing the biggest segment of the overall market.

The audio capability for the 1001-series monitors was previously discussed because, as mentioned, a major competitor had already brought a monitor with audio capability to the market. But Manuel Scriba, the segment manager for the 1001-product line, found the market too small to justify adding the audio feature, but the telephone call from RisingSun changed everything. As Scriba recalled, "Suddenly the program, named Shogun, would help us to meet our financial, market share,

493

customer relationship and competitive business goals," he said. "First and foremost," he commented, "a new program had to fulfill our business goals. That's what it is all about—the business goals."

As for the financial goal, the order was large enough to cover the development and research cost associated with the program and make the desired contribution to the company's bottom line. But it was more than that. International Instruments, Inc. was focused on market share and customer relationship as key strategic goals. It was clear that Shogun would support the achievement of these strategic goals. "We had excellent customer relationships before this program, and if an important customer wanted to have the new feature—and the program was financially viable—what else could we do but satisfy them?" asked Scriba. "On top of that," he added, "Shogun would also provide gain in market share for this monitor product line. Not only could we increase our market share, but we also could attain our competitive goal, which was to preempt our competitor from gaining more market share in Japan."

Scriba proposed the new program to Robin Weiland, vice president of International Instruments, Inc. "We had the chance to increase our market share," recalled Weiland. "It was tough, but feasible. So the question was no longer, 'does it make sense?,' but rather, 'can we get this done in only 11 months?.' The next steps were to assign a program manager and set up the program as soon as possible."

THE FOCAL POINT FOR BUSINESS RESULTS

Melanie Lehr came from a strategic marketing position and was new to program management. As she remembered, "I was new to the company and new to program management, but I knew the company pretty well from my former jobs. This program was about to become a great challenge for me and the company. But I was glad to have the support from the program management office."

The program management office, a knowledge base for the program management activities throughout the company, provided not only standardized but also flexible processes and tools for each of the programs. "We don't expect naturally-born program managers," said Bob Mitchell, head of the PMO. "Program managers are made; they are trained on the job and in classrooms. Through the years, we developed a skill-set map for program managers. A program manager's task is strategic—the focal point for business success. They must have strategic skills. They lead all kinds of people—some easy to lead, some difficult. Therefore, they need leadership skills. They face tough times, requiring tough character. Hence, we expect them to have a special set of intrapersonal skills. Programs cost a lot of money and require many resources, so program managers need to have financial skills. Since we develop highly technical products, program managers should have working-level technical skills. Finally, the customer must be understood inside and out, requiring program managers to have customer skills. In summary, we require six sets of skills from our program managers."

Like every program manager, Lehr was held accountable for business results. Therefore, her job was one of the most critical in the company. "She did not have all of the skills we require," said Mitchell. She was, however, an experienced engineer with an MBA, eight years of experience in design engineering and strategic marketing, and was involved with many programs as a member of extended teams and as a project manager. She had all of the required skills except for the program management set. "We planned to promote Melanie to program manager in the long term," Mitchell added. "But when Shogun was added to the product road map, we did not have a seasoned person available, so we assigned the program to her. New program managers need as much help as possible. Mentoring from others who have had the same challenges is the best way to teach them. We solicited the help of an excellent mentor, which provided the program management skill set Melanie was missing. Standardized processes and tools also helped her master the challenge. This example demonstrates that you do not need to reinvent the wheel. And Melanie did a great job."

"What helped me most," Lehr recalled, "was the mentoring part. Marcel Greenhill was a program manager for 20-plus years and did all kinds of programs. His mentoring and support taught me all of the small tricks and tips that helped during the start-up phase, which was the toughest for me. Setting up a program required tremendous effort because it involved multiple projects and disciplines in the company, such as engineering, marketing, finance, production, and sales. Cross-discipline development was really important for achieving the business goals."

INTERDEPENDENT PROJECTS

Preparations were undertaken to assemble the program team. A program requires a lot of different people with different backgrounds, making it difficult to coordinate them within one team. The usual program management approach at International Instruments, Inc. was to define two layers—a PCT and several extended teams. The PCT for Shogun consisted of 12 members. They represented the functions that had major involvement, such as marketing, mechanical engineering, manufacturing, purchasing, and finance. Additional required functions for the program were also represented—like systems, engineering hardware and software, quality control, promotion for the new product introduction support. Each member of the PCT led his or her extended team.

Lehr commented, "It wasn't like running 12 different projects and assembling them at various stages throughout the process. We had to breakdown the whole program into multiple interdependent projects. Each project manager managed a project within his or her own function or discipline. I, on the other hand, managed across projects—coordinating them to make sure that functional objectives were in tune with the business objectives. Essentially, my job was cross-project and cross-disciplinary."

The projects on the Shogun program were all interrelated and dependent on one another. For instance, if marketing could not finalize the required specifications with the customer, engineering could not start to develop required features, quality could not be controlled, and manufacturing would be delayed.

Managing the interdependencies between the projects was a real challenge for Lehr. "I knew that I would not be able to manage all of the projects in detail and take a close look at everything," she said. "So I needed to make sure that each project could function individually. At the same time, everybody needed to be aware of the consequences of their actions for the other project teams. Projects have delays, no matter how well you plan ahead. The only question is whether or not the team has enough of a chance to 'extinguish' the fire before it affects the other projects and finally the whole program. Transparency is important."

BUSINESS GOALS

Shogun was started with a specific set of goals—profitability, market share, development and product cost, product performance, and time. Development cost was estimated as $5.1 million, while the performance was set by the customer's requirements. Product cost required more detailed input for the exact configuration of features. The time frame was a tough constraint because the product needed to be delivered in 11 months.

COORDINATED PLANNING

Jumping into action, Lehr worked out a rough plan with the PCT. Each project was addressed with the initial budget and timeline identified. Cross-project interdependencies were identified in a very rough manner.

These outlines were then given to the PCT members, who, because they represented their functional projects, had the best insights. Each of the projects was then broken down into detail by the responsible project managers. They added their insights on scheduling and resources and experience of how to complete a program of this scope.

"I was given the outline for the broader marketing function," remembered Christian Foyer, who was the marketing project manager. "I had to address steps like customer validation and continuous screening for market requirements. The way to achieve these goals was up to me and my extended team. Melanie only needed to know our results, deliverables, and risks. I sat down with my team and we developed our project work breakdown structure (WBS) to deliver our part of the program. My extended team could, thus, focus on the marketing jobs, while I supervised the work and ensured alignment with the other projects."

After the project managers had created the detailed WBS for their individual projects, the PCT, under the guidance of Lehr, assembled a complete WBS for the program. The WBS outlined the specific tasks for each of the functions in respect to other projects, or more precisely, cross-project dependencies. Only those work packages that had an impact on other projects were considered important on the program level. More detailed outline and scheduling were the responsibility of the PCT members.

"This was not always easy," Foyen commented. "I had created the WBS for my team and, although it was a tough time frame, we managed to get all of the parts completed in time—at least on the outline. But when we sat down in the program core team and worked through the WBS, Melanie required some changes. These were mostly earlier deadlines because there were other projects that needed my input sooner. So we accelerated some tasks and left others with more time. But it made sense—not for the marketing function—but for the overall program. This understanding became crucial. We constantly reassessed the program progress in the program core team and adjusted for problems."

The PCT then agreed on important milestones that were tracked. Tools like the program strike zone and the program dashboard enabled Lehr to effectively track the progress of Shogun and all of its projects. The dashboard provided an overview of the current status, and the program strike zone showed whenever a program-level milestone was significantly delayed. "Communication is the key for everything," Lehr stated. "But you always have a mentality of 'bad news never travels up.' Tools like the program strike zone and program dashboard helped us to have an objective measurement and information system that everybody could agree on."

COORDINATED MANAGEMENT

The program involved nearly every function within the company and tight interdependency between all of the projects was a given. Communications became a must-do on a regular basis. "We had regularly scheduled meetings and communication patterns, as well as more informal and driven-by-demand communication," recalled Lehr. "If there was a problem, there needed to be communication. Nothing else mattered."

The PCT met once a week to discuss program progress and issues. If changes were needed, the whole team had to agree on them. Alignment of the project teams was important, especially with the short time frame. "The whole process is not always straightforward, but it helps a lot if the members know each other very well and communicate across the functions to solve smaller problems on their own," commented Lehr.

In addition to the PCT meetings, some of the functions needed to work together more closely and more often. As an example, the engineering function required input from marketing for prioritizing the features. Prioritization of features drove prioritization of work for the design team.

The PCT members met with their project teams as required. The engineering project manager, Gregory Wolfe, recalled his approach. "I had the largest project team comprised of eight engineers," he said. "To keep them updated, we met every week directly after the program core team meeting. In doing that, I ensured that they had the most recent information and that they felt more involved with the program. Of course, I did not need to deliver all of the information. Other than the business view of the program, my project was focused solely on the technical work—to get the product engineered in time and in the desired cost range. But there was also a motivational aspect to keep everybody focused on the program vision and strategy."

Communication in the program did not only flow from the top down. In International Instruments, Inc., programs are critical to the company's business strategy and the achievement of strategic business goals. This importance was reflected in the program reviews for senior management, which were held on a monthly basis. Several vice presidents attended, headed by Robin Weiland, and spent the whole day listening to program managers' reports, dissecting issues, and removing obstacles—occasionally killing programs that were not accomplishing their strategic goals. The program managers' reports were based on the information contained in the program strike zone and dashboard.

Lehr commented, "What senior management did for the program was what I did for each of the projects. In the end, management is responsible for the program. So they need to know what's going on—not every single step, but the major milestones. So I kept them updated—just like the program core team and the functional representatives kept me updated. It was just on a broader scale. The closer the program came to its desired launch date, the more often there were management reviews. Management wanted to increase control. That was tough on one hand, because it sometimes took time that I could have spent on other things. On the other hand, it ensured their support."

ALIGNING EXECUTION WITH STRATEGIC OBJECTIVES

Since the program was initiated by RisingSun's demand for an additional feature, customer focus was very strong. Foyer stated, "Besides the strict validation processes in the beginning of the program, we ensured that each major step and each feature we planned to implement was validated and approved by the customer."

But RisingSun was not the only customer in that sense. "The 1001-series monitors had additional capabilities in event logging, etc.," Foyer said. "This was partially demanded by RisingSun but also determined within my department to be a major value-add to customers in general. So we looked at both RisingSun's specifications and the overall market demand. We wanted to have a competitive product for the whole market."

At the midpoint of the program, Wolfe realized there was a problem for the hardware project team. "In order to implement all of the desired features, we weren't going to be able to meet our deadline," he said. "We pushed hard, but final feature requests from marketing were too late, and the additional quality testing required would take us too long. At that point we needed to make a decision."

Wolfe deferred the question to Foyer, who commented, "This was a tough problem for us. RisingSun wanted to have a portability feature in the 1001 series, but we needed to eliminate it in order to meet the schedule deadline. For us, it was either the portability feature or the deadline, and it was hard to determine."

Foyer faced a decision. "To find a solution, we examined the marketplace and called some customers to ask if they would need the portability feature," he said. "In the end, it was determined that the market didn't really need the feature, so we talked to RisingSun who agreed it was desired by not essential."

FLEXIBILITY

Flexibility was really important for the program. "It was a complex program with an aggressive schedule," Lehr commented. "You can't just develop a plan and wait for the results. We tried to plan as well as we could, but there were a considerable number of changes that occurred. We sat together in the PCT and asked, 'What can we do? What does it take to do implement a change, and how much would it cost?' We were flexible because we needed to be. Having a team that worked so well together was really the key for everything."

Midway through the program, it became obvious that the team would need more people and more resources to accomplish the program in the desired time frame. Thus, management decided to support the program by having engineers from the company's Japanese engineering group work on Shogun. "There were problems in the beginning," Lehr remembered. "One of the engineers was very junior and didn't speak English. We had a politically sensitive issue, but senior management helped us by moving the engineer to a program with a less aggressive schedule."

FINISHING WITH FLYING COLORS

Bringing the product to launch was still a challenge due to considerable pressure from upper management. The team had an opportunity to prove whether the communication and, more importantly, the informal networks were working. "The end was brutal," recalled Lehr. "We needed to get everything out as quickly as possible, and although I did as much as I could, I couldn't have managed on my own. The project teams worked well together, and everybody headed in the same direction. We wanted to meet our goals. The program was finished within the 11-month target, with higher product performance, and within the given development budget. That was the easier, more tactical part of the goals. We had to wait an additional nine months after the product launch to see if we had achieved what we strategically planned in the business case: increase profits, achieve gains in market share, and preempt a competitor. We did."

CLOSING

Many program management aspects in the Shogun story are quickly recognizable. The teams' dedication to coordinated management, focus on business goals and bottom line, understanding of cross-project dependencies, and effectively utilizing horizontal- and vertical-management techniques are things that each of us have observed in programs. Our experience tells us that when these elements are present, a successful program is likely to result.

Appendix B

ConSoul Software

Andrea Hayes-Martinelli and Dragan Z. Milosevic

"Wait a minute" said Bali Balebi, the Silverbow program manager, while passionately waving his hands. "Do I understand you correctly that senior management is saying that my program must hit the release date, and if that requires dropping the two automation features, it is okay?"

" Yes," responded Christine Smiley, the PMO director, "you understood me. But please calm down, we need cool heads now."

Balebi continued, "So, first we add the automation features despite the program team telling us that the planned program duration of 21 months would only allow for the 8 original features. Now, we are in the integrate phase three months before we get to deployment, and I'm being asked to drop the automation features because we're a month behind schedule? I'm sorry if I'm having a tough time keeping a cool head, but we can't do that."

"They are not asking you, they are directing you to remove the features in order to get back on schedule; the delivery date is crucial," replied Smiley. "Again, please calm down and tell me why we can't drop the automation features. Give me a logical argument that I can take back to senior management. I can't just go back to them and say we can't remove the features because the program manager is passionately against it."

"Okay," said Balebi. "Two reasons, first I have already made an announcement to our lead customers that the automation features will be included in the next release. Second, the features have already been integrated with the other features in the release. It may take longer to back them out and redo the integration than to just continue with the integration as is."

"Oh, now I understand the problem," responded Smiley. "Let me talk to Matt Short (vice president of enterprise software), and you sit down with your team and review all possible options to make the original delivery date, both with and without the automation features. Also call the lead customers who we know plan to purchase the new release. Tell them about the possible delay of the two automation features until the next release and see how they react."

COMPANY CONTEXT

Before we see what program actions were taken on the Silverbow program, we need to evaluate ConSoul Software. Program decisions need to be put into perspective of the corporate context in which they exist, as the company environment heavily influences program decisions and strategy.

Overview

ConSoul Software was a late entry in the facilities and construction software industry, which was dominated by two primary competitors. Matt Short, vice president of enterprise software at ConSoul, recognized that, although the competition had a foothold in the industry, their products were not user-friendly. Short described the competitive environment. "The competition had tremendous advantages," he said. "Their products were just packed with features. The problem was, their products were extremely cumbersome to use and required large amounts of user training." ConSoul's strategy to enter the industry and to continually gain market share was to provide customized solutions that focused on ease of use and required little or no user training. "It had to be that easy, or no one would buy our products," Short said.

ConSoul is now an award-winning software product development enterprise, which produces integrated-software solutions for accounting, payroll, fixed asset management, human resources, customer relations, and e-commerce applications. The company specializes in developing integrated- and customized-software solutions for more than 20,000 customers in the facilities development and construction industry.

This case describes how ConSoul struggled with the management of a program—Silverbow—in its attempt to release an upgrade and grab more market share in a dominated market and the challenges it faced.

History

ConSoul Software was founded in 1978 as a start-up company. Its mission is to empower its customers to succeed by providing them with extraordinary software and services. It does so by focusing on four primary business postulates:

- *Providing best-in-class products:* By developing customized software solutions packaged as office suites, ConSoul gains both repeat business and a continually increasing customer base.
- *Implementing a rapid software release cycle:* ConSoul delivers software products with new features and functions every six months for most product lines.
- *Developing ease of use and integrated solutions:* Compared to its competitor's products, ConSoul's software solutions focus on ease of use, which results in little or no user training and tight integration with customers' existing business systems.
- *Providing competitive pricing:* A unique bundling strategy and integrated-feature set allows ConSoul to provide its customers with a total solution at a fair and competitive price.

Business Strategy

The strategy that ConSoul employed is best described by what Snow and Hambrick call a customer-intimacy strategy.[1] A customer-intimacy strategy has proven to be an effective way for companies to enter markets that are already dominated by a few competitors. Competitive advantage can be gained by delivering what a specific customer wants, not what the market as a whole may want. This type of organization demonstrates superior aptitude in advisory services and relationship management.[2] Its business structure delegates decision making to employees who are close to the customers and its management systems are geared toward creating results for carefully selected and nurtured clients. The corporate and program culture embraces specific, rather than general, solutions and thrives on deep and lasting client relationships.

ConSoul Software has three principle strategic objectives, listed in priority order:

1. Maximize customer satisfaction
2. Provide high product and service quality
3. Maintain competitive time-to-market delivery

These strategic objectives guide the work of the company and its program teams, as demonstrated in the following sections.

Strategic Program Planning

As stated earlier, a program manager within a business unit is charged with delivering an entire software solution through multiple software releases to ConSoul's customers. Program release planning is accomplished and communicated through the use of a program road map tool. An example of a program road map for the enterprise solutions business unit is shown in Figure B.1.

The road map represents the software releases planned for each program over a two-year time frame within each of the four product lines. Over the two-year time frame, each program consists of multiple release projects. It should be noted that even though the releases are shown as sequential events on the program road map, work is performed in a concurrent manner with overlapping activities and resources.

	Q1'07	Q2'07	Q3'07	Q4'07	Q1'08	Q2'08	Q3'08	Q4'08
Financial Releases		▲ 1.0		▲ 2.0		▲ 3.0		▲ 4.0
Operations Releases	▲ 2.0		▲ 3.0		▲ 4.0		▲ 5.0	
Estimating Releases		▲ 1.5			▲ 2.0		▲ 2.5	
Corp. Solutions Releases		▲ 1.0		▲ 2.0		▲ 3.0		▲ 4.0

Figure B.1   Program Road map.

Organizational Structure

ConSoul Software is composed of two business units—enterprise software solutions and office software solutions—each led by a corporate vice president. The business units are supported by centralized finance, marketing, and manufacturing organizations, as illustrated in Figure B.2.

Each product development business unit is organized as a matrix structure and is led by three functional directors: director of program management, director of engineering, and director of technical communications. Each business unit is further segmented by product lines. For example, as shown in Figure B.2, the enterprise solution business unit is segmented into financial, operations, estimating, and corporate solutions product lines. A program manager is assigned to each product line and is charged with delivering the entire product solution through multiple software releases. Each release is organized as a separate project and is managed by a technical project manager.

The software development and software test/quality assurance (QA) resources report directly to the engineering functional organization and are loaned to the product line program managers and release project managers as needed. Representatives from other centralized support organizations are also assigned to the product line programs to represent their respective functions.

URGENT SILVERBOW TEAM MEETING

After Balebi's conversation with Smiley, he called an urgent meeting with the Silverbow core team. The agenda was short—plan for options to finish the program. Balebi briefly explained, "You know that the program plan shows we'll be a month late if we include all the ten features. Senior management has sent us a message that we must hit the release date, and if that requires dropping the two automation features, it is Okay." He continued, "Remember that senior management requested that we add the automation features, despite us telling them that our schedule wouldn't support it. Now, they have reversed their position. The trouble is we have already made an announcement to our lead customers that we have ten features in the next release, including the two automation features. Dropping them may significantly lower our sales.

Christine instructed us to put our heads together and review all possible options to meet the original delivery date, with and without the

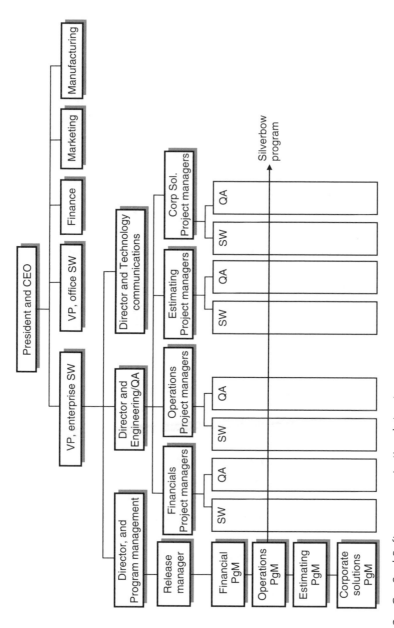

Figure B.2 ConSoul Software organizational structure.

automation features. We will call the lead customers who we know plan to buy the new release as well, and see what kind of damage the removal of the automation features may cause.

Now, let's review our program from top to bottom, which may help us consider all the possible options. If you have any questions about senior management's request, you can ask them as we go through the review."

THE SILVERBOW PROGRAM

Silverbow is a code name for a program that was funded and executed in the office solutions business unit of ConSoul Software. The Silverbow program supports ConSoul's operations segment and provides a comprehensive set of document control capabilities. The program consists of six projects, each structured as a software release package delivered to customers in consecutive release cycles. Total duration of the program was 22 months. It was estimated that company revenue would increase 53 percent in the first year and 46 percent in the second year due to software licensing agreements associated with the Silverbow program. Table B.1 summarizes the characteristics of the Silverbow program.

Table B.1   Silverbow program characteristics.

Silverbow Program Characteristics	
Industry	Facilities and construction software
Product description	Software upgrade (project management software)
Program description	Provide solutions for common document control processes (handling of meeting minutes, daily reports), submittal packages, notices, issues)
Product novelty	Derivative
Technological Uncertainty	Low technology
System scope	Full system
Pace	Fast and competitive
Business unit	Operations
Customer	External
Strategic goal	Extension of current product line for increased life cycle revenue
Program size/duration	20 people/22 months

Program Structure

The Silverbow program utilized a PCT structure, which is common for all ConSoul product-development programs. The PCT structure was highly matrixed, with many interfaces between the engineering, test and quality assurance, technical communications, and marketing functional organizations. The PCT was led by the program manager and consists of the current release project manager, the subsequent release project manager, the engineering manager, and the software test/QA manager. Project managers report directly to the director of engineering and indirectly to the program manager for the duration of their involvement in the program. The program manager, in turn, reports directly to the director of program management.

Each project was segmented into multiple subteams that corresponded to primary product features such as drawing logs, meeting minutes and reports, as shown in Figure B.3. Each feature had an engineering lead, a test and QA lead, and a technical communications lead, as well as

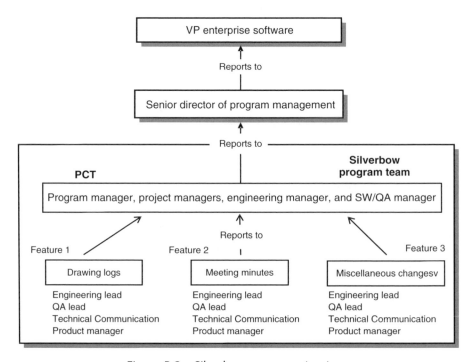

Figure B.3   Silverbow program structure.

Figure B.4   The ConSoul PLC phases.

a marketing product manager who managed the technical design and development work for his or her respective function on the project.

The PCT is responsible for the product and business results derived from the delivery of the family of releases that constitute the program. The project teams are responsible for designing, coding, and QA testing of the various features that comprise the software release.

The Program Process

Silverbow was one of the first programs to use the new product development process adopted by ConSoul. The process is broken into the following elements: program phases, kitting, schedule management, cost management, risk management, resource management, and communications.

Program phases: As illustrated in Figure B.4, the PLC consisted of seven phases: predefine, define, plan, design, integrate, deploy, and evaluate.

Generally, a program is started by creating what ConSoul calls the story, which describes the end product in terms of the usage model. The usage model is a description of the interaction between the user and the program product that identifies the product's benefit to the user.[2] From there, use cases (sequence of interactions between the user and the product), primary features for each use case, high-level resources, and timeline estimates, and a release plan are developed. A feasibility analysis is then performed.

Once feasibility has been proven, each project—and the program as a whole—is scoped, functional specifications are developed, and development and resource allocation plans are created and presented to executive management for approval. If approved, the program moves into the design phase and execution of the project plans begins.

Throughout the design phase, customer change requests are evaluated and approved changes are added to the program and project plans. During the final integration phase, the software code is integrated, software test and QA is performed, release documentation is generated, and the

Table B.2   PLC phases and activities.

Phases	Major Actions
Predefine	Perform feasibility analysis with marketing, product development, technology: perform financial analysis; get high level estimates
Define	Identify product definition team; create research plan; perform customer task and needs analysis; risk management; develop product definition; create initial estimate
Plan	Develop the functional specifications; resource allocation; develop plans to support all the business needs; present plan to marketing and make sure that this is what they wanted before going into design phase; negotiate with marketing; go/no-go decision
Design	Execute projects; do change requests from customers for better design solutions: integrate code; system test
Integrate	Generate release documents; finalize products; assess project
Deploy	Duplicate software disks; update internal systems; new order fulfillment; release fulfillment
Evaluate	Program retrospective; customer evaluation; overall evaluation of phase assessments

current release product is evaluated for deployment. Once a product reaches deployment, focus shifts to manufacturing, order fulfillment, and customer support.

The predefine, define, and plan phases are executed only once for a given program. The design, integrate, and deploy phases are executed once for each project and, therefore, multiple times throughout the life cycle of the program.

Once the final project is deployed, the program enters the evaluation phase, in which a program retrospective is performed and customer feedback information is evaluated. Table B.2 summarizes the primary activities associated with each phase of the ConSoul PLC.

Kitting: Project scope was managed by a process referred to as kitting. The idea was to segment software deliverables into five-day work packages. At the end of five days, or one cycle, each subteam within a project released a deliverable for each of the primary features. Each deliverable consisted of one element of the product that was functional and fully tested. This process of iterative software development is known as Agile development.[3]

The kitting process was used for the first time on the Silverbow program and proved to be a challenge for the project managers because it did not lend itself to the linear process of historical project management methods. Tracy Brooks, the release project manager, described the problem to Balebi. "Kitting doesn't address anything about logical units of work of a software project," he said. "Software development is not manufacturing, as the kitting idea is. A software process doesn't chunk up the same way a manufacturing process does."

The source of the problem was having two planning systems in place, which were incompatible. Per the first one, the program and each of the projects were planned by major milestones corresponding to large deliverables and phase completion. Per the second one, small deliverables of five-day increments (kits) were scoped. As a result, kitting performance and milestone performance were not aligned, leaving the performance-to-schedule metric ambiguous. As changes to the program scope were approved, the feature changes were either incorporated into existing kitting plans or new software kits were developed. Balebi was able to determine that the dichotomy and lack of knowledge of how to coordinate the two planning systems were a major reason for the program delay of one month.

Schedule management: The schedule was developed based on time esti-mates from the engineering and software test/QA functions, not as a program team. High-level time estimates were first developed in the pre-define and define phases, then detailed estimates were developed during the kit planning process in the plan phase.

Milestones were based on major project deliverables and standard program decision checkpoints and monitored every two weeks by the PCT. Example milestones included software detailed design complete, user-interface complete, code complete, code freeze, and product ship release.

Cost management: Cost of the program was expressed in terms of the number of people on the team and the number of hours to complete their tasks, not on a budgeted dollar amount. ConSoul sets budgets based on business solutions, rather than individual programs. The program manager, therefore, does not know the cost of the program in dollar value. Cost is managed through the program resource management process.

Balebi believes this is an area of potential improvement. "I think the current program cost-management process can get us in trouble," he said. "Nobody realizes how much money it actually takes to develop a product."

Risk management: Risk management was a shared process on the Silverbow program, both by phase and by team. During the program predefine phase, the marketing function was responsible for identifying potential risk events, evaluating the risk level, and developing necessary risk management plans. During the other phases of the PLC, the PCT drove the risk management process.

Risks contained to a single project were managed by the project manager and his or her project team, while risks that had potential affects on other projects, or program success as a whole, were managed by the PCT.

Each risk event was quantified according to the following two variables: probability of occurrence and severity of impact. Priority-one risks were the highest risk events, while priorities two and three were relatively lower. All priority-one risk events had to be communicated to all internal stakeholders, and a mitigation or avoidance plan had to be created. Priority-two and three risk events were monitored only on a recurring basis.

Resource management: Resource management was one of the primary responsibilities of Silverbow's program manager; the program manager also focused on ensuring that the project teams, as well as the PCT, were fully staffed with capable people to complete the work. Once a program plan is approved at ConSoul, resources are committed to the program by the various functional managers. The Silverbow program was fully resourced according to the program plan as it entered the design phase of the PLC.

However, resources were a primary constraint on the program. Because the program manager did not have a monetary budget to use for managing program constraints, the majority of scope, schedule, and quality issues were addressed by changes in program resources. For example, the program manager had to continually adjust resources to accommodate the additional features requested by customers or other stakeholders and address any quality issues associated with the current release or previous releases already operational in the field.

Communications: The PCT met once a week, generally on Monday mornings, to review program and project status, current issues, potential risk events, and planned activities for the coming week. Project status was focused on the kitting progress for each of the feature teams on the project. A review of each of the feature team's five-day kitting progress and deliverables was conducted. Program status focused primarily on progress toward major program milestones, resolution of current problems, and review of risk mitigation or avoidance plans for each of the priority-one risks.

Each project team conducted their own team meetings once per week at the beginning of the program, then two to three times per week during the later stages of the design and integration phases. The focus of the project team meetings was on the technical details associated with each feature team.

Communication of program status to internal stakeholders was accomplished by the program manager in the form of a formal status report once every two weeks. The report was sent directly to the senior management team and posted on the program website for all other interested parties.

Program Metrics and Tools

The Silverbow Program employed standard metrics and tools for all ConSoul programs and focused on measurement of the following processes: scope management, schedule management, risk management, and quality management. As mentioned in the previous section, program cost is not measured.

Scope management: Program and project scope was measured by the number of kits associated with each feature in a project. Performance was measured by the number of kits completed versus kits planned and required a mitigation plan for variance recovery. Wall charts were used to manage scope and performance by tracking the number of kits completed by marking the kits off as they were finished, then monitoring the kitting progress against the original plan.

Schedule management: Performance to schedule was focused on progress toward completion of the key program milestone dates. As mentioned earlier, there was no correlation between completion of program milestones and completion of the five-day kitting activities. A standard Gantt chart was utilized to manage the program schedule.

Risk management: Risk management metrics included the total number of risk events for each priority level (high, medium, or low) and progress toward mitigation or avoidance of each high-priority risk. Timely development of a risk mitigation or avoidance plan for each priority-one risk and the completion date were closely monitored. A risk-management spreadsheet was used to track progress. Key fields in the spreadsheet included risk entry date, risk description, root cause, severity of impact, probability of occurrence, priority level, mitigation/avoidance plan, targeted completion date, and status.

Quality management: Software quality was measured by the number of defects recorded, resolved, and outstanding for each release. Software quality is the primary indicator used for the ship release decision. ConSoul utilizes an in-house developed tracking database to manage software defects on all of its development programs. The tool categorizes defects by severity level and resolution owner.

Table B.3 summarizes the program metrics and tools used for the Silverbow program.

Table B.3   Program metrics and tools.

Program Processes	*Metrics*	*Tools*
Scope and Performance Management	Number of kits completed versus planned	Kitting and wall chart
Schedule Management	Performance toward meeting key program milestones	Microsoft Project
Risk Management	Total number of risks per priority level (high, medium, low)	Risk management spreadsheet
	Time to develop risk mitigation plan for high priority risks	
Quality Management	Number of defects recorded	In-house tracking database
	Number of defects resolved	
	Number of defects outstanding	

Alignment Between Business Strategy and Program Execution

As stated earlier, ConSoul Software uses a customer-intimacy strategy to compete and gain competitive advantage in the facilities and construction software industry. Each business unit focuses on delivering particular features with specific, rather than general, solutions and does so with a lot of attention paid to maintaining the highest quality, customer involvement, and satisfaction possible.

All elements of the program management discipline explained previously are aligned with the three principle strategic objectives: maximizing customer satisfaction, providing high-quality products and services, and maintaining competitive time-to-market delivery schedules. Each element of program execution and its alignment to strategy is shown in summary form in Table B.4 and described in detail thereafter.

Table B.4   Program elements aligned to strategy.

Program Management Elements	Alignment
Program Organization	Matrix structure with many interfaces between engineers, software testers, technical communications, and product managers
Program Process	Milestones are deliverable driven. Change requests often come from marketing and customers. Risk is based upon uncertainty toward meeting strategic goals
Program Metrics	Measure program details toward achievement of strategic goals. Focus on meeting major program milestones, quality of the product, management of customer requested changes and management of risk
Program Tools	Scope, performance tracking, and risk tools are important. Schedule tools are standard and flexible, whereas there is no specific tool for cost
Program Culture	Customer driven, collaborative, wide-open communication, high degree of respect, pride, and teamwork

Program organization: The Silverbow program organization is aligned to the business strategy by the utilization of a matrix structure that enables a high degree of cross-discipline interaction and collaboration. Interdependencies between engineering, software test/QA, technical communications, and product managers are focused on maintaining customer satisfaction and delivering quality products. The team was made as small as possible to facilitate open communication.

Product managers were the representatives and voices of customers, and their responsibility was to make sure that the product matched the customers' needs. The software test/QA team was responsible for assuring the highest-quality software possible. Mike Billard, the engineering lead for the Minutes Meeting feature, told Balebi, "We do set dates, but we are not hard set on achieving the dates because customer satisfaction and quality factors take precedence. We would drop everything if a customer had a problem with one of our products in order to fix it for them."

Program process: The program process adopted by ConSoul is highly focused on satisfying their customers. Customers are encouraged to be involved in the process by defining their needs and wants in the predefine phase—evaluating preproduction releases during the design phase, participating in the integration process, evaluating the product under operational conditions, and providing feedback to the program teams.

Designing the software is accomplished through a process of iteration, allowing for continual improvement, maturity of the product over time, and an opportunity for customer change requests to be evaluated and incorporated.

Program metrics and tools: The program metrics and tools utilized were highly aligned with the three primary strategic elements by focusing on the measurement and tracking of software quality, management of program and project risks to success, feature content, and performance to committed schedule.

Program culture: The program culture was highly influenced by the business strategy of the company. Therefore, there is excellent alignment between business strategy and program execution. Strategic objectives are clear, concise, and measurable, and the program strategy, organization, processes, metrics, and tools all support attainment of the business objectives. The cross-discipline program structure provides clear lines of responsibility and authority, and the PLC process has well-defined phases and phase-exit criteria. Some of the culture's other elements include the following:

- Cross-discipline collaboration and open communication. There are no walls between engineering, software test/QA, program and project managers, or high-level executives.
- The organization is driven by the customers.
- The status of employees' work is tied to how well they know the product.
- The environment encourages a high degree of respect, pride, and teamwork.

WHAT ARE THE SILVERBOW PROGRAM OPTIONS?

Team Silverbow spent nine hours in two days planning for review of the options. They rechecked the status, reviewed their strategy, organization, process, metrics and tools, how they were applied, and how the relevant decisions were made. Multiple iterations of what-if analyses were performed, resource reallocation was evaluated, pro and con analyses of each option were carried out, and rough comparisons established. Eventually, three final options emerged. Balebi will present the following options to senior managers:

Option 1:

Strategy: Stay on the current course. Keep the scope, all features intact; shoot for the same deadline, using the same plan; and pursue quality as goal 1.

Pros:	We know the game.
	The team feels comfortable playing the way it knows.
	The goals are reachable.
Cons:	We will most likely be late by one month.
	We will lose sales.
	Team members may start finding new jobs as the program is nearing the end.

Option 2: Keep all features intact and hit the announced release date.

Strategy: Crash schedule, fast-track, and outsource. First, in some parts of the program, we will crash the schedule by adding more resources, retaining activity dependencies and shortening critical path activities. Second, we will fast-track some parts of the program by breaking activity dependencies and creating new ones to overlap and speed up the schedule. Third, we will outsource the software testing to an external, top-notch company to accelerate the schedule.

  Pros: It will be a feat.
            Releasing product as announced.
            Making customers happy.

  Cons: We may not be able to deliver on time with the quality
              desired.
            Outsourcing was never tried before.
            Fast-tracking is a new approach for us.

Option 3: Drop the two features and finish on time.

Strategy: Drop the two features. That enables the elimination of time to integrate the two new features, making it much easier to deliver on time. Also, to pay attention to the quality of the remaining features.

  Pros: Will deliver on time with good quality.
            Will release as announced.
            It is still the best product in the market.

  Cons: We may lose about 20 percent of the product sales
              (phone survey of lead customers).
            Because we failed on our product feature promises,
              we may lose some customers permanently.
            We may have a tarnished reputation.

MEETING WITH EXECUTIVES

After getting word from Smiley that the Silverbow program manager was ready to present a set of options, senior managers agreed to see Balebi for a brief review. The senior managers were organized as a Product Approval Committee (PAC) who have the authority to approve all product programs, perform gate reviews, and make major strategic program decisions. The PAC included vice presidents of marketing, enterprise software, office software, finance, manufacturing, and the director of program management. The PAC was headed by the vice president of marketing, John Biffin. When Balebi entered the conference room, the PAC members signaled that they were ready to start the meeting.

"Bali, Christine told us that you have some new options to offer," stated John Biffin.

"Yes, I have three options—some old, some new. Bear with the details, although I will be brief," replied Balebi. Balebi took seven minutes to present his three options, not forgetting to mention the source of his data, whenever he could. Finally, he wrapped up, with what he believed was his punch line. "After you decide which option to pursue, the program team will make it happen."

Biffin's response took Balebi by surprise. "Actually, Balebi, we won't choose the option, you will. We saw you today for one reason only. We wanted to make sure that you have a reasonable plan, and you do, and tell you that we will give you whatever resources you need. Choose your option and go for it. But you should know, we will hold you accountable."

Stunned, Balebi tried to think about how to respond. But it was too late to say anything because the PAC members had already adjourned the meeting. Balebi hurried out and was ready to reconvene his team members to decide which option to pursue.

CLOSING

The Consoul software example demonstrates some very good practices of the program management discipline, as well as a couple of opportunities for improvement. This example shows how business strategy drives the program management practices, structure, methods, metrics and tools. In particular, how business strategy influences trade-off decisions on a program.

This example also shows the impact of not utilizing consistent scheduling and budgeting processes to manage both projects and programs.

REFERENCES

1. Snow, C. C. and D. C. Hambirck, "Measuring organizational strategies: some theoretical and methodological problems," *Academy of Management Review*, Vol. 5, No. 4 (1980).

2. Treacy, Michael and Fred Wiersema, *The Discipline of Market Leaders: Choose Your Customers, Narrow Your Focus, Dominate Your Market.* Reading, MA: Addison Wesley, 1995.

3. Schwaber, Ken. *Agile Project Management with Scrum.* Redmond, WA: Microsoft Press, 2004.

Appendix C

Planet Orbit

Peerasit Patanakul and Dragan Z. Milosevic

INTRODUCTION

With $467 million in total budget and 144 months in duration, Planet Orbit is an ambitious program. Its objective is to build a spacecraft with a photometer for identifying terrestrial planets in the universe. Scientists believe that this program will eventually help them understand the extent of life on other planets and across much of the universe. "It represents a breakthrough in science that has the potential to change mankind's views about his position and place in the universe," according to Eric Anderson, Planet Orbit program manager.

Next week the PDR (preliminary design review) will be held. The review committee is expected to be tough, and Anderson knows from experience that the program has to be in excellent shape to be granted approval to move from definition to the design phase. He believes the program is progressing well technically, but he is aware of some interpersonal and human relations issues that may be of concern to the stakeholders. Additionally, Anderson is concerned that the latest schedule has some serious disconnects with senior management's delivery expectations. However, he feels that the PCT can push the team hard enough to make up any schedule shortfalls that they may encounter during the development phase.

Anderson is excited to have the Planet Orbit program finally ready to advance to the next phase of development. It has been three years since the program was initially approved. He still feels that the objective of the program fits well with the mission of the space agency—exploring life in the universe.

Anderson has been through the status of the program several times already in preparation for the PDR meeting with the committee. However, he believes it may be useful to recap with some of his key team members one more time to ensure all issues and concerns have been properly addressed. He reviewed his status documentation before Charles Wright's arrival to his office. Wright is the project manager of the local site for the Planet Orbit Program. When Wright arrived, they reviewed the program again in comprehensive detail.

PROGRAM BACKGROUND

The Planet Orbit program is a special space mission program sponsored by the space agency. The assigned program manager, Anderson, is a senior scientist with the agency. He had initiated the idea for the Planet Orbit program and had proposed it to the Space Agency multiple times since 1992. "Our idea is to use a transit photometer that will help increase the understanding of life in the universe and other planets like Earth," Anderson described. In several instances in the past, the program cost estimates and proposed scientific methodology were compared to much larger and more complicated missions. As a result, with each selection cycle, the program was forced to respond to new demands and requests. In particular, the program had to redo its costing several times to validate its accuracy and had to prove the viability of its chosen data collection technique with ground-based demonstrations. Additionally, they had to demonstrate that the research center had the capability for the end-to-end management of the program. It was not until December 2001, after persistent attempts to prove the idea for almost a decade, that the program was eventually selected.

PROGRAM ORGANIZATION AND MANAGEMENT

Because the breadth, depth, and complexity of the program are so large, the agency determined that all of the program components (program management, ground segment management, and flight segment management) were too much a single program manager to manage individually. Therefore, in June 2002, the space agency selected the spacecraft laboratory as the mission management partner in an attempt to reduce risk. Additionally, the Technology Corporation was selected as the industrial partner of the program for development of the hardware. Under this arrangement, the research center was still responsible for roughly 70 percent of the direct cost from instrument development and delivery and science management.

The structure of Planet Orbit is a program form of organizational structure involving multiple subteams in three organizations (Figure C.1). The structure is atypical because the program is shared by two centers research center and spacecraft laboratory. The program has a PMO, which is an entity created for a specific period of time that is dedicated specifically to the Planet Orbit program. The PMO provides a variety of managerial and administrative support that is pertinent to managing a major and complex program such as Planet Orbit.

Team members were selected by the PCT, which represented the functional project managers reporting to Anderson. The criteria for selection of the team members included experience, competency, dedication, and enthusiasm. External hiring for the program included employees from the spacecraft laboratory and the Technology Corporation who were also picked by the Planet Orbit team.

The Planet Orbit team size varied across program phases as the demand for personnel fluctuated. At the end of the definition phase, the team at the research center was composed of 25 full-time employees; the team at spacecraft laboratory totaled 12 full-time employees and 5 part-time members; the Technology Corporation's team included 80 full-time employees. Because the program is geographically dispersed across three different organizations, communication among the members is critical. The major means of communication include phone (frequent), e-mail (frequent), face-to-face meetings (frequent), and teleconference (weekly). There are also the casual and informal communication channels, including periodic social gatherings, picnics, and barbeques.

A formal team building exercise (referred to as four-dimensional (4-D) training) was held somewhat late after the start of the program. The 4-D training, provided by an external organization, helped the team members get to know one another, form cohesiveness within the team, and create "smoothness and harmony" across the organizations involved.

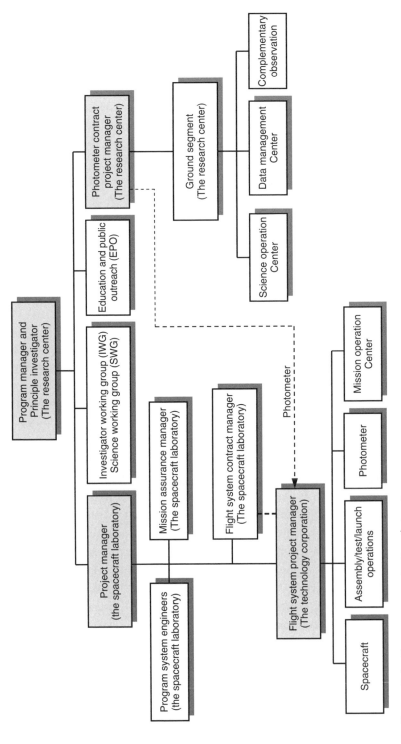

Figure C.1 Program organizational structure.

This became a vehicle for cross-organizational and, cross-site team building and integration of the research center, spacecraft laboratory, and Technology Corporation personnel. However, this was not as effective as originally planned. Many team members felt that 4-D training, although a good initiative, came too late to significantly benefit the team.

As program manager, Anderson is responsible for the execution of the entire mission and the scientific integrity of the investigation. The responsibility for day-to-day management of the program is delegated by the program manager to the project team at the spacecraft laboratory, where the project manager resides. The Planet Orbit's Educational and Public Outreach (EPO) Organization, the Investigator Working Group (IWG) and the Science Working Group (SWG), all report directly to the program manager. In general, the research center and its program manager are responsible for the development and test of the photometer and the overall ground segment. The spacecraft laboratory and its project manager, Wright, are responsible for providing the flight system, which includes the spacecraft, flight segment assembly/test/launch operations (ATLO), and flight segment checkout. Technology Corporation is a subcontractor to the spacecraft laboratory, and is responsible for developing, building, integrating, and testing the flight system and mission operations center, in addition to launching the flight segment.

For the Planet Orbit program, top management of the agency is involved during the gate approvals and program reviews. Three formal review teams are established to evaluate the program. These include the program formal review (PFR) team—the standing review board picked from various key organizations; the independent assessment team (IAT)—selected from the agency; and the systems management office assessment (SMOA) team—selected from the spacecraft laboratory. The PFR has 21 members, the IAT has 6–9 members, and SMOA has 2–4 members.

Review teams are used as the mechanism to evaluate whether the program is in line with the agency mission. Generally, the review teams make recommendations (so-called "findings") and evaluate the readiness of the program to proceed to the next phase. Reviews help the program manager and team manage the program and stay in line with senior management's expectations and the formal objectives for the program. The team must address any significant plan deviations and inform the review panels to ensure that they agree the team has properly addressed and resolved their concerns. This mechanism is used during all gate reviews.

PROGRAM STRATEGY

The program objectives must align directly with the mission and strategy documented by the space agency. The objective of the Planet Orbit Program is to build a spacecraft with a photometer for space exploration. Its scientific goal is to conduct a census of extraterrestrial solar planets by using a photometer in a heliocentric orbit. The first priority or strategic focus of the program is science goals, followed by the clear mandate to stay within a number of constraints (schedule, cost, and technical integrity). The science focus is clearly associated with agency value and is reinforced by the program manager to the core team and other team members throughout the program.

Product definition: The major products of the program include a photometer onboard a spacecraft and its associated ground system. The photometer is an instrument for measuring the luminous flux of a light source to observe the periodic dimming in starlight caused by planetary transit. When a planet passes in front of its parent star, it blocks a small fraction of the light from that star. If the dimming is truly caused by a planet, then the transits must be repeatable. Measuring three transits—all with a consistent period, duration, and change in brightness—provides a rigorous method for discovering and confirming planets. Simply speaking, the program aims to build the instrument that will find the terrestrial planets in the habitable zone of other stars. This includes developing and testing the photometer and the overall ground segment, developing the flight system, and flight segment in-orbit checkout system.

Program value: The program is very well aligned with the agency's strategic plan. The value of the program is to contribute to the answer to the fundamental questions of "Does life in any form, however simple or complex, carbon-based or other, exist elsewhere than on Earth?" and "Are there Earth-like planets beyond our solar system?;" and to provide exciting scientific results of great visceral interest to the general public about exploration.

Success criteria: Time, cost, and performance are the major criteria determining the success of the program. The agency specified that the program cannot drop below the minimum performance it promises and must launch within a certain time. With a target launch date of next year, the program will be initiated on board an expendable launch vehicle used to launch the spacecraft with the photometer, and the end of the baseline mission is five years after lauch. In terms of cost, the program cannot exceed either the development cost cap or the funding profile.

PROGRAM SCOPE

The Planet Orbit program is built around the following three major deliverables: The research center is responsible for developing and testing the photometer and the overall ground segment. The work done in the laboratory includes the development of the flight system (spacecraft, flight segment, and ATLO) and flight segment on-orbit checkout system. The Technology Corporation is responsible for developing, building, integrating, and testing the flight system. The major customer who will benefit from the program is the space agency, the parent organization of the research center and the spacecraft laboratory.

PROGRAM PROCESSES

The Planet Orbit program has six major PLC phases. The first phase is advance studies (prephase A). The objective of this phase is to produce a broad spectrum of ideas and alternatives for missions from which new programs can be selected. Second, is the preliminary analysis (phase A). In this phase, the team has to determine the feasibility and desirability of a suggested new major system and its compatibility with the agency's strategic plans. Third is the definition phase (phase B), which has an objective to define the program in sufficient detail to establish an initial baseline that is capable of meeting mission needs. Next is the design phase (phase C). In this phase, the team completes the detailed design of the system. Then, the program goes to the development phase (phase D) to build the subsystems and integrate them to create the system, while developing confidence that it will be able to meet the system requirements, deploy the system, and ensure that it is ready for operations. The last phase in the life cycle is operation (phase E). In this phase, the team has to make sure that the system actually meets the initial need and then dispose of the system in a responsible manner. Figure C.2 summarizes the PLC and the major milestones, including the timeline.

Major milestones include mission concept review (MCR), mission definition review (MDR), system definition review (SDR), preliminary design

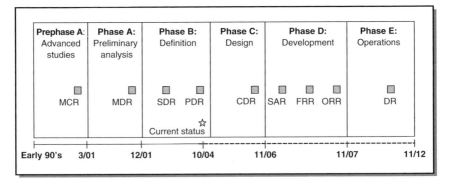

Figure C.2   PLC and major milestones.

review (PDR), critical design review (CDR), system acceptance review (SAR), flight readiness review (FRR), operational readiness review (ORR), and decommissioning review (DR). These phases and milestones are consistent and standard to the agency. There is definition and understanding of the decisions that need to be made at each milestone review. Planet Orbit is currently scheduled to pass the PDR stage. The first review team is the PFR, which is a standing review board. Their job is to verify or recommend that the program move forward on the basis of the completeness of its requirements and the understanding of the agency's requirements. The second review team is the IAT, whose job is to make the recommendation of whether the program should be confirmed. Next, the SMOA team evaluates the aspects of the program, based on the interest of the chief engineering office. After the program achieves PDR approval, it must meet a major constraint set by the agency that the program must then launch within 36 months, including one month of commissioning. This means that phases C and D cannot exceed more than 36 months, collectively.

The tactical management of the Planet Orbit program includes a formal WBS, roughly seven levels, in which specific team leaders are responsible for certain levels. Wright primarily has control over level 1 (program level), 2 (segment level), 3 (system level), and 4 (subsystem level); whereas, the project managers in the photometer, flight segments, and functional teams have been responsible for the lower levels (assembly, subassembly, and parts). Each level contains numerous activities. The spacecraft, as an example, has approximately 4000 activities. The WBS was used as the baseline to develop the program schedule and cost.

The schedule was structured to conform to the following boundary conditions: The funding profile for fiscal year 2005 through fiscal year 2007; the start of phase C and phase D to be after the preliminary design review and confirmation review; and phase C and phase D must have a duration of 36 months or less. The cost-estimation process involved triangulation of various methods such as a bottom-up, top-down process, independent cost modeling, and comparison with other similar programs (analogous estimating).

While a mitigation plan is developed for risks, the team has a mechanism with the mission assurance organization that tracks issues and problems separately. The agency has set a formal standard on risk mitigation (risk management policy document) in which the risks are rated based on their probability of occurrence and severity. Red, green, and blue charts are used to capture the varying degree of risks, which is discussed in the quarterly and monthly management reviews. Risks generically are

Table C.1   Controls requiring program manager and higher-level approval.

Controlled Parameter	*Controlled by Enterprise Associate Administrator*	*Controlled by Lead Center Director*	*Controlled by Program Manager*	*Change Request Thresholds*
Funding by year			Yes	Change in approved program budget
Level-1 requirements (Technical Requirements)	Unless controlled by program manager		Yes	Anything beyond agreed upon de-scope options
Program Objectives	Unless controlled by program manager		Yes	Anything beyond agreed upon de-scope options
Program Plan		Yes	Yes	Changes in scope, schedule, or budget

owned by the program, and they are assigned to a responsible individual who resolves and reports up the management chain. Planet Orbit is a fairly low-risk mission because it has a high-technology readiness level and does not involve on-board human life.

The program is tightly monitored and controlled. All technical performance, cost, or schedule parameters that require approval by the agency administrator, the associate administrator, the lead center director, or program manager are identified in Table C.1. These controlled parameters are well documented and shared with all team members in progress meetings and reports.

PROGRAM METRICS AND TOOLS

Several project management applications are used in the program. A critical path schedule and Gantt chart are used for scheduling and are based on the WBS. The number of activities in the schedule is in the thousands and disaggregated through the system of hierarchical scheduling, based on WBS system levels. Standard enterprise-management tools are used for budget and expense tracking. The team is *required* to do an earned-value analysis starting with phase C for the Technology Corporation's contract. Off-the-shelf commercial products, such as Live-Link, Project, and Doors are used to collaborate among the three major organizations. However, all team members have not used them universally.

Program performance is measured and tracked very carefully. In particular, aggregate actual-to-plan metrics for the master schedule and total budget are required by the agency.

READY FOR THE REVIEW

Anderson and Wright completed their review of the program documentation and concluded that the program was proceeding well. There were a couple of issues that they discussed, which had been the source of team frustration in the past.

The first concern was that cross-site coordination was not going as well as expected, creating several time delays and inefficiencies. Early cross-organizational training turned out to be relatively ineffective. Implementation of some selected software tools, such as Live-Link, helped the situation considerably, but they decided that more effective management by the program manager and core team leaders was essential.

The second key issue related to the schedule constraints imposed by the agency. For the most part, the completion within 36 months was, in essence, dictated by senior management. The program team accepted this date without the appropriate analysis and consolidation of schedule to support it. Additional schedule analysis, since the last milestone review, indicated that the target completion date may not be achievable and will need to be appropriately discussed at the upcoming PDR meeting.

As they concluded, Anderson and Wright recapped where they thought they stood with the program overall. Anderson summarized as follows:

- We have a strong team with high morale.
- We have competent personnel in all key positions who are very experienced and knowledgeable in their respective fields.
- We have clearly defined goals, focus, and strategy.
- The program aligns well to the strategic goals of the agency, which helps in terms of management support.
- The program is quite stable in terms of its scope and science.
- The program baselines are set.
- The program schedule is a challenge but still manageable.
- We have a strong program management team.
- We know what we have to focus on in the future to be successful.

He concluded, "I think we are very well prepared for the upcoming committee review. I'll review what we discussed today with the other core team members, and I think we are ready to proceed."

CLOSING

The planet orbit example demonstrates how the program management model can scale as a viable management approach for small, multi-discipline development efforts to very large and complex efforts involving multiple organizations. However, as program size and complexity grow, the capabilities and experience of the program manager becomes more important. Fundamentally, this program is sound, however, it is evident that cross-project and cross-organization collaboration and synergy is breaking down. This shows the importance of leadership and other "soft skills" on the part of program managers.

Appendix D

General Public Hospital

Peerasit Patanakul and Dragan Z. Milosevic

INTRODUCTION

The time is 6:30 A.M., midweek in the month of February. Julia Skown is the program manager for the Time Keeping System (TKS) Program. She is in a conference room looking at the agenda for a program review meeting to assess the status on the TKS program, which is scheduled at 8:00 A.M. In the room with her are Stacey Cook, a payroll specialist, and Tom Black, an IT project manager. They are members of the PCT and employees of General Public Hospital (GPH), and it is obvious that all three are nervous regarding the potential outcome for their program.

There is good reason for them to be nervous. The purpose of the meeting today is a program baseline review (go/kill/hold) with senior management. *Go* means the program is ready to proceed, *kill* means the program will be terminated, and *hold* means the program needs to be reworked and brought back for review. The team is concerned that senior management may decide that the TKS program must be put on hold. This decision will mean that the program will be sent back for rework and lose its priority for available resources. In other words, it may be put on the back burner.

PROGRAM BACKGROUND

The IT organization for GPH took the lead in championing the TKS program. Operating under the CIO, the group provides various types of services and solutions for both internal (doctors, nurses, students) and external (patients) customers. Some of those services and solutions include the registration system, employees' clock in/out system, payment system for patients, and computing power for medical research, which is among the best in the United States. The vision of the hospital is to be a national and international leader in health care, education, research, and technology development. To help accomplish the vision, strategic goals for the IT group provide the basis for the programs. Generally, the uniqueness of the IT organization is that it operates in one of the few academic medical centers in the United States that employs a "best-of-breed" strategy. Therefore, the group achieves high-quality solutions by using best-in-breed commercial products, rather than developing their own and integrating the solutions in a patchwork fashion. John Menegy, CIO, says, "We work with individual vendors to find the best solutions, and then we successfully integrate the solutions with our existing products."

There are more than 300 employees in three major groups under the applications division director of the IT organization. These three groups include Academic and Research Applications Support, University Applications, and Health Care Applications. The TKS program resides in the Academic and Research Applications Support group as depicted in Figure D.1. TKS is a formal program that has a program manager and a team of skilled and knowledgeable members assigned. All the team members are motivated and committed to the program.

POTENTIAL PROGRAM SCOPE CHANGE

The objective of TKS was to implement a new time-keeping system for GPH. It was a one-million dollar program and was anticipated to take 18 months to complete. The motivation for the program came when the vendor that provided services for the current time-keeping system sold its system to another company. As a result, GPH was forced to make a change because support for the current system would be discontinued by the new vendor. Management had to either purchase a new system from the new vendor, or seek other vendors for time-keeping systems. The first option appeared to offer fewer risks (familiarity with the previous

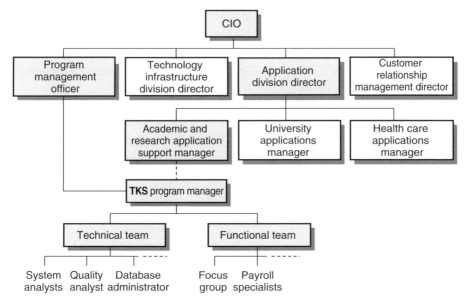

Figure D.1   Organizational chart and TKS program team structure.

generation of the system) and enabled GPH to keep the front-end system for its employees (for example, tax system, clock in, clock out).

Skown, Cook, and Black came to the office early to do last minute preparations before the management review. Still concerned about the outcome of the review meeting, they discussed the major issues.

Skown: Stacey, let me ask you this. Frankly, what do you think about our program so far?

Cook: I think we are doing great. There have been problems here and there, but I think we took care of them.

Skown: Okay. But Stacey, you are the team member representing our customers. If you anticipate any problems that we can prevent from happening, please let us know. We really want the new system to function properly.

Cook: Don't worry about this, Julia. I will tell you if I see anything wrong. I know that the quality of the new system is our top priority. Plus, I don't want a system that cannot pay people accurately. That's why I am here. Thank you, though, for involving me in this program and valuing my opinions so far.

The discussion turned to adding a new transaction-inquiry feature to the TKS system, which will make the system much more user-friendly—a feature that currently is not in the program plan. However, the new feature imposes more technical difficulty than what the team had expected and will cause an increase in the program cost and schedule.

Skown: So, the question here is whether we will move on to the next phase without this feature, or do we want to spend more time adding it to the new system? How long will it take us to add this feature, Tom?

Black: On the technical side, it will take us at least two months to do, meaning that the total program delay will be two months.

Skown: And you said that we have to buy new hardware because the current system is not compatible, correct?

Black: Correct. But we already budgeted for most of the hardware; it's just a couple of pieces that we didn't know we needed at the beginning of the program.

Skown: How much will this cost us?

Black: Our vendor told me that this additional hardware will cost about $100,000.

Skown: $100,000 is a lot of money. What is their lead time?

Black: Well, it will take about three weeks. I already included this when I told you that it will take us at least two months to add this feature.

Skown: All right. Do our existing testing and implementation plans account for this new feature?

Black: Mostly, yes. But you never know what to expect with new hardware. It is a program risk but probably only a medium severity. Our vendor is familiar with this new hardware and should be able to help us in a timely manner if we encounter any problems.

Skown: Okay. So, in a nutshell, if we add this new feature, we will have about a two-month delay and be $100,000 over the planned budget. This means that instead of us implementing the new system at the beginning of July, we will have to wait until the beginning of September. Plus, we will end up well over budget.

Cook: But if we do not add this feature now, we will be on schedule, right?

Skown: Yes, if everything goes according to our plan. Our options are to add this feature later when we do a future system upgrade or spend more time and cost now.

Black: Julia, what do you think we should recommend to the review committee?

Skown: First, I don't think they will let me go into a lot of discussion on the topic. They pretty much know what is going on in TKS because as you know I talked to them a few times in the last month. Second, there are not many options to consider. Third, I will tell them to consult the focus group because they represent our customers and other stakeholders. We have arranged to have the focus group stand by so they can come in five minutes into the review, assuming the committee agrees that they should participate.

Cook and Black headed to get coffee. With just a few minutes remaining before the review meeting, Skown used the time to review the various elements of the TKS program.

TKS PROGRAM TEAM

The structure of the TKS program was a dedicated, autonomous team, including representatives from several functional groups (for example, hardware engineers an intern and five technical services personnel). The technical team consisted of three system analysts, a quality analyst, a project manager, and a database administrator. Customer representatives, such as payroll specialists were also part of the team. Also representing the customers was a focus group that was chosen by the functional sponsors. Most team members were selected based on their skills and knowledge and were motivated and worked well as a group. In addition, the TKS program also had strong support from the PMO which is responsible for developing, implementing, and continuously improving the program management processes and tools to improve program performance, and to assist the IT organization in achieving its business objectives.

PROGRAM STRATEGY

The health care industry places a large emphasis on the quality of the products and services it develops. It is translated into the business strategy of GPH, in which it primarily focuses on maintaining a stable, limited line of products or services; it offers best quality with low cost; and it delivers what its customers want. Influenced by the business strategy, program management elements are directed to satisfy customer needs through the delivery of these quality products and the services. The first priority of TKS is to install the system and make it work correctly with all other existing systems. The major constraints are cost and schedule because the program has limited resources and a strict go-live date, previously announced to all employees. The following section lists the program objectives, product definition, valve proposition, and its success criteria.

Objectives: The objective of the TKS Program is to implement a new time-keeping system for the organization. The target customers are payroll specialists and other users of the new system.

Product definition: The product is a time-keeping system that stores all employee-related transactions. More precisely, TKS has to integrate employee data, leave accrual balances, labor schedules, and modified pay rules into a new system. The scope of the program is to replace the current time-keeping system. This includes developing new interfaces, training 500 time keepers, and deploying the application to all personal computing devices. Even though it is an off-the-shelf product, the team has to work on the details with the vendor, especially the product interfaces with the existing hardware, which required some modifications. The inputs from payroll specialists are vital for getting initial set up and configuration of the system. Resources are required from the payroll office, field technical services, network applications, Unix administration and database administration, and a focus group consisting of timekeepers and managers from a cross-section of the hospital.

Value proposition: The advantage of the new solution is that other products available are not nearly as mature or advanced as the one that the company is employing. TKS will give GPH the best solution available today. In addition, the new system has the advantage over the previous product generation in terms of reporting capability, scheduling capability, robustness, and various types of features. From a cost perspective, the new system helps reduce the number of part-time employees due to the efficiencies gained from the new generation of product.

Success criteria: The primary success measure of the program is the accuracy of the system. Other success criteria include the following: User-friendly interface, July go-live date, and $1 million total cost. These success measures were recognized up front and are well articulated in the program. To meet these success factors, the team understands that the program must have clear goals and scope, a high level of communication, effective system testing, and a high level of stakeholder support and buy in. In addition, the effective use of program management practices are understood as factors contributing to program success.

PROGRAM PROCESS AND TOOLS

Closely following the standard process the PMO created, TKS has four major phases to complete: integration and request, program planning, program execution, and implementation and support. There are also three major decision checkpoints: functional review, baseline review, and customer approval (see Figure D.2). The TKS program is currently at the end of stage two—the program planning stage. This standard process is referred to as the program management solution development life cycle (PMSDLC). It was influenced by the organization's strategy of having a stable, standardized, and tested process with quality and customer focus. To ensure the use of the process, management demanded

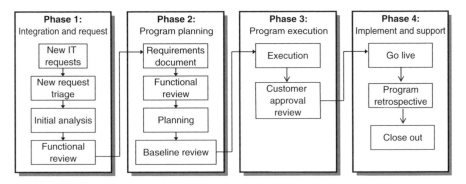

Figure D.2   Program phases.

a two-day PMSDLC training class in which all senior management, program managers, and major team members were required to attend. Program and project tools focus on quality assurance, cost, schedule, and performance to success criteria.

Not all programs within the IT organization are required to go through all the standard phases and milestones. The PLC is scalable upon the type of the programs with keen emphasis on achieving quality with minimum cost. Maintenance programs, which are those that require minimum efforts (a couple of weeks to a couple of months in duration) and are initiated to fix problems in existing products or services, go through an accelerated life cycle. Minor programs are those that require in-depth planning because they involve some degree of risk impact on the organization. A minor program normally lasts from two months to a year in duration and will pass through all phases of the PMSDLC but require fewer steps in Phase 1. A major program like TKS is required to go through the entire PLC. A major program normally takes from six months to three years in duration.

The scope of the TKS program was defined based on business requirements. The program schedule was developed on the basis of standard hour estimates of previous programs. A Gantt chart was created to support the development of the program schedule. The program manager worked with the technical sponsor, the vendor, and the PCT to determine the budget. This included the cost of hardware, software, consulting fees, licensing fees, resource hours, and training. Even though TKS is considered a low-risk program because it involved few new technologies, the team follows the standard risk management process. The program team documents the probability of multiple risks that may occur in the program in terms of dates and/or dollars. The program team created the risk list and developed contingency plans to address and resolve each risk as needed.

In terms of quality, the program has a quality analyst whose responsibility is to ensure that the program will satisfy the needs for which it was undertaken. Activities such as measuring, examining, and testing were performed to determine whether the results conformed to the requirements. Actions are taken to bring defective or nonconforming items into compliance to the specification requirements. Checklists are also used to track quality.

In the TKS program, the means of communication include program meetings, program review presentations, e-mails, and the phone.

The program manager met with the program steering committee to achieve program approval. After approval, the program manager reports the program progress back to the committee periodically through formal meetings that are held once a quarter. The program manager meets with the program team daily, while involving the customers on a regular basis. The program manager holds regularly scheduled meetings and documents the minutes from these meetings with action items for ongoing follow-up by the PCT. In addition, the program has an integration process to make sure that the various elements of the program are properly coordinated. It is done through an iterative process of multiple meetings and consistent communications on a regular basis.

PROGRAM METRICS

The significant program metrics tracked on the TKS program include milestone completion, risk mitigation, budget tracking, and the number of key program scope changes. Program schedule is tracked using milestones and tasks performed to the schedule baseline. Program cost is tracked by expenditure, account balance, hardware and software cost, consulting fee, training, and staff cost. This cost metric is tracked separately by the program manager in the program reviews and is not shared with all team members. The program risks are tracked by the mitigated risks versus exposure.

PROGRAM CULTURE

In general, the organizational culture is diverse, complex, formal, and relatively inflexible to change. Quality and customer focus are valued highly by all employees.

GPH has a wide array of employee types—physicians, nurses, administrators, engineers, and technicians—which makes it diverse. To control this diversity in styles and personalities, GPH opted for a more rigid and hierarchal form of organizational structure. In addition, most employees in the organization are reluctant to change, especially as it pertains to new technology.

Because the time-keeping system is an emotional subject for everybody in the organization, the TKS program team has to be sensitive to the needs of the customers and convince them to accept the new change. Quality and accuracy is key in the minds of the customers. The program has consistent, regular, and open communication, which contribute to team members and customer representatives understanding their roles and responsibilities.

THE PROGRAM BASELINE REVIEW MEETING

The door to the management conference room opened and Skown and her team were invited to join the program review committee. The committee consisted of John Menegy (GPH CIO), John Bacon (customer relationship director), and Art Counter (payrol-director).

John Menegy: This is the baseline review meeting in which we will decide whether to approve the TKS program to move into Phase 4 of development. Since we have discussed the status of your program with you several times over the past few weeks, I think we understand what's going on with the program. What is the agenda for today's meeting?

Skown: Here is the agenda:

- Summary of program status (program manager)
- Schedule update
- Financial update
- Risks-mitigation activities
- Resource status
- Key-customer events and interactions
- Presentation of the goal accomplishment
- Transaction inquiry feature assessment

Menegy: I don't think we need to cover all agenda items. As I stated, we pretty much know your status. But we want to hear if there is anything new to add.

Skown: Well, adding the new transaction inquiry feature we were asked to assess will add two months to the schedule and $100,000 in cost. It is a lot, we know. But we need to ask our customers and other stakeholders if they believe the new feature is necessary.

Bacon: You mean to ask our focus group?

Skown: Yes, I think they are the right people to ask. They are the ones who clock in and clock out everyday. Plus, they are the representatives from their functional departments.

Black: Quality is our focus. All in all, we want the new system to work accurately and be easy to use and maintain. And we all know that the payroll specialists and the focus group are our customers. I think we need feedback from the focus group.

Menegy: How long will it take for them to get here?

Skown: Five minutes, they are waiting for our call. All five of them.

Menegy: Wow, you've done your preparation. John and Art, do you agree that we bring the focus group in here?

Bacon and Counter: Yes, agreed.

The focus group joined the meeting. For several minutes, the committee talked to each member of the focus group. They all believed that it was in the best interest of GPH to add the new feature now and proceed with the program execution. However, the committee members were well aware of the significant adverse ramifications to a slip in the schedule for completion of the program. The benefits of adding the new feature versus the costs of delaying the introduction was weighed carefully. Menegy consulted with the other members of the committee and announced to the team that they will need to deliberate with the committee and will communicate their decision in half an hour.

With that announcement, Skown excused herself to go check her e-mail. When she got to her office, she stood for a few minutes before opening her e-mail account and thought to herself, *Half an hour—that feels like an eternity! I think the information from the focus group will save the program. At worst, they will come back with a hold decision but no kill decision.* After responding to a few e-mails from her program team, Julia returned to the meeting to hear the executive team's decision.

Final Thoughts on Program Management

The key message that we convey in this book is that program management is directly linked to the success of the business. Several factors regarding the business define when program management should be used.

When senior management has critical developmental efforts that require a tight connection between the strategic objectives of the business and execution, the program management model is recommended. This situation will require a leader that possesses broad and comprehensive business knowledge, excellent people skills, and a solid-grounding technically. Generally, these efforts will be characterized by high levels of complexity and significant cross-discipline dependency.

The traditional interpretation of project management is most appropriate for those business situations that are not directly tied to business results and that are less dynamic and complex. Project management fundamentals may be best defined and described via PMBOK, the international and defecto U.S. standard. According to the standard, project management is about management of a single, individual project, with a primary focus to accomplish the triple-constraint goals (time, cost, and quality). More precisely, project management is tactical in nature, focused on execution success. The project manager manages vertically within a single project—assuring that work effort generates desired deliverables on time, within budget, and at require performance levels. It is clear that in-depth project management and functionally specific technical competencies are dominant.

Project management provides outstanding systematic procedures that organizations use to produce triple-constraint deliverables in less dynamic and complex business situations. But, such focus may no longer be suitable for organizations to win in today's marketplace. The demands have changed, as organizations face more development efforts in significantly dynamic and complex business situations. Such changes necessitate new best practices. In this case, one that puts strategically aligned program management in the center of the company's competitive stage.

We define program management as the coordinated management of interdependent projects over a finite period of time to achieve a set of business results. If we elaborate on its basics, it is obvious that program management is much different from project management. Program management is strategic in nature and focused on the business success of the company. Alignment relates to the execution with business strategy. Successful delivery of the right product, service, or infrastructure capability is vital, and it must be at the right time. Unlike the project manager who manages vertically, the program manager manages horizontally across the functional projects involved with a program. Also, the program manager assures that the cross-project work effort remains feasible from a business standpoint.

But these are no more than details, as important as they may be. On a fundamental level, company leaders need to understand the degree of strategic importance any large development effort has on the success of the business and decide whether they need to think and manage in a project management or program management mode. If the focus is on improved business results, the choice is clear.

Index